Forecasting and Risk Management Techniques for Electricity Markets II

Forecasting and Risk Management Techniques for Electricity Markets II

Guest Editor

Yuji Yamada

MDPI

Basel • Beijing • Wuhan • Barcelona • Belgrade • Novi Sad • Cluj • Manchester

Guest Editor
Yuji Yamada
Institute of Business Sciences
University of Tsukuba
Tokyo
Japan

Editorial Office
MDPI AG
Grosspeteranlage 5
4052 Basel, Switzerland

This is a reprint of the Special Issue, published open access by the journal *Energies* (ISSN 1996-1073), freely accessible at: https://www.mdpi.com/journal/energies/special_issues/HO0MEWIM2U.

For citation purposes, cite each article independently as indicated on the article page online and as indicated below:

Lastname, A.A.; Lastname, B.B. Article Title. *Journal Name* **Year**, *Volume Number*, Page Range.

ISBN 978-3-7258-4825-6 (Hbk)
ISBN 978-3-7258-4826-3 (PDF)
https://doi.org/10.3390/books978-3-7258-4826-3

Contents

About the Editor

Yuji Yamada

Yuji Yamada (Professor, Ph.D. in Engineering). Yuji Yamada was born in Matsumoto City, Nagano, Japan, in 1969. He received his B.S. and M.S. degrees from the University of Chiba and the Tokyo Institute of Technology in 1993 and 1995, respectively. In 1998, he earned his Ph.D. in Engineering from the Tokyo Institute of Technology, where his doctoral thesis focused on global optimization for robust control under the supervision of Dr. Shinji Hara. From 1998 to 2001, he was a postdoctoral scholar in Control and Dynamical Systems at the California Institute of Technology (Caltech), working with Dr. John Doyle. In 2001, he also served as a course lecturer at Caltech, teaching a Ph.D. course entitled Stochastic Simulation in Finance, with advisory support from Prof. Peter Bossaerts. In 2002, he joined the University of Tsukuba as an associate professor and has been a full professor at the Institute of Business Sciences since 2013. He served as Dean of the Institute from April 2018 to March 2021 and as Chair of the Master's Program in Systems Management from April 2015 to March 2017. Since August 2023, he has served as President of the Japanese Association of Financial Econometrics and Engineering (JAFEE). His research interests span finance, optimization, data science, and control theory, with a particular focus on applications in electricity and financial markets. He has led several research projects on renewable energy and electricity trading, including the Grant-in-Aid for Scientific Research (A) projects 16H01833 and 20H00285, from the Japan Society for the Promotion of Science (JSPS).

Preface

The electricity market is undergoing rapid transformation as it adapts to the increasing penetration of renewable energy sources and the transition toward decentralized power systems. These changes have introduced new layers of uncertainty in both electricity supply and demand, as well as heightened volatility in market pricing. To address these challenges, market design must evolve to accommodate flexible trading systems, data-driven forecasting, and advanced financial instruments for risk mitigation.

Building on this foundation, Forecasting and Risk Management Techniques for Electricity Markets II continues the exploration begun in the first volume, further examining how electricity markets can respond to this shifting landscape. While the initial volume laid the groundwork by identifying critical sources of uncertainty and emerging risks, the current Reprint expands the discussion to include recent developments in forecasting techniques, market simulations, and financial-engineering approaches. The growing relevance of decentralized energy resources, peer-to-peer trading, and climate-linked risks underscores the urgency of developing robust decision-making tools.

This Reprint presents a diverse collection of contributions that reflect the interdisciplinary nature of modern electricity market research. Topics include probabilistic forecasting methods for load and price, agent-based simulations for decentralized coordination, blockchain-based market mechanisms, and the design of energy and weather derivatives to hedge against forecast and volume risks. These studies offer valuable insights for energy producers, grid operators, financial engineers, and policymakers striving to build resilient and sustainable electricity markets. In doing so, this Reprint demonstrates how perspectives from economics, engineering, finance, and data science can be integrated to address evolving market needs.

We hope that this Reprint will serve not only as a valuable resource for researchers and practitioners, but also as a stimulus for further exploration into the design of electricity markets that are both adaptive and resilient in the face of accelerating change. Special thanks go to all the authors and contributors who made this volume possible, and to Ms. Shary Song, Section Managing Editor, for her dedicated support in organizing both the Special Issue and this Reprint. Last but not least, I would like to express my heartfelt gratitude to my wife, Yumiko, my son, Yukei, and my daughter, Sakura, for their unwavering support in my life.

Yuji Yamada
Guest Editor

energies

MDPI

Article

Improving the Efficiency of Hedge Trading Using Higher-Order Standardized Weather Derivatives for Wind Power

Takuji Matsumoto [1,*] and Yuji Yamada [2,*]

1 Faculty of Transdisciplinary Sciences for Innovation, Kanazawa University, Kanazawa 920-1192, Japan
2 Faculty of Business Sciences, University of Tsukuba, Tokyo 112-0012, Japan
* Correspondence: mtakuji@staff.kanazawa-u.ac.jp (T.M.); yuji@gssm.otsuka.tsukuba.ac.jp (Y.Y.)

Abstract: Since the future output of wind power generation is uncertain due to weather conditions, there is an increasing need to manage the risks associated with wind power businesses, which have been increasingly implemented in recent years. This study introduces multiple weather derivatives of wind speed and temperature and examines their effectiveness in reducing (hedging) the fluctuation risk of future cash flows attributed to wind power generation. Given the diversification of hedgers and hedging needs, we propose new standardized derivatives with higher-order monomial payoff functions, such as "wind speed cubic derivatives" and "wind speed and temperature cross-derivatives," to minimize the cash flow variance and develop a market-trading scheme to practically use these derivatives in wind power businesses. In particular, while demonstrating the importance of standardizing weather derivatives regarding market liquidity and efficiency, we propose a strategy to narrow down the required number (or volume) of traded instruments and improve trading efficiency by utilizing the least absolute shrinkage and selection operator (LASSO) regression. Empirical analysis reveals that higher-order, multivariate standardized derivatives can not only enhance the out-of-sample hedge effect but also help reduce trading volume. The results suggest that diversification of hedging instruments increases transaction flexibility and helps wind power generators find more efficient portfolios, which can be generalized to risk management practices in other businesses.

Keywords: efficient frontier; generalized additive model; LASSO regression; optimal hedging; weather derivatives; wind power

1. Introduction

In the context of achieving carbon neutrality goals, the penetration of wind power, a typical clean energy source, has been steadily increasing worldwide. According to the International Energy Agency [1], the global wind power capacity is increasing at an accelerating rate and this trend is expected to continue. Small wind turbines, which can be installed at residential sites, have especially attracted attention in recent years, and their market is also expected to grow steadily [2]. With these forces in play, the number and types of players in the wind power markets are also expected to increase worldwide. Then, the need to manage the risk of fluctuations in wind power generation revenues becomes increasingly vital for wind power businesses.

With regard to risk management in wind power generation businesses, in addition to avoiding the risk of loss through accurate forecasting, "hedging" approaches play important roles, which involve the various use of financial transactions/contracts to compensate for inevitable loss risks. Risk management through forecasting approaches is often discussed in the context of avoiding imbalance risk due to high-cost imbalance charges imposed on forecast errors; for example, optimizing market transactions as proposed in [3,4]. However, since it is extremely difficult to accurately forecast and compensate for losses, particularly against the long-term ahead horizon, financial hedging transactions may be used to control such cash flow fluctuation risks. Recently, a method combining both approaches has been proposed, such as the use of derivatives based on prediction errors to hedge against losses derived from prediction errors [5]. Thus, for effective risk management in the electricity business, in addition to improving forecast accuracy, risk "hedging" to compensate for financial losses is vitally essential [6], and a variety of hedging products exist in the electricity market [7]; the focus of this study is also on the latter issue: the concept of hedging.

For hedging "price" risk in the wind power business, financial instruments, such as power futures and bilateral contracts, and support measures, including feed-in tariffs (FIT), can be used. However, few financial instruments are available for hedging "volume" risk in practice. For example, even if a wind power generator has a contract to supply a certain amount of electricity at a fixed price, it may incur losses when electricity cannot be generated owing to poor wind conditions. However, few instruments have been devised to hedge these losses, leaving a room or need for research in developing practical and effective trading schemes to manage the volume risks.

As one of the few instruments for treating these volume risks, Wind Power Futures (WPF) have been listed on the European Energy Exchange (EEX; note that this was delisted in 2020) and Nasdaq. WPFs are contracts with a weighted average of area-wide wind power generation in Germany/Austria as the underlying index. As the underlying index is the amount of electricity generated, the WPF may be effective for large-scale wind power generators in the target area. However, more specific and refined products (derivatives or forwards/futures) may be required for individual power producers, including small-scale wind farms or even households (prosumers) with small wind power generators operating in a specific region. For instance, a more effective tool can be a flexible instrument tailored to the specific weather conditions of the generator's business area using only observed weather values as the underlying index. These weather derivatives can help ensure transparency as a financial instrument since the generator's individual situation will not affect the underlying index. Here, we propose an approach for using such "weather" derivatives, while emphasizing transparency and flexibility in product design.

Several studies have examined derivatives for wind power, including not only WPF [8,9] but also the European put-type quanto options [10], wind put barrier options [11], and wind call options [12]. Typically, these derivatives have been studied in the context of pricing rather than how effectively they hedge risk. Papers focusing on the hedge effect of wind derivatives, although very limited, include [13], which analyzed the effectiveness of the WPF, and [14], which showed the effectiveness of option-type derivatives on wind speed. In contrast, Yamada [5] proposed a unique derivative for hedging wind power risk and demonstrated its effectiveness based on non-parametric regression using the generalized additive model (GAM) [15]. The implication presented there is the fact that the hedge effect can be enhanced by estimating arbitrary smooth functions as derivative payoffs rather than using linear or conventional option-type functions (see [16] for the original idea and a genealogy of prior works on this). Subsequently, this hedging approach was extended to multivariate models [17] and also to retailers and solar power producers [18,19] and demonstrated that it also has a high hedge effect for them. In these studies, hedgers use non-parametric regression techniques to estimate smooth payoff functions and obtain optimal weather derivatives. However, derivatives estimated by non-parametric regression require bilateral contracts between insurers and hedgers in practice because the optimal payoffs are hedger-specific. Therefore, there remained room to consider the use of market principles from the perspective of improving the trading efficiency. Therefore, while the derivatives proposed in this present study have the same motivation as these series of studies in terms of constructing arbitrary payoff functions, we propose a new unique "market trading scheme" in which weather derivatives of various types and dimensions are introduced in advance and traded by multiple power utilities (hedgers). In other words, instead of trading (contracting) arbitrary payoff functions, by introducing various types (several to a dozen) of "standardized derivatives," hedgers with different loss functions (dependent on generation output patterns) can combine the necessary instruments and trade them so that every hedger can replicate their own optimal payoff functions. This would also lead to improve trading efficiency through liquid trading by multiple players (details in Section 2).

For specific modeling, this study draws inspiration from previous research on wind power forecasting models. In some previous studies, wind power generation can be generally modeled in the form of a cubic function of wind speed, and temperature is also often taken into account as the next most influential weather variable after wind speed (see Appendix A). Taking this into account, we apply their insights to design weather derivatives based on wind speed and temperature for wind power businesses. In addition to estimating non-parametric payoffs using a two-dimensional tensor product spline function [20] that can capture the complex interaction of wind speed and temperature on wind power generation, and evaluating the hedge effect of previous methods, we introduce new composite and standardized derivatives, such as the "wind speed cubic derivatives" and "wind speed and temperature cross-derivatives," and demonstrate the effectiveness of derivative portfolios replicated with them.

As noted above, our primary methodological motivation is to design and replicate optimal nonlinear payoffs for effective hedging, and there are relevant studies in this context as well. For example, Oum et al. [21] and Oum and Oren [22] constructed optimal nonlinear payoff functions on electricity prices for volume (demand) risk hedging in electric utilities. Brick and Roncoroni [23] also derived nonlinear payoffs for temperature using a similar approach. These results share analogous ideas with ours that optimal (and effective) payoff functions are searched based on smooth nonlinear shapes and then are replicated using standardized instruments for hedging volume risks. In particular, their methods use the approach proposed by Carr and Madan [24], which showed that any twice-differentiable non-linear payoff function can be replicated by a risk-less asset (the bond), a single risky asset (the stock), and European options on the same underlying asset with any strikes. However, their approach requires infinitely many standardized derivatives (i.e., puts/calls with any strikes) in principle and may not be easy to achieve a sufficient approximation using a restricted number of standardized derivatives (e.g., Oum and Oren [22] present a calculation case where dozens of different options are used simultaneously, despite being univariate). Instead, our approach is to consolidate product types needed to replicate the payoffs by creating new monomial-type derivatives and to propose a trading scheme involving such derivatives rather than being limited to the idea of using existing option-type derivatives. In this respect, this study also illustrates new implications for the methodological motivations and perspectives of previous studies.

Moreover, while our recent studies have addressed the design of refined derivatives for wind power [25,26], this study provides a new perspective by demonstrating the usefulness of our proposed approach not only with respect to its hedge effect but also "trading efficiency." Simply trying to increase the hedge effect by using various types of derivatives or combinations of products might not be original as an idea in hedge modeling. However, as a new methodological contribution in hedging decision making, we propose the strategy to estimate a hedging model for a standardized derivatives portfolio with the least absolute shrinkage and selection operator (LASSO) regression, which can simultaneously achieve both the hedge effect and the small transaction volume. In particular, we show that adjusting the regularization parameters of LASSO can reduce the number of product types that should be traded by exchanges/hedgers while maintaining hedge effect. We also find that as the variety of derivatives offered by an insurer or exchange for trading increases, hedgers can find more efficient derivatives portfolios using LASSO regression trading strategies. Note that the hedging problem using a statistical approach, especially the most straightforward ordinary least squares (OLS) method, has had several applications since Ederington [27] proposed the method of minimum variance hedging in the futures market, and it has been applied to various empirical studies in the context of cross hedging, etc. (see, e.g., the papers [28,29] and the literature review therein.). Similar studies for the electricity market include [30], which used OLS to estimate a hedging model consisting of temperature (degree days), water flow, and gas prices against electricity spot prices. Other related studies in terms of using LASSO regression include those that applied it to portfolio selection for index tracking in the stock market and examined its performance [31,32]. However, to the best of our knowledge, the idea of using LASSO regression to enhance trading efficiency in hedging problems has not been proposed before (perhaps this is because most existing research on hedging takes as its starting point the idea of using the existing limited types of hedge-related instruments.), and this study provides a new perspective in this context as well. Thus, this study presents multi-layered novel ideas, not only proposing a form of trading market for multiple derivative products in anticipation of future changes, such as the decentralization of the electricity market environment, but also presenting a LASSO regression-based trading strategy at the same time as a solution to the problem of trading costs inherent in the diversification of financial products.

The remainder of this work is structured as follows. In Section 2, while developing our market trading scheme, we also discuss the effects of trading multiple derivatives and then formulate specific hedging models. In Section 3, we use empirical data to verify the estimation results of the model and hedge effect, and in particular we reveal the mechanism that the LASSO regression allows us to construct efficient portfolios out of using multiple derivatives. Finally, Section 4 presents our conclusions.

2. Methods

2.1. Market Trading Model

We first develop a market model and discuss possible trading schemes for the proposed weather derivatives. The overall picture of the transaction envisioned in this study is illustrated in Figure 1. Assume that a wind power generator sells the electricity to an off-taker under a power purchase agreement (PPA). The generator enters into a derivative contract with an insurance company (or purchases derivatives on an exchange market) in advance to hedge the risk of fluctuations in future electricity sales. Then, at delivery time t, the generator delivers the volume of power output V_t to the off-taker and receives (or pays if negative) a cash flow $g(V_t)$ each time as a settlement fee, where the function $g(\cdot)$ is predetermined in the PPA. At the same time, due to the pre-contracted weather derivatives, the power generator receives $payoff(M_t)$ that depends on the meteorological index (calculated from the observed values) M_t (M_t may include wind speed, temperature, etc., and the function $payoff(\cdot)$ is predetermined when the derivatives contract), and the generator's loss is compensated to some extent. In this manner, the power generator can mitigate the risk of cash flow fluctuations.

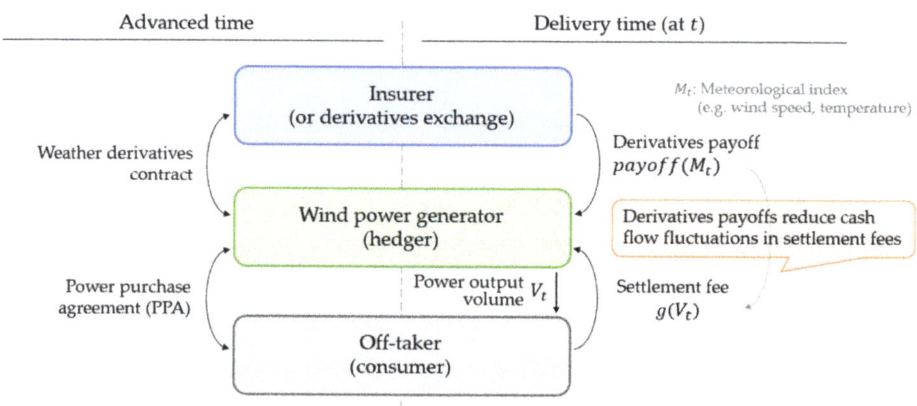

Figure 1. Overall picture of the transactions assumed in this study (from derivatives contracts to settlements).

In designing such weather derivatives, this study deals with two types of derivatives: non-parametric (made-to-order) and standardized derivatives. Figures 2 and 3 provide a conceptual comparison of both derivative schemes, focusing on the market trading models and payoff functions (with the problem to be solved), respectively. Non-parametric derivatives use a bilateral contractual form where each hedger seeks an individual optimal derivative payoff, and both the seller and buyer agree with the transaction. Meanwhile, in the derivative trading scheme of our interest, the seller (i.e., an insurance company) can offer derivatives with standardized payoff functions (i.e., monomials of the weather indexes as the payoff functions) to any buyer, whereas buyers may agree to purchase them and include them in their portfolio. The buyer should choose the types and trading volumes of derivatives based on their own generation patterns and their correlation with the weather indices used. This standardization will allow hedgers (buyers) to trade only those instruments that are necessary according to their own risk tolerance. In addition, unlike non-parametric derivatives where payoffs are determined on a made-to-order basis, standardized derivatives allow multiple players to trade individual products. Similarly, if a sufficient number of sellers sell standardized derivatives, the market principle will help in better and fairer price discovery, and simultaneously many players may participate; this can increase the liquidity of the market. Furthermore, this study also proposes a way to use LASSO regression for decision-making on how to select effective derivative products to be traded in the market (or which and how much products hedgers should trade from the available products) in such a standardized derivatives market trading scheme, as will be discussed in detail in Section 2.5.

Additionally, instead of setting the insurance company as a potential seller, a more efficient trading scheme can be assuming that baseload power generators (who sell baseload power generation in the wholesale market) are the sellers of derivatives (risk underwriters). This is because when the derivatives' payout is higher (i.e., when wind generation is lower), the baseload generator's revenue from the wholesale power price also tends to be higher because of the reduced power supply, thereby conveniently offsetting cash flows. (For example, Ketterer [33] showed logically and empirically that an increase (decrease) in the supply of wind power affects the decrease (increase) in electricity prices.) In this transaction scheme, an insurance company can serve as an intermediary rather than a simple seller. Indeed, studies have considered similar schemes [34] for electricity, temperature, and/or solar radiation derivatives. Here, we focus on standardized schemes that are highly applicable to such transactions and that enhance liquidity market trading.

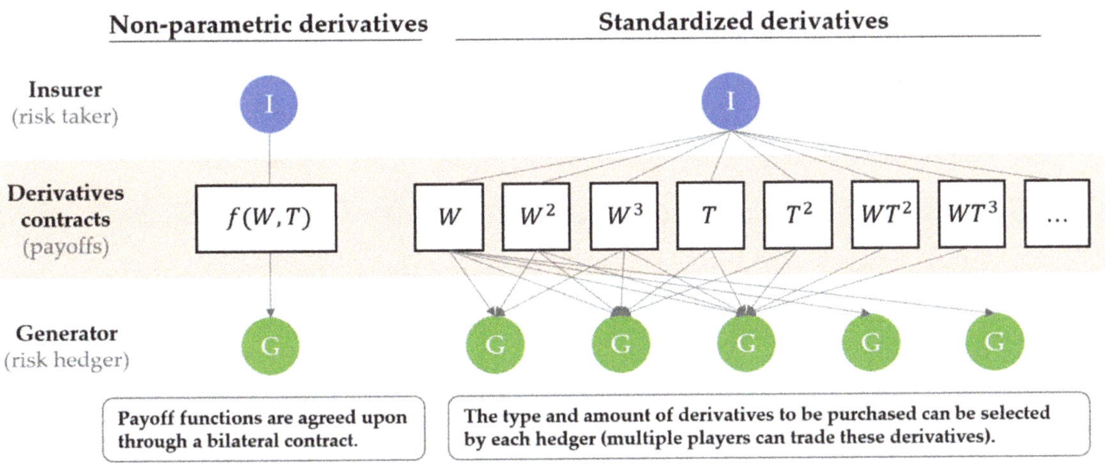

Figure 2. Market trading model for wind derivatives. Note: "W" represents wind speed index and "T" represents temperature index.

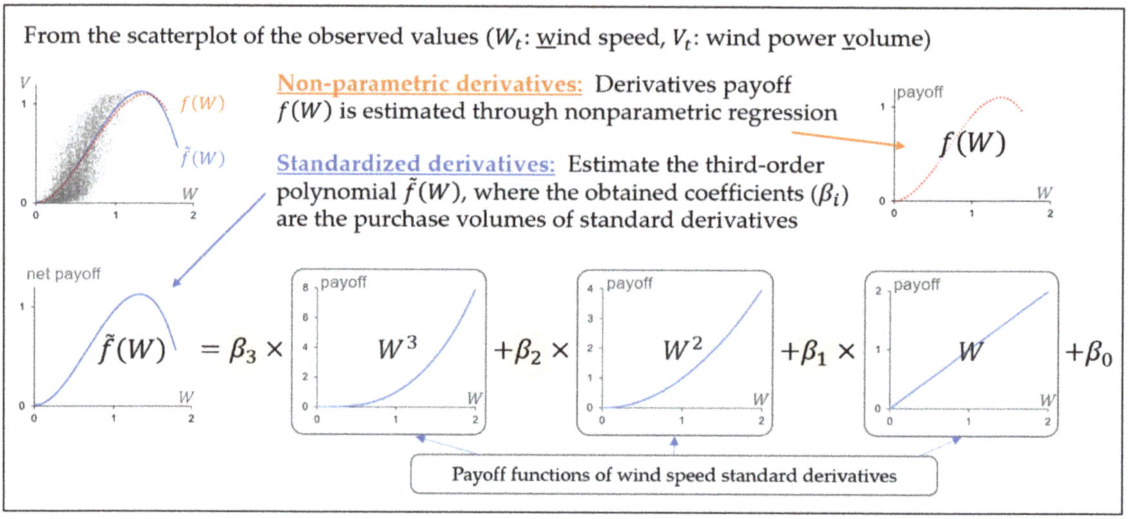

Figure 3. Conceptual diagram of methods for determining standardized (non-parametric) derivatives' trading volume (payoff functions). Note: this conceptual diagram is an example of using only wind speed derivatives.

Note non-parametric derivatives are based on smoothing splines whereas standardized derivatives are based on (normal) polynomials, as will be discussed in detail in the next section. The difference is that the former is expressed using a smoothing spline function consisting of highly flexible piecewise (local) polynomials and is estimated by a GAM, a type of non-parametric regression (more precisely, it is expressed as a linear sum of multiple basis functions; see, e.g., [19] for the specific formulas). On the other hand, the latter is represented by a limited number of monomial-type functions. In other words, the former derivatives with payoffs represented by myriad polynomials do not fit into standardization, and to distinguish them from monomial-type standardized derivatives, we refer to them as non-parametric derivatives for convenience.

2.2. Minimum Variance Hedging Problem

In this study, we assume that the hedged risk is only the volume of wind power generated, and we do not consider price risk. Hedging volume risk in electric business is a challenging issue because it is generally difficult to completely hedge [7], and this is why several recent studies, especially in the renewable energy business, have often addressed the hedging problem for volume risk only [5,14,17,35]. These studies suggest that even if price risk is excluded (i.e., even if price risk can be completely hedged), weather derivatives are still effective because the volume risk of renewable energy is heavily linked to weather-related indices. With this in mind, we particularly consider the problem of mitigating cash flow fluctuations derived from power generation variations using derivative payoffs on wind speed and temperature. For example, if the hedger (wind power producer that generates V_t at time t) has a fixed-rate and fixed-quantity supply obligation contract in advance, where any shortage or surplus from the supply obligation amount v is settled at a fixed unit price θ, the profit and loss function to be hedged (i.e., $g(V_t)$ in Figure 1) is $\theta(V_t - v)$. Note that this situation of assuming only volume risk corresponds to a power producer that has concluded a PPA with a general contract structure called "pay-as-produced," and PPAs have been rapidly increasing in recent years as a new alternative to the conventional FITs [36].

Then, the minimum variance hedging problem defined below can be modeled without loss of generality by simply setting the hedge target to the generation volume V_t. In this hedging problem (for a detailed exposition, see [34]), the hedger's primary aim is to minimize the variance of the net cash flows of the electricity sales revenue (wind power generation volume) V_t and the derivative payoff, $payoff(W_t,\ T_t)$, composed of wind speed index W_t and temperature index T_t at each point in time:

$$\text{minimize}:\ \text{Var}[V_t - payoff(W_t,\ T_t)] \tag{1}$$

where the term $payoff(W_t,\ T_t)$ refers to the payoff function for non-parametric derivatives or the net payoff of the derivative portfolio in our standardized derivative scheme. The latter corresponds to the problem of finding the "contract volume" of standardized derivatives (detailed in the next subsections). Note that the units of each term in Equation (1) are currencies (e.g., EUR), but in this study, which compares among hedging models (to be constructed in the next and subsequent sections), the conversion of units and scales is not considered because it does not affect the results of the analyses. For example, when the unit price of electricity generation is θ', the payoff function of either model estimated is simply multiplied by θ', and the same results (e.g., hedge effect, which will be discussed later) will be obtained. Note also that Equation (1) has a degree of freedom with respect to the payoff's location (first-order moment); therefore, it is possible to impose constraints on the mean value. For example, if a zero-mean constraint is imposed on the payoff itself (i.e., $\overline{payoff(W_t,\ T_t)} = 0$), it can be designed as a derivative with zero expected payoff without requiring an initial payment (or a premium) at contract time.

2.3. Non-Parametric Derivatives

In the case of the minimum variance hedging problem with non-parametric derivatives, the problem of optimizing the derivatives' "payoff function" $payoff(W_t,\ T_t) := f(W_t,\ T_t)$ in Equation (1) corresponds to estimating the following GAM:

$$V_t = f(W_t,\ T_t) + \varepsilon_t \tag{2}$$

where f is the two-dimensional tensor-product spline function [20] and ε_t is the residual with a mean of 0. Because the estimation of GAM involves minimizing the sum of squares of the residual ε_t under the smoothing constraint on f, it is synonymous with the variance minimization in Equation (1). Note that in this study, the GAM is estimated using the function gam() in the R package "mgcv" [20].

It is possible to design a "prediction error derivative" (i.e., a derivative with weather prediction errors as the underlying index) by removing the time-dependent trend from the underlying weather index (see, for example, [18]). However, we take an approach that does not remove the time-dependent trend from the underlying index, prioritizing the intuitive and straightforward cross-trend estimation of wind speed and temperature (e.g., the estimated payoff function shown in Section 3.1).

2.4. Standardized Derivatives

Next, in the case of minimum variance hedging problem with standardized derivatives, the problem is to optimize "contract/trading volume" of standardized derivatives based on the idea for approximating the payoff functions of non-parametric derivatives, $f(W_t,\ T_t)$, using simple multivariate polynomials denoted by $\widetilde{f}(W_t,\ T_t)$. We consider the following model:

$$V_t = \widetilde{f}(W_t,\ T_t) + \varepsilon_t, \quad \widetilde{f}(W_t,\ T_t) := \sum_{0\le i\le I,\ 0\le j\le J} \beta_{ij} {W_t}^i {T_t}^j \tag{3}$$

where ${W_t}^i {T_t}^j (i = 0, \ldots I;\ j = 0, \ldots, J)$ are the payoff functions for different standardized derivatives. By estimating Equation (3) using OLS, hedgers can estimate the optimal (in the context of minimum variance hedging) contract volume β_{ij} for each standardized derivative. Note that when $i, j = 0$ is a constant, the term can be regarded as a discount bond (with purchase unit β_{00}).

To consider the sufficient orders of derivatives, previous findings on wind power models are helpful (see Appendix A). Wind power output is generally modeled using a third-order formula for the wind speed [37]. In addition, wind turbines tend to reduce their power output when the temperature becomes too hot or too cold owing to a decrease in air pressure or freezing of components. This suggests an inverse U-shaped relationship with respect to temperature (see, e.g., https://www.windturbinestar.com/wind-turbine-annual-output.html, accessed on 15 June 2022). Based on the above considerations, this study adopts $I = 3$ and $J = 2$ in the hedging model in Equation (3). Note that the validity of assuming a nonlinear model was also verified by Ramsey's RESET test [38] using our empirical data (see Appendix B).

Note also that because the average of the observed values with respect to W_t and T_t is positive, simply using the payoffs of second- or higher-order derivatives (i.e., when i or j is greater than 2) as powers of the observed values may cause unnecessary large variations in ${W_t}^i$ or ${T_t}^j$. To avoid this problem, we design the initial (i.e., first-order) underlying indexes by subtracting the historical average from the observed value in advance; that is, $W_t := W_{obs,\,t} - \overline{W_{obs,\,t}}$, $T_t := T_{obs,\,t} - \overline{T_{obs,\,t}}$. This helps us in making derivatives more manageable and in increasing trading efficiency. The specific mechanism is discussed later (particularly in detail in Appendix D) using empirical data.

A recent study [26] has also shown that when designing hourly wind derivatives, wind speed derivatives at not only the same hour but also an hour before and after are effective. To apply this to our standardized derivative design, we set the wind speed index at each hour as a three-hour average value, including an hour before (lagged) and after (forward). That is, together with the above correction, the wind speed index used here is defined as follows: $W_t := (1/3)\sum_{\tau=t-1}^{\tau=t+1}\left(W_{obs,\,\tau} - \overline{W_{obs,\,\tau}}\right)$.

2.5. Hedge Trading Strategy Using LASSO Regression

The more incorporated the standardized derivatives are, the better replicated the nonlinear payoffs can be. However, using too many of them may reduce the out-of-sample period fit (i.e., hedge effect) due to overfitting. In addition, unnecessarily trading multiple derivatives may lead to trading inefficiencies in the form of increased product management costs or monitoring costs in the derivatives market. Similarly, reducing trading volume may be more desirable for efficient hedging strategies (see Appendix C for this discussion). Therefore, the strategy for estimating Equation (3) with the LASSO regression [39] may be more effective and efficient. It should be noted that, when dealing with conventional hedging problems, the standard OLS regressions were often applied for estimating hedging models since only a limited number of (existing) hedging instruments were mainly considered in the previous study. However, in this study, which introduces a market scheme for multiple derivatives that can be customized for a variety of hedgers, the use of LASSO regression should be particularly effective to ensure robustness and to improve trading efficiency (choice of trading instruments).

In LASSO regressions, the residual sum of squares with the L1 norm constraint on the regression coefficient vector is minimized. That is, each coefficient β is obtained as a solution to the following optimization problem using the regularization parameter λ:

$$\text{minimize}: \sum_{t=1}^{T}\left(V_t - \sum_{0\le i\le I,\,0\le j\le J}\beta_{ij}{W_t}^i{T_t}^j\right)^2 + \lambda\sum_{0\le i\le I,\,0\le j\le J}\left|\beta_{ij}\right| \tag{4}$$

Solving (4) allows us to reduce or eliminate less relevant variables [39]. That is, the LASSO regression will help sellers of derivatives (insurers or exchanges) identify products that are inherently in high need. On the other side for the buyers, it would be more efficient if they could select only the necessary (effective) derivatives while considering their risk tolerance and reduce the volume of each derivative. Note that when estimating the LASSO regression, it is desirable to standardize variables (i.e., "z-score" normalization) to provide equivalent penalties for variables with different units; this is also followed in this study. We estimate the LASSO regressions using the R package "glmnet," where standardization of variables before model estimation and rescaling of coefficients after estimation are automatically performed as default settings [40,41].

For the regularization parameter λ, some values may be specified in advance. However, the cross-validation (CV) approach is an effective rational selection method. The most common way is to select the λ that minimizes the mean CV error (λ_{min}); another method is to use the "one-standard error rule" (1SE rule: λ_{1se}). Hastie et al. [40] defined the 1SE rule as a method to choose the most parsimonious model (i.e., the largest λ), whose error is no more than one standard error above the error of the minimized error model, and suggested using the 1SE rule to select the best model. That is, the 1SE rule allows us to select the simplest model with the same accuracy as the best one [42]. Here, the adoption of this 1SE rule may help us in estimating a more efficient derivatives portfolio; that is, a portfolio with sufficient hedge effect while narrowing down the required number (or volume) of trading instruments as much as possible. Hence, our empirical analysis tests the effectiveness of trading strategies with LASSO regression by using both λ_{min} and λ_{1se}, as well as by varying λ continuously.

3. Results

Next, we present the results of estimating the payoff function and calculating the hedge effect. Here, we use the following observed values of wind power generation and weather in Denmark, where wind power is very prevalent.

- Wind power generation [MWh]: actual power generation in Eastern Denmark (DK2) (downloaded from https://www.nordpoolgroup.com/en/Market-data1/, accessed on 11 January 2021)
- Wind speed [m/s] and temperature [°C]: observed values at Copenhagen Airport (downloaded from http://rp5.ru/metar.php?metar=EKCH, accessed on 11 January 2021)

The estimation (in-sample) and validation periods (out-of-sample) are 2019 and 2020, respectively, using hourly data for all 24 h days from January 1 to December 31. The estimated trends and trading volumes (coefficients) were obtained from the in-sample period data only, whereas the hedge effects were calculated using the out-of-sample period data. Note that this study used data through 2020, which were available at the time the data were obtained, but as of the time of this paper's submission, the data, including the most recent period, were no longer available to the public on that site.

Considering that sufficient historical data are often not available in practical situations, we estimated the hedge models from only one-year of in-sample data and calculated the hedge effect for the following year; this also helps in checking the robustness of the proposed method. Specifically, the hedge effect uses the variance reduction ratio (VRR) as 1 − VRR. VRR is defined as follows:

$$\mathrm{VRR} := \mathrm{Var}[V_t - payoff(W_t,\ T_t)]/\mathrm{Var}[V_t] \tag{5}$$

3.1. Estimated Trend (Non-Parametric Derivatives)

The trend of the estimated non-parametric derivative (i.e., $f(W_t, T_t)$ in Equation (2)) is shown in Figure 4. As mentioned in Section 2.3, a cubic form in the wind speed direction and a near-quadratic form in the temperature direction were observed. For example, the sensitivity of wind power generation to wind speed was low when the wind speed was low, then increased as the wind speed increased, and finally decreased when the wind speed became too high. Meanwhile, the effect of temperature peaked at approximately 5–10 °C, and the power generation decreased at both higher and lower temperatures.

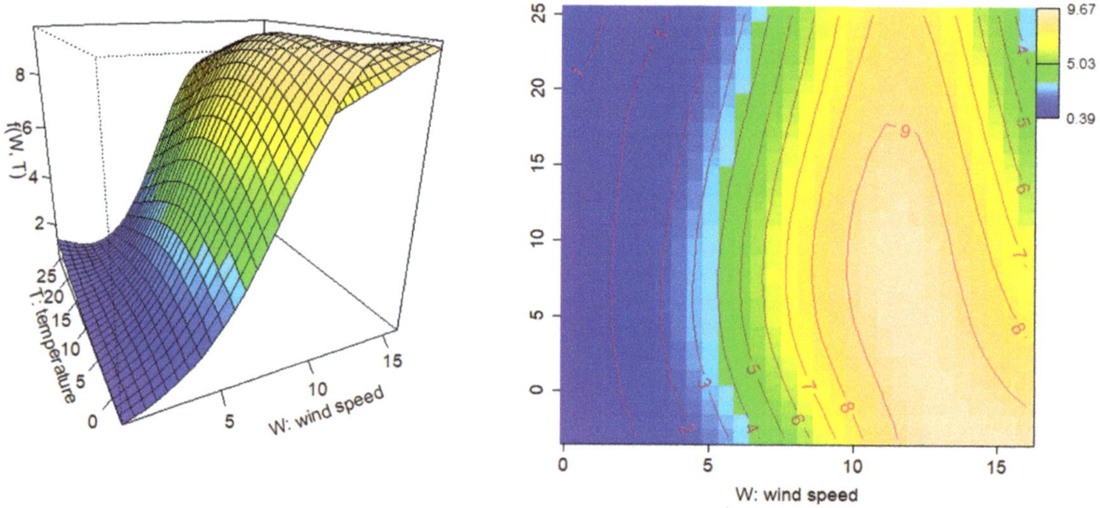

Figure 4. Estimated trends of non-parametric derivatives (left: perspective, right: contour).

3.2. Measurement of Hedge Effects

Figure 5 plots the cumulative hedge effect, and the hedging products included in each hedging model are shown in Table 1 (note that the operators "+" and "*" are not strictly defined but are used for convenience to correspond intuitively with the long regression equation). The blue (orange) dots indicate the hedge effect of standardized (non-parametric) derivatives. For reference, the light blue and light orange dots are plotted together, showing the case where wind speeds before (lag) and after (forward) one hour are not included in the underlying indexes, as described in Section 2.4. Evidently, including wind speeds at the before and after hours in the underlying index is effective for improving the hedge effects; almost 3% improvements are observed in all hedging models.

Next, for both non-parametric and standardized types, the hedge effects increasd as the number of derivative combinations increased (i.e., the number of non-linear degrees of freedom increased). The hedge effect of using only "wind speed forwards" (W1) was 65.9%, while including "wind speed cubic derivatives" (W3) improved the hedge effect to 68.1% (68.3% for wind speed non-parametric derivatives). For the wind-speed squared derivative (W2), no improvement was observed in the hedge effect when combined with wind speed forwards without wind speed cubic derivatives, indicating that both derivatives are meaningful only when used in combination. The effect of temperature standardized derivatives was similar. The hedge effect can be increased to 69.3% when the "temperature squared derivative" is included (69.8% when temperature non-parametric derivatives are used) and 70% when all "wind speed and temperature cross-derivatives" are included (70.5% when the two-dimensional non-parametric derivative of wind speed and temperature is used).

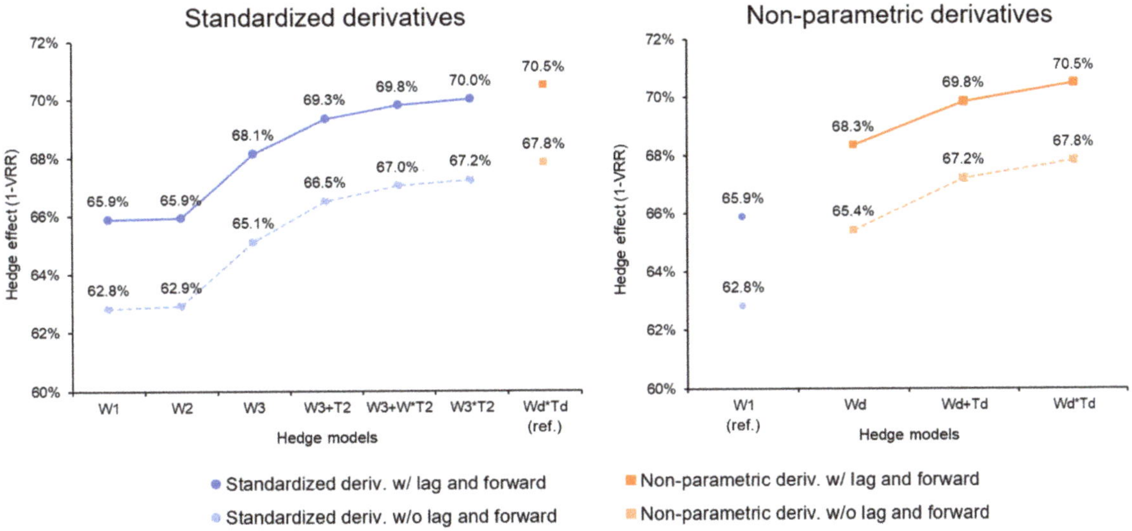

Figure 5. Hedge effects of standardized derivatives and non-parametric derivatives.

Table 1. Components of hedging instruments by hedging model.

		Standardized Derivatives Model						Non-Parametric Derivatives Model		
		W1	**W2**	**W3**	**W3+T2**	**W3+W*T2**	**W3*T2**	**Wd**	**Wd+Td**	**Wd*Td**
Standardized derivatives	W	✓	✓	✓	✓	✓	✓			
	W^2		✓	✓	✓	✓	✓			
	W^3			✓	✓	✓	✓			
	$T + T^2$				✓	✓	✓			
	$WT + WT^2$					✓	✓			
	$W^2T + W^2T^2 + W^3T + W^3T^2$						✓			
Non-parametric derivatives	$f_W(W)$							✓	✓	
	$f_T(T)$								✓	
	$f_{te}(W,\ T)$									✓

For confirmation, the actual wind power generation versus payoffs of the derivative portfolio in the W3*T2 model is plotted in Figure 6. Because the payoff of the derivatives may be thought of as the predicted wind power generation (assuming that wind speed and temperature were observed), Figure 6 can be viewed as a fitting plot considering Equation (3) as a forecast model. Notably, the fluctuations of the two were generally consistent, except when power generation was extremely high/low. The wind speed and temperature indices are based on data observed at only one location (Copenhagen Airport); the model's accuracy may improve by refining the indices.

The results of the above hedge effects are similar, even when viewed by month, as shown in Table 2. Notably, the effect of the "wind speed cubic derivative" improves the hedge effect in most months, and in some months, by more than 10% (e.g., 10.6% in March and 64.2% in August) compared to the case where only wind speed forward was used. While the hedge effect decreased in a few months, this decrease was limited.

The improvement in the hedge effect of "wind speed and temperature cross-derivatives" is particularly pronounced in the summer months. This may be due to the temperature variable successfully explaining the relationship between the continuous decrease in generation output (generation efficiency) with increasing temperature during the summer months, and hence the increased effectiveness of the temperature (cross) derivatives. In particular, August had the strongest non-linearity and interaction; the hedge effect improved by 114.6% when the temperature cross derivatives were included. Thus, the temperature-related derivatives do affect the reduction of wind turbine output (cash flow loss) due to high temperatures. Therefore, applying temperature derivatives or wind speed/temperature cross-derivatives especially for the summer months may provide useful hedges.

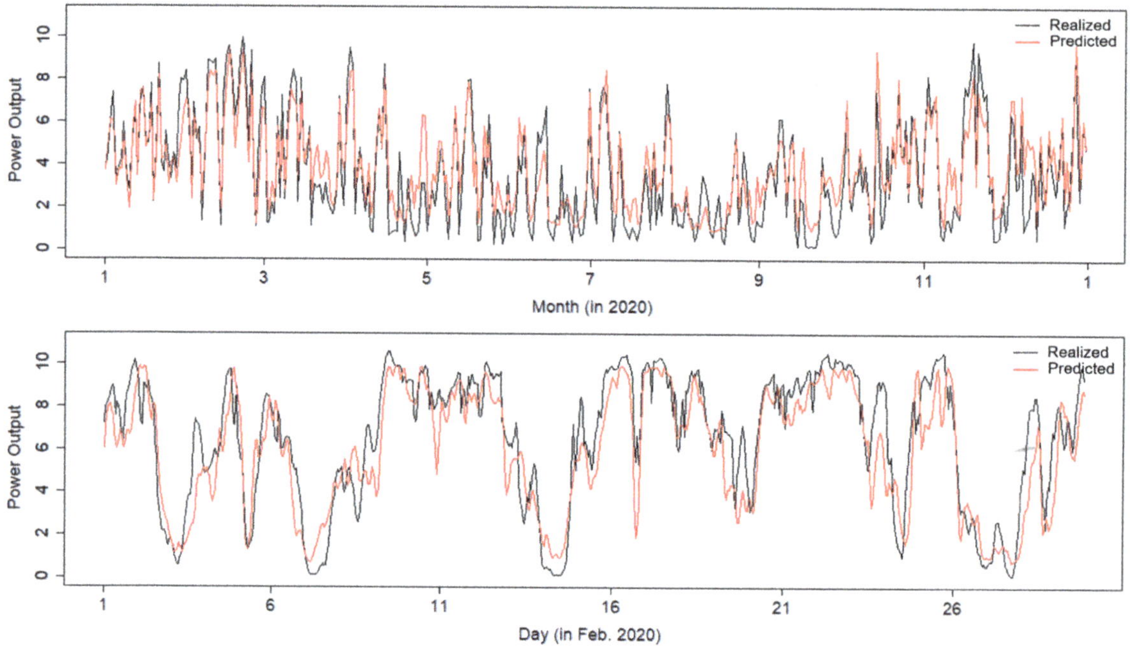

Figure 6. Realized versus predicted values of wind power generation in the out-of-sample period (Upper: 2020, Lower: February 2020). Note: the upper graph plots daily averaged values for visibility. The realized data are the raw wind production data downloaded from https://www.nordpoolgroup.com/en/Market-data1/ (accessed on 11 January 2021).

Table 2. Monthly hedge effect and improvement rate against the W1 model.

		Monthly Hedge Effect												All Period
		Jan	Feb	Mar	Apr	May	Jun	Jul	Aug	Sep	Oct	Nov	Dec	
Hedge effect (1-VRR)	W1	**68.6%**	76.1%	65.4%	59.1%	58.7%	40.9%	66.3%	15.4%	63.5%	62.8%	72.8%	**60.9%**	65.9%
	W3	67.5%	79.4%	**72.3%**	**59.9%**	59.5%	43.9%	69.1%	25.3%	68.8%	62.5%	75.9%	60.3%	68.1%
	W3*T2	68.6%	**80.7%**	71.5%	59.2%	**61.1%**	**46.9%**	**74.0%**	**33.0%**	**68.8%**	**63.8%**	**76.0%**	59.8%	**69.3%**
Improvement rate (to "W1")	W3	−1.6%	4.4%	**10.6%**	**1.4%**	1.4%	7.2%	4.3%	64.2%	8.3%	−0.4%	4.2%	**−1.1%**	3.4%
	W3*T2	**0.0%**	**6.1%**	9.3%	0.2%	**4.0%**	**14.7%**	**11.6%**	**114.6%**	**8.4%**	**1.7%**	**4.4%**	−1.8%	**5.2%**

Note: improvement rate is the ratio of the increase in the hedge effect of each hedging model relative to that of the W1 model; the largest value among the models in each month for both hedge effect and improvement rate is shown in bold.

3.3. *Trading Efficiency Using LASSO Regression*

Finally, we tested the effectiveness of the "trading product selection" strategy of the LASSO regression described in Section 2.3. This strategy seems particularly effective when one intends to trade multiple standardized derivatives. Therefore, here, we take the "W3*T2" model (with an original hedge effect of 70.0%) as an example and estimate this model with a LASSO regression instead of the normal OLS. The regularization parameter λ in the LASSO regression is treated in two ways, λ_{min} and λ_{1se}, as described in Section 2.5.

The contract/trading volumes (i.e., absolute values of the coefficients of the explanatory variables) for each derivative estimated by this method, including other hedging models using standardized derivatives, are shown in Table 3. The hedge effects shown in Figure 5 are also described at the bottom of the table. First, comparing the W3*T2 model with the models estimated using the LASSO regression, the hedge effect decreased slightly (by one percentage point) to 69.3% when using Lasso-λ_{1se} and remained virtually unchanged from the original 70.0% when using Lasso-λ_{min}. However, there are notable differences in the product selection and volumes of trading, as discussed below.

First, regarding the product selection, we see that the trading volume of W^3T was zero when Lasso-λ_{1se} is used. This is because the L1 regularization term in LASSO, explained in Section 2.5, reduced the variable's coefficient (trading volume of the instrument) with the lowest contribution to 0. Note that increasing λ just a bit more makes it possible to further reduce the volume of multiple instruments traded to zero, which is discussed in detail later with a figure at the very end of this section.

Table 3. Derivatives contract volume by hedging model, including LASSO regression-based trading strategies.

	W1	W2	W3	W3+T2	W3+W*T2	W3*T2	lasso.λ_{min}	lasso.λ_{1se}	Stdev of Payoffs
Contract volumes of each derivative (absolute values of coefficients)									
W	0.98465	0.97895	1.13905	1.11814	1.16114	1.18486	1.18426	1.05063	2.3
I(W^2)	-	0.00378	0.06930	0.06541	0.06286	0.05867	0.05820	0.03367	8.3
I(W^3)	-	-	0.01406	0.01369	0.01376	0.01460	0.01455	0.00728	62.3
T	-	-	-	0.03324	0.03283	0.04237	0.04244	0.03022	6.4
I(T^2)	-	-	-	0.00569	0.00613	0.00694	0.00697	0.00579	46.3
I(W * T)	-	-	-	-	0.00355	0.00246	0.00229	0.00080	13.8
I(W * T^2)	-	-	-	-	0.00113	0.00180	0.00177	0.00006	128.2
I(W^2 * T)	-	-	-	-	-	0.00258	0.00262	0.00051	56.3
I(W^2 * T^2)	-	-	-	-	-	0.00016	0.00017	0.00011	428.9
I(W^3 * T)	-	-	-	-	-	0.00015	0.00017	.	361.8
I(W^3 * T^2)	-	-	-	-	-	0.00004	0.00004	0.00003	2651.4
Contract volumes of derivatives portfolio									
Simple sum	0.98465	0.98273	1.22241	1.23617	1.28140	1.31463	1.31348	1.12909	
Weighted sum	2.22805	2.24632	4.02585	4.40015	4.69229	5.31285	5.30554	3.74113	
Hedge effects of derivatives portfolio									
1-VRR	0.65892	0.65929	0.68146	0.69343	0.69797	0.70006	0.70009	0.69348	

In assessing trading volume, we examine not only the "simple sum" of the total trading volume of the derivatives portfolio but also a "weighted sum" weighted by the standard deviation of the payoffs of the individual derivatives. Note that the "weighted sum" is a useful measure of overall trading inefficiency, considering that hedge trading-related costs, such as the risk premium that a seller may demand, can depend on the instrument's volatility (see Appendix C for more details). In both cases, the total volumes of the derivatives portfolio with LASSO are significantly smaller than those with regular OLS. For instance, the weighted sum using Lasso-λ_{1se} decreased from 5.31 to 3.74, a reduction of approximately 30%.

To verify the effectiveness of the LASSO regression to control trading volume, including the other hedging models (i.e., from W1 to W3+W*T2 models), we plot the relationship between contract volume and the hedge effect using the hedging model in Figure 7; here, the weighted sum in Table 3 is used as the trading volume on the vertical axis. First, as shown by the blue line, the trading volume increased with each additional product type, and simultaneously the hedge effect also increased. This result is consistent with practical intuition, as it suggests a trade-off exists between the hedge effect and its cost. However, the results of the trading strategies in the LASSO regression are more interesting. In Figure 7, they are drawn as dashed curves extending from the plot of each hedging model (i.e., the blue points where the trading volume is estimated by OLS, corresponding to $\lambda = 0$) to the lower left as λ gradually increases. Notably, the trading strategy estimated by LASSO for the W3*T2 model is generally below the line drawn by other models. That is, the strategy of using LASSO to estimate the trading volume of all derivatives (including higher-order derivatives) creates a "Pareto optimal" portfolio. Furthermore, the "Pareto optimal" frontier extends downward as the number of product types increases. This suggests that the availability of multiple instruments (i.e., trading multiple hedge products) not only increases the hedge effect but also contributes to reducing the hedger's trading volume. For example, a portfolio with the W3 model (using wind speed derivatives only) would be an effective solution to increase the hedge effect as much as possible in a situation where no temperature-related derivatives were available. However, if temperature derivatives can be traded, one can construct a more efficient mixed portfolio in which only the trading volume can be noticeably reduced with the same hedge effect.

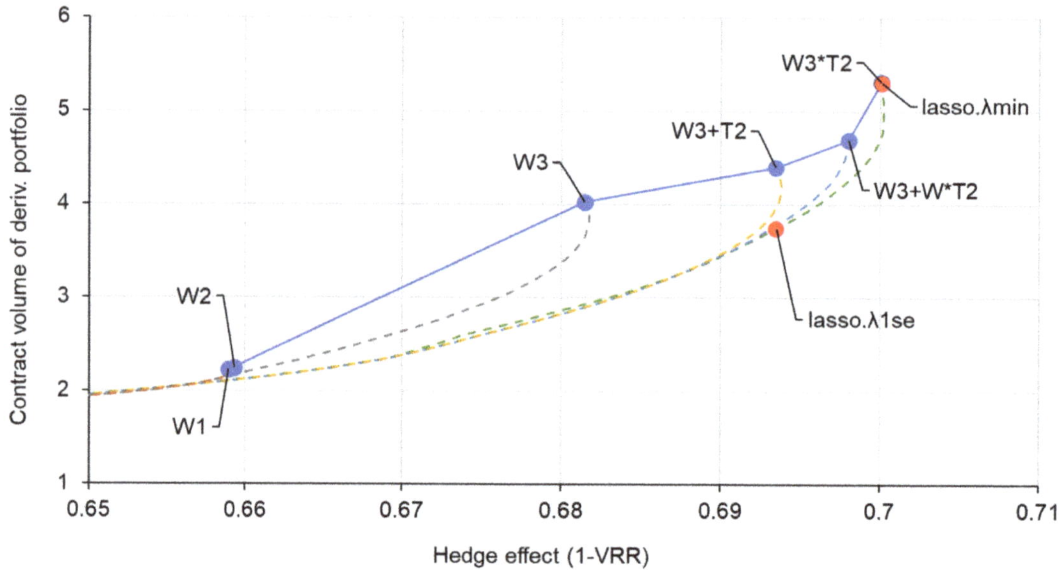

Figure 7. Relationship between contract volume and hedge effect by hedging model. Note: the term "lasso" refers to the W3*T2 model estimated by LASSO regression. Hedge effect is measured in out-of-sample periods. Note: the blue dots represent the estimation results from OLS and the dashed curves represent the estimation results obtained by varying λ in the LASSO regression.

Note that the above analysis results are for the case where the hedging model is estimated using one year of in-sample data for 2019, as stated at the beginning of Section 3. Here, in order to verify the robustness of the results, as reference scenarios, we also calculated the results when using (A) two years of data for 2018–2019 and (B) half a year of data for July-December 2019, as the in-sample period. The pairs of {hedge effect, contract volume (weighted sum)} for the Lasso-λ_{1se} strategy were calculated to be {0.692, 3.819} for Case (A) and {0.682, 3.528} for Case (B). In both cases, the hedge effects were smaller than that of the benchmark case using one year of in-sample data, i.e., 0.693. Even when viewed in a two-dimensional plot that takes into account the trading volume, points A and B are inside the efficient frontier shown in Figure 7. The reason for this may be that taking too long of a learning period has the disadvantage of increasing the impact of changes in generation patterns due to renewal or expansion of generation facilities, while taking too short of a learning period makes robust estimation difficult. In general, for one-year ahead hedging, the use of the most recent one-year in-sample period, as adopted for the benchmark case, is considered appropriate from the perspective of ensuring a hedge effect.

For reference, Figure 8 shows the path diagram of the coefficients when the W3*T2 model is estimated by the LASSO regression with respect to the regularization parameter λ. Notably, increasing λ generally reduces the absolute value of each coefficient (i.e., the volume of each derivative contract). Further, each coefficient shrinks to zero as λ increases owing to the L1-norm penalty of the LASSO regression. If large transaction costs arise depending on the types of products traded (e.g., monitoring costs of price changes by product), then the decision of increasing λ can be made with reference to this path diagram to reduce the number of product types. For example, from the red vertical line in Figure 7, which represents λ_{1se}, some product types can be excluded from the portfolio by adjusting λ to slightly larger values. In any case, if λ is set too large, the coefficients (derivative volume) converge to zero and the hedge effect shrinks. Overall, to construct an efficient portfolio without impairing the hedge effect, the use of the 1SE rule λ is well balanced, as demonstrated above, and may provide an excellent criterion in the hedger's decision-making process for the trading strategy.

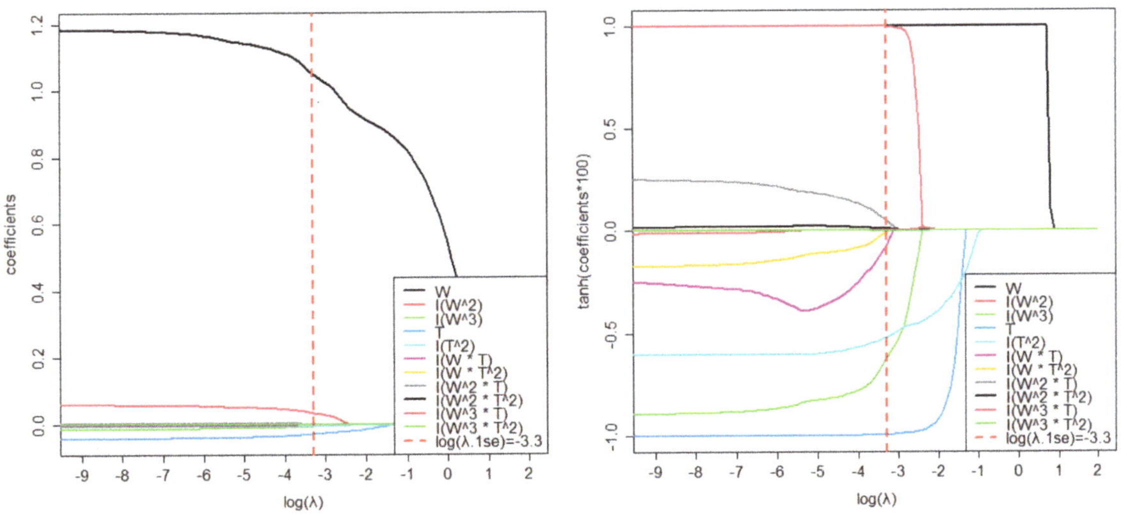

Figure 8. Relationship between the regularization parameter λ and the coefficients in LASSO regression. The graph on the right transforms the scale of the left graph using the hyperbolic tangent function ($tanh(100x)$) for ease of viewing.

4. Conclusions

This study demonstrated the effectiveness of hedging strategies using portfolios of high-order standardized weather derivatives applied to the wind power generation business. Our main contributions are as follows. First, we proposed new high-order monomial-type standardized derivatives, such as wind speed cubic derivatives and wind speed/temperature cross-derivatives, and demonstrated their effectiveness. Second, while developing a specific market trading scheme, we also showed the practical benefits of standardizing weather derivatives. Third, we proposed a "trading product selection" strategy using LASSO regression and demonstrated its effectiveness in terms of both increasing the efficiency of hedge trading and ensuring the hedge effect. Finally, we visualized the latent mixed trend in wind power generation relative to temperature and wind speed, and we noted the implications in the context of wind power forecasting models.

Essentially, we have shown that the hedge effect can be enhanced by incorporating various underlying hedge instruments and increasing the non-linearity of the portfolio. In particular, our new implication is that constructing a standardized derivatives portfolio is significant not only in hedge effect but also in hedge efficiency. By providing a variety of standardized hedging instruments, insurance companies (or exchanges) can give hedgers more flexible options for building their derivative portfolios. Estimating hedging models using LASSO regressions can help insurers/exchanges make decisions to identify efficient instruments. In addition, if hedgers adopt a trading strategy of LASSO regression, they can find and construct more Pareto-optimal portfolios, which help deal with the trade-off of reducing trading volume (transaction costs) and improving the hedge effect. Importantly, this mechanism may have generalizable implications for risk management practices in other businesses.

The primary idea provided by this study is in the original market scheme using new standardized derivatives for granular risk hedging, which would be increasingly needed in future decentralized electricity markets. Although relatively simple statistical methods such as OLS and LASSO regression were used for the methodology, our approach should have an advantage from the standpoint of practicality and applicability to the real world, i.e., a market scheme on methodologies that are as simple and straightforward as possible are superior. As a first step to examine the potential use of new higher-order standard derivatives, this study deals only with volume risk, but it would be interesting to examine the possibility of simultaneous hedging with price risk, which should be addressed in the future. In addition, the perspective of this research can be versatile, for example, in building a derivatives market for the distributed solar power generation business. The research questions for such applications may include verifying the effectiveness at daily granularity and further refining the hedging model by taking seasonality into account. Furthermore, from the perspective of further reducing transaction costs, methods other than LASSO regression could be used. LASSO regression has the advantage of reducing the number of hedge products due to its ability to shrink the coefficients of variables to zero, but if the costs associated with transaction volume (e.g., risk premiums) are more pronounced than the costs associated with the number of products (e.g., monitoring costs), other regularized regression methods, such as Ridge regression, might be superior. Further investigation of such questions may also be a subject for future work.

Derivatives with finer granularity and more variety, such as those addressed here, may have more demand once infrastructure, such as peer-to-peer power trading platforms, is in place. For instance, a fully digital trading platform will most likely reduce transaction costs. Moreover, a platform based on promising blockchain technology can enable immediate settlement or payment. (For example, the EnergyWeb blockchain has been developed for the energy sector and uses the "Proof-of-Authority" consensus algorithm. This blockchain has transaction rates hundreds to thousands of times higher than the "Proof-of-Work" conventional algorithm used by Bitcoin and other crypto applications, and it allows for instantaneous settlement [43].) Granular derivatives can be particularly useful for small players who wish to minimize fluctuations in their digital wallets. In any case, with the diversification of power resources, the increasing variety of hedging needs is an emerging trend. Then, naturally, hedging products should also be diversified. Given such future changes, the market design concept of diversification of hedging instruments and efficient trading scheme, as presented in this study, could be a potential basic idea to realize the decentralized power trading that will be needed in the future. Design methodologies, standardization schemes, and trading strategies for refined weather derivatives, such as those proposed here, may have useful implications in the electricity market, where transaction granularity is becoming increasingly finer in space and time.

Author Contributions: Conceptualization, T.M. and Y.Y.; methodology, T.M. and Y.Y.; software, T.M.; validation, T.M.; data curation, T.M.; writing—original draft preparation, T.M.; writing—review and editing, Y.Y.; visualization, T.M.; supervision, Y.Y.; project administration, Y.Y.; funding acquisition, T.M. and Y.Y. All authors have read and agreed to the published version of the manuscript.

Funding: Grant-in-Aid for Scientific Research (A) 20H00285, Grant-in-Aid for Challenging Research (Exploratory) 19K22024, and Grant-in-Aid for Young Scientists 21K14374 from the Japan Society for the Promotion of Science (JSPS).

Institutional Review Board Statement: Not applicable.

Informed Consent Statement: Not applicable.

Data Availability Statement: All data used in this study were obtained from the following websites: https://www.nordpoolgroup.com/en/Market-data1 (accessed on 11 January 2021); http://rp5.ru/metar.php?metar=EKCH (accessed on 11 January 2021). Note that the publication of the data on the former website has been terminated as of the time of submission of this paper, but the data for the period covered in this study are available from the first author.

Acknowledgments: This paper is an extension of the one presented at ICEER 2022, and their proceedings are to appear in "Energy Reports" (see Matsumoto and Yamada (2023)). This work was supported by Grant-in-Aid for Scientific Research (A) 20H00285, Grant-in-Aid for Challenging Research (Exploratory) 19K22024, and Grant-in-Aid for Young Scientists 21K14374 from the Japan Society for the Promotion of Science (JSPS).

Conflicts of Interest: The authors declare no conflict of interest.

Appendix A. Research on Wind Power Output Forecasting

To develop an elaborate hedging model for wind power projects, we briefly reviewed studies on wind power forecasting. Broadly, wind power forecasting methods can be classified into time-series and exogenous variable models [44]. In recent years, the latter have emerged due to advances in computing technology and artificial intelligence, including several models that use machine learning (ML) methods such as artificial neural networks and deep learning [45]. Wind power generation is strongly correlated with wind speed, where intensity non-linearities exist. The relationship between wind turbine and wind speed has been analyzed in detail, focusing on technical characteristics; for example, in [46–48]. In general, the wind power curve is often described in the form of a cube of wind speed [37]. Marčiukaitis et al. [49] also compared and validated various models of non-linear regression with wind speed as an explanatory variable. Temperature is the next most commonly used explanatory variable after wind speed [45]. Some studies have demonstrated the effect of using temperature to improve accuracy using ML methods [50–52]; however, these studies do not clarify the structural effects of temperature and wind speed (including their interaction) on wind-power generation. Therefore, we intuitively illustrate the interaction between temperature and wind speed on wind power generation using a three-dimensional plot (Figure 4). This can provide useful insights in the context of wind power forecasting models.

Appendix B. Verification of Non-Linearity by Ramsey's RESET Test

This study, which proposes the concept of trading schemes and strategies for high order standard derivatives, does not enter into the problem of optimizing the order of polynomial equations (3). That is, it is not the purpose of this study to determine whether third or second order is optimal as a power of wind speed or temperature. However, to verify the assumptions of the nonlinear models, Ramsey's RESET test [38], a common test for linear regression, was performed on several models. In this test, the order of diagnostic variables for the linear function (null hypothesis) can be specified (see, for example, [53] as an empirical study), and the test was performed using the R package "lmtest" [54]. The results suggest the presence of strong nonlinearities, as shown in Table A1. In other words, it is worthwhile to use derivatives with higher-order monomial payoffs in combination with wind and temperature futures that have linear payoffs.

Table A1. The result of Ramsey's RESET test.

Model	Diagnostic Variables	F-Statistic	p-Value
V ~ W	squares, cubes	264.97	$< 2.2 \times 10^{16}$
V ~ T	squares	305.21	$< 2.2 \times 10^{16}$
V ~ W + T	squares, cubes	294.66	$< 2.2 \times 10^{16}$

Note: this table is calculated using the function resettest() of the R package "lmtest" for the in-sample period data for Denmark presented in Section 3. "Model" is the "formula" of the original linear model and "Diagnostic variables" are the "power" of the fitted values of the linear model [54]. As the table shows, the p-values are essentially zero, meaning that the linear models of the null hypothesis are strongly rejected. For more information on this test, see, for example, "Chapter 9" in [55].

Appendix C. Hedge Trading Related Costs That Can Depend on Trading Volume

In general, an increase in hedging volume may result in an additional cost for a hedger from the following reasons. First, the bid/ask spread (BAS) is often used as a proxy for measuring the direct cost of transactions for market instruments [56]. Since a hedger must pay a BAS-equivalent transaction cost per trading (see, e.g., [57,58]), an increase in the trading volume leads to a rise in such a direct transaction cost. Another potential cost (which, in some cases, may be greater than the BAS) is the "risk premium" that the seller of a derivative contract adds to the selling price. The price of weather derivatives or derivatives (futures) in electricity markets is often formulated as the expected price of the underlying asset plus a risk premium (see, for example, [59] for weather derivatives and [60] for electricity futures), and the risk premium is interpreted as compensation for the seller's risk burden (see, for example, [61,62]). That is, for the buyer, it corresponds to the cost of lowering his exposure [63]. Thus, curtailing the trading volume reduces this invisible cost corresponding to the risk premium.

In addition, both the BAS and the risk premium tend to increase as the volatility of the asset increases. For example, it has been demonstrated that there is a positive correlation between volatility and BAS [64,65] and volatility and risk premium [66] in various futures markets. In other words, the values calculated in Section 3.3, where the trading volume is weighted by the standard deviation of the payoffs of each instrument, have a certain rationality as an indicator of hedge trading related costs.

Appendix D. The Case of Using Derivatives with No Average Value Correction for the Underlying Index

In Section 2.4, we noted that when designing higher-order standardized derivatives, it is desirable to subtract the mean from the first-order underlying (we refer to this case as a "corrected case"); here, we show the reason for this. First, if such a correction is not made (an "uncorrected case"), the volatility of higher-order derivatives will be unnecessarily large, which will worsen the efficiency of the transaction. Figure 7 in Section 3.3, shows the relationship between trading volume ("weighted sum" considering standard deviation) and hedge effect for the "corrected" derivatives; the corresponding plot for the "uncorrected" case is shown in Figure A1 below. While the absence of the correction has little effect on the hedge effects as they are perfectly consistent in the ordinary least squares (OLS) estimation case, the standard deviation-weighted trading volumes (trading costs) are several times larger than those shown in Figure 7. Although the least absolute shrinkage and selection operator (LASSO) regression strategy is effective in reducing trading volume in this uncorrected case as well (the volume is reduced to approximately 1/3 using the 1SE rule), the volume can be reduced much more when using corrected derivatives. Specifically, even with the 1SE rule, trading volume decreased to 3.74 in the corrected case, while it only decreased to 7.68 in the uncorrected case.

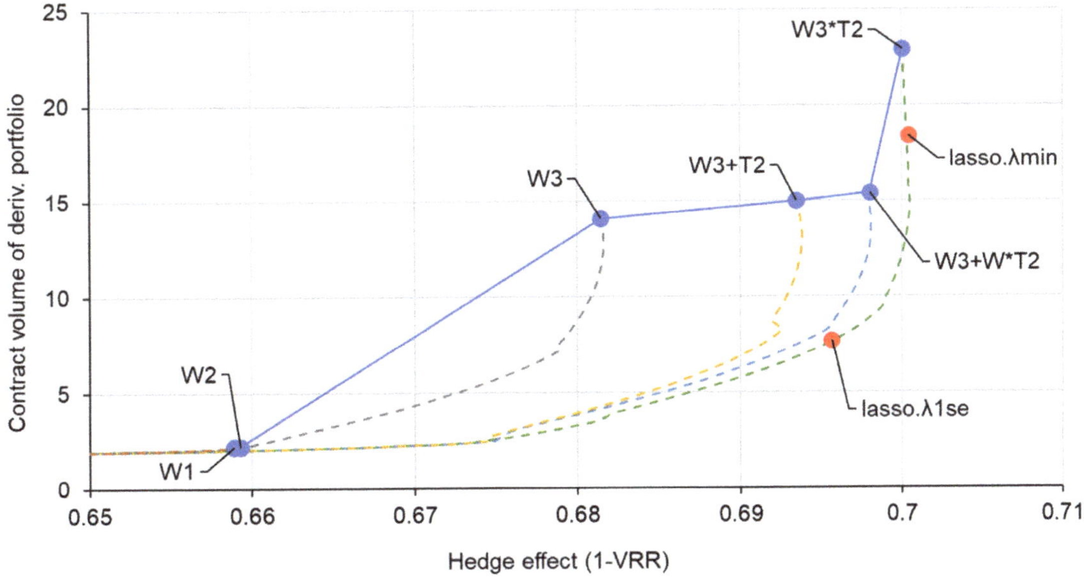

Figure A1. Relationship between contract volume and hedge effect by hedging model (cases in which no mean value correction was made for the underlying indexes). Note: the term "lasso" refers to the W3*T2 model estimated by LASSO regression. Hedge effect is measured in out-of-sample periods. Note: the blue dots represent the estimation results from OLS and the dashed curves represent the estimation results obtained by varying λ in the LASSO regression.

The path diagram of the estimated coefficients (derivative volume) from the LASSO regression for the uncorrected case is also shown in Figure A2 (for comparison, see Figure 8 for the corrected case). Note that the coefficients are not stable with respect to the moving λ. For example, the hedger's position in wind futures has changed significantly from a sell position to zero position and then to a buy position as λ increases. Further, the paths of the other coefficients are also generally not smooth. This is irrational from the perspective of efficient decision making in trading practice. In addition, assuming, for example, that this derivative supported continuous trading, such an unstable volume path can lead to unnecessary rebalancing of trading positions. This is not desirable from a transaction-cost perspective.

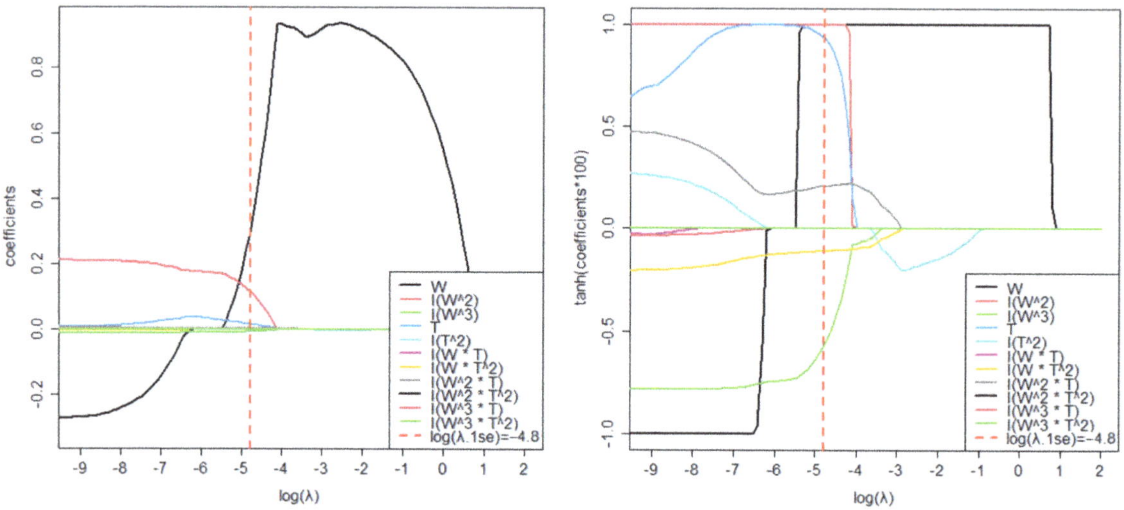

Figure A2. Relationship between the regularization parameter λ and the coefficients in LASSO regression (i.e., cases with no mean value correction for the underlying indexes). The graph on the right transforms the scale of the left graph using the hyperbolic tangent function ($tanh(100x)$) for ease of viewing.

References

1. IEA. Renewables 2021 Analysis and Forecast to 2026. Available online: https://www.iea.org/reports/renewables-2021 (accessed on 15 June 2022).
2. Market Research Future. Small Wind Power Market. Available online: https://www.marketresearchfuture.com/reports/small-wind-power-market-4568 (accessed on 15 June 2022).
3. Botterud, A.; Wang, J.; Bessa, R.; Keko, H.; Miranda, V. Risk management and optimal bidding for a wind power producer. In *IEEE PES General Meeting*; IEEE: Minneapolis, MN, USA, 2010; pp. 1–8. [CrossRef]
4. Bathurst, G.; Weatherill, J.; Strbac, G. Trading wind generation in short term energy markets. *IEEE Trans. Power Syst.* **2002**, *17*, 782–789. [CrossRef]
5. Yamada, Y. Optimal Hedging of Prediction Errors Using Prediction Errors. *Asia-Pacific Financ. Mark.* **2008**, *15*, 67–95. [CrossRef]
6. Eydeland, A.; Wolyniec, K. *Energy and Power Risk Management: New Developments in Modeling, Pricing, and Hedging*; John Wiley & Sons: Hoboken, NJ, USA, 2002; Volume 97.
7. Deng, S.; Oren, S. Electricity derivatives and risk management. *Energy* **2006**, *31*, 940–953. [CrossRef]
8. Benth, F.E.; Pircalabu, A. A non-Gaussian Ornstein–Uhlenbeck model for pricing wind power futures. *Appl. Math. Financ.* **2018**, *25*, 36–65. [CrossRef]
9. Gersema, G.; Wozabal, D. An equilibrium pricing model for wind power futures. *Energy Econ.* **2017**, *65*, 64–74. [CrossRef]
10. Benth, F.E.; Di Persio, L.; Lavagnini, S. Stochastic Modeling of Wind Derivatives in Energy Markets. *Risks* **2018**, *6*, 56. [CrossRef]
11. Rodríguez, Y.; Pérez-Uribe, M.; Contreras, J. Wind Put Barrier Options Pricing Based on the Nordix Index. *Energies* **2021**, *14*, 1177. [CrossRef]
12. Kanamura, T.; Homann, L.; Prokopczuk, M. Pricing analysis of wind power derivatives for renewable energy risk management. *Appl. Energy* **2021**, *304*, 117827. [CrossRef]
13. Christensen, T.S.; Pircalabu, A. On the spatial hedging effectiveness of German wind power futures for wind power generators. *J. Energy Mark.* **2018**, *11*, 71–96. [CrossRef]
14. Masala, G.; Micocci, M.; Rizk, A. Hedging Wind Power Risk Exposure through Weather Derivatives. *Energies* **2022**, *15*, 1343. [CrossRef]
15. Hastie, T.; Tibshirani, R. *Generalized Additive Models*; Chapman & Hall: Boca Raton, FL, USA, 1990.
16. Yamada, Y. Valuation and hedging of weather derivatives on monthly average temperature. *J. Risk* **2007**, *10*, 101–125. [CrossRef]
17. Matsumoto, T.; Yamada, Y. Cross Hedging Using Prediction Error Weather Derivatives for Loss of Solar Output Prediction Errors in Electricity Market. *Asia-Pacific Financ. Mark.* **2018**, *26*, 211–227. [CrossRef]
18. Matsumoto, T.; Yamada, Y. Simultaneous hedging strategy for price and volume risks in electricity businesses using energy and weather derivatives. *Energy Econ.* **2021**, *95*, 105101. [CrossRef]
19. Matsumoto, T.; Yamada, Y. Customized yet Standardized Temperature Derivatives: A Non-Parametric Approach with Suitable Basis Selection for Ensuring Robustness. *Energies* **2021**, *14*, 3351. [CrossRef]

20. Wood, S.N. *Generalized Additive Models: An Introduction with R*; Chapman and Hall: New York, NY, USA, 2017.
21. Oum, Y.; Oren, S.; Deng, S. Hedging quantity risks with standard power options in a competitive wholesale electricity market. *Nav. Res. Logist.* **2006**, *53*, 697–712. [CrossRef]
22. Oum, Y.; Oren, S.S. Optimal Static Hedging of Volumetric Risk in a Competitive Wholesale Electricity Market. *Decis. Anal.* **2010**, *7*, 107–122. [CrossRef]
23. Brik, R.I.; Roncoroni, A. Static mitigation of volumetric risk. *J. Energy Mark.* **2016**, *9*, 111–150. [CrossRef]
24. Carr, P.; Madan, D. Optimal positioning in derivative securities. *Quant. Financ.* **2001**, *1*, 19–37. [CrossRef]
25. Matsumoto, T.; Yamada, Y. Multivariate Weather Derivatives for Wind Power Risk Management: Standardization Scheme and Trading Strategy. *Energy Rep.* **2023**, *in press*.
26. Yamada, Y.; Matsumoto, T. *Construction of Mixed Derivatives Strategy for Wind Power Producers*; University of Tsukuba: Tokyo, Japan, 2023; *manuscript in preparation*.
27. Ederington, L.H. The hedging performance of the new futures markets. *J. Financ.* **1979**, *34*, 157–170. [CrossRef]
28. Zainudin, A.D.; Mohamad, A. Cross hedging with stock index futures. *Q. Rev. Econ. Finance* **2021**, *82*, 128–144. [CrossRef]
29. Ong, T.S.; Tan, W.F.; Teh, B.H. Hedging effectiveness of crude palm oil futures market in Malaysia. *World Appl. Sci. J.* **2012**, *19*, 556–565.
30. Woo, C.-K.; Horowitz, I.; Hoang, K. Cross hedging and forward-contract pricing of electricity. *Energy Econ.* **2001**, *23*, 1–15. [CrossRef]
31. Sant'Anna, L.R.; Caldeira, J.F.; Filomena, T.P. Lasso-based index tracking and statistical arbitrage long-short strategies. *N. Am. J. Econ. Finance* **2020**, *51*, 101055. [CrossRef]
32. Chen, Q.-A.; Hu, Q.; Yang, H.; Qi, K. A kind of new time-weighted nonnegative lasso index-tracking model and its application. *N. Am. J. Econ. Financ.* **2022**, *59*, 101603. [CrossRef]
33. Ketterer, J.C. The impact of wind power generation on the electricity price in Germany. *Energy Econ.* **2014**, *44*, 270–280. [CrossRef]
34. Yamada, Y.; Matsumoto, T. Going for Derivatives or Forwards? Minimizing Cashflow Fluctuations of Electricity Transactions on Power Markets. *Energies* **2021**, *14*, 7311. [CrossRef]
35. Mosquera-López, S.; Uribe, J.M. Pricing the risk due to weather conditions in small variable renewable energy projects. *Appl. Energy* **2022**, *322*, 119476. [CrossRef]
36. RE-Source. Risk Mitigation for Corporate Renewable PPAs. 2017. Available online: https://resource-platform.eu/wp-content/uploads/files/statements/RE-Source%203.pdf (accessed on 3 February 2022).
37. Ko, W.; Hur, D.; Park, J.-K. Correction of wind power forecasting by considering wind speed forecast error. *J. Int. Counc. Electr. Eng.* **2015**, *5*, 47–50. [CrossRef]
38. Ramsey, J.B. Tests for Specification Errors in Classical Linear Least-Squares Regression Analysis. *J. R. Stat. Soc. Ser. B (Methodol.)* **1969**, *31*, 350–371. [CrossRef]
39. Tibshirani, R. Regression Shrinkage and Selection Via the Lasso. *J. R. Stat. Soc. Ser. B* **1996**, *58*, 267–288. [CrossRef]
40. Hastie, T.; Tibshirani, R.; Friedman, J.H.; Friedman, J.H. *The Elements of Statistical Learning: Data Mining, Inference, and Prediction*; Springer: New York, NY, USA, 2009; Volume 2, pp. 1–758.
41. Friedman, J.; Hastie, T.; Tibshirani, R.; Narasimhan, B.; Tay, K.; Simon, N.; Qian, J. Package 'glmnet'. CRAN R Repositary. Available online: https://cran.r-project.org/web/packages/glmnet/index.html (accessed on 15 June 2022).
42. Krstajic, D.; Buturovic, L.J.; Leahy, D.E.; Thomas, S. Cross-validation pitfalls when selecting and assessing regression and classification models. *J. Cheminform.* **2014**, *6*, 10. [CrossRef] [PubMed]
43. Alladi, T.; Chamola, V.; Rodrigues, J.J.P.C.; Kozlov, S.A. Blockchain in Smart Grids: A Review on Different Use Cases. *Sensors* **2019**, *19*, 4862. [CrossRef] [PubMed]
44. Foley, A.M.; Leahy, P.G.; Marvuglia, A.; McKeogh, E.J. Current methods and advances in forecasting of wind power generation. *Renew. Energy* **2012**, *37*, 1–8. [CrossRef]
45. Hanifi, S.; Liu, X.; Lin, Z.; Lotfian, S. A Critical Review of Wind Power Forecasting Methods—Past, Present and Future. *Energies* **2020**, *13*, 3764. [CrossRef]
46. Chaurasiya, P.K.; Kumar, V.K.; Warudkar, V.; Ahmed, S. Evaluation of wind energy potential and estimation of wind turbine characteristics for two different sites. *Int. J. Ambient Energy* **2021**, *42*, 1409–1419. [CrossRef]
47. Chaurasiya, P.K.; Ahmed, S.; Warudkar, V. Wind characteristics observation using Doppler-SODAR for wind energy applications. *Resour.-Effic. Technol.* **2017**, *3*, 495–505.
48. Werapun, W.; Tirawanichakul, Y.; Waewsak, J. Wind shear coefficients and their effect on energy production. *Energy Procedia* **2017**, *138*, 1061–1066. [CrossRef]
49. Marčiukaitis, M.; Žutautaitė, I.; Martišauskas, L.; Jokšas, B.; Gecevičius, G.; Sfetsos, A. Non-linear regression model for wind turbine power curve. *Renew. Energy* **2017**, *113*, 732–741. [CrossRef]
50. Aksoy, B.; Selbaş, R. Estimation of Wind Turbine Energy Production Value by Using Machine Learning Algorithms and Development of Implementation Program. *Energy Sources Part A Recover. Util. Environ. Eff.* **2019**, *43*, 692–704. [CrossRef]
51. Bilal, B.; Ndongo, M.; Adjallah, K.H.; Sava, A.; Kébé, C.M.; Ndiaye, P.A.; Sambou, V. Wind turbine power output prediction model design based on artificial neural networks and climatic spatiotemporal data. In Proceedings of the 2018 IEEE International Conference on Industrial Technology, Buenos Aires, Argentina, 26–28 February 2020; pp. 1085–1092.

52. De Giorgi, M.G.; Ficarella, A.; Tarantino, M. Assessment of the benefits of numerical weather predictions in wind power forecasting based on statistical methods. *Energy* **2011**, *36*, 3968–3978. [CrossRef]
53. Vaona, A. Spatial autocorrelation or model misspecification? The help from RESET and the curse of small samples. *Lett. Spat. Resour. Sci.* **2009**, *2*, 53–59. [CrossRef]
54. Hothorn, T.; Zeileis, A.; Farebrother, R.W.; Cummins, C.; Millo, G.; Mitchell, D.; Zeileis, M.A. Package 'lmtest'. Testing Linear Regression Models. Available online: https://cran.r-project.org/web/packages/lmtest/lmtest.pdf (accessed on 3 February 2023).
55. Wooldridge, J.M. *Introductory Econometrics: A Modern Approach*; Cengage learning: Boston, MA, USA, 2015.
56. Demsetz, H. The Cost of Transacting. *Q. J. Econ.* **1968**, *82*, 33–53. [CrossRef]
57. Lobo, M.S.; Fazel, M.; Boyd, S. Portfolio optimization with linear and fixed transaction costs. *Ann. Oper. Res.* **2006**, *152*, 341–365. [CrossRef]
58. Georgiev, K.; Kim, Y.S.; Stoyanov, S. Periodic portfolio revision with transaction costs. *Math. Methods Oper. Res. (ZOR)* **2015**, *81*, 337–359. [CrossRef]
59. Chen, F.Y.; Yano, C.A. Improving Supply Chain Performance and Managing Risk Under Weather-Related Demand Uncertainty. *Manag. Sci.* **2010**, *56*, 1380–1397. [CrossRef]
60. Bunn, D.W.; Chen, D. The forward premium in electricity futures. *J. Empir. Financ.* **2013**, *23*, 173–186. [CrossRef]
61. Longstaff, F.A.; Wang, A.W. Electricity Forward Prices: A High-Frequency Empirical Analysis. *J. Financ.* **2004**, *59*, 1877–1900. [CrossRef]
62. Gagliardini, P.; Ossola, E.; Scaillet, O. Time-Varying Risk Premium in Large Cross-Sectional Equity Data Sets. *Econometrica* **2016**, *84*, 985–1046. [CrossRef]
63. Woodard, J. *Options and the Volatility Risk Premium*; Pearson Education: London, UK, 2011.
64. Batchelor, R.A.; Alizadeh, A.H.; Visvikis, I.D. The relation between bid–ask spreads and price volatility in forward markets. *Deriv. Use Trading Regul.* **2005**, *11*, 105–125. [CrossRef]
65. Bryant, H.L.; Haigh, M.S. Bid–ask spreads in commodity futures markets. *Appl. Financ. Econ.* **2004**, *14*, 923–936. [CrossRef]
66. Yoo, J.; Maddala, G.S. Risk premia and price volatility in futures markets. *J. Futur. Mark.* **1991**, *11*, 165–177. [CrossRef]

energies

MDPI

Article

P2P Electricity Trading Considering User Preferences for Renewable Energy and Demand-Side Shifts

Daishi Sagawa [1,*], Kenji Tanaka [1], Fumiaki Ishida [2], Hideya Saito [2], Naoya Takenaga [3] and Kosuke Saegusa [3]

[1] School of Engineering, The University of Tokyo, Tokyo 113-8656, Japan
[2] The Kansai Electric Power Co., Inc., Osaka 530-8270, Japan
[3] BIPROGY Inc., Tokyo 135-8560, Japan
* Correspondence: sagawa@ioe.t.u-tokyo.ac.jp

Abstract: In the global trend towards decarbonization, peer-to-peer (P2P) energy trading is garnering increasing attention. Furthermore, energy management on the demand side plays a crucial role in decarbonization efforts. The authors have previously developed an automated bidding agent that considers user preferences for renewable energy (RE), assuming users own electric vehicles (EVs). In this study, we expand upon this work by considering users who own not only EVs but also heat pump water heaters, and we develop an automated bidding agent that takes into account their preferences for RE. We propose a method to control the start time and presence of daytime operation shifts for heat pump water heaters, leveraging their daytime operation shift function. Demonstration experiments were conducted to effectively control devices such as EVs and heat pumps using the agent. The results of the experiments revealed that by controlling the daytime operation of heat pumps with our method, the RE utilization rate can be improved compared to scenarios without daytime operation shifts. Furthermore, we developed a simulator to verify the outcomes under different scenarios of demand-side resource ownership rates, demonstrating that higher ownership rates of EVs and heat pumps enable more effective utilization of renewable energy, and that this effect is further enhanced through P2P trading. Based on these findings, we recommend promoting the adoption of demand-side resources such as EVs and heat pumps and encouraging P2P energy trading to maximize the utilization of renewable energy in future energy systems.

Keywords: P2P energy trading; electric vehicle; heat pump water heater; optimization

1. Introduction

1.1. Background

Renewable energy (RE) plays a crucial role in the global decarbonization trend, contributing to the large-scale deployment of clean energy sources. In the transition towards decarbonized energy systems, it is important to focus not only on diversifying power generation, including RE, but also on demand-side energy management, such as energy storage and energy efficiency improvements [1]. Consequently, this study emphasizes demand-side controls that utilize energy storage systems and heat pumps on the demand side.

Peer-to-peer (P2P) energy trading has emerged as a promising solution for enabling dynamic demand-side energy management and coordinating a multitude of distributed energy resources. Various studies have examined different aspects of P2P energy trading, including optimizing trading strategies and market design [2,3], harnessing blockchain technology for efficiency, transparency, and security [4–8], and demonstrating the effectiveness of these systems through real-world case studies [9–13]. For instance, Liu et al. [2] and AlSkaif et al. [3] proposed innovative P2P energy trading platforms to optimize demand response schemes and trading preferences, respectively. These platforms aim to improve the overall efficiency of energy management in residential systems. Blockchain technology has been extensively explored to facilitate secure and transparent P2P energy trading, as evidenced by the work of Dang et al. [6] for big industrial energy users and Shukla et al. [8] for smart grid applications. Real-world case studies further substantiate the potential of P2P energy trading systems. For example, the Brooklyn Microgrid project, discussed by Mengelkamp et al. [10], illustrates how blockchain technology can be effectively implemented to create a decentralized microgrid energy market.

Not only the technical aspects but also the social acceptability and institutional aspects of energy management and decentralization are studied. Studies [14,15] focus on the role of blockchain technology in the energy sector and institutional development for distributed energy systems. Study [16] examines business models and policy interactions for microgrids, finding natural gas technologies as the most robust option. Study [17] evaluates peer-to-peer energy sharing mechanisms, with supply and demand ratio mechanisms outperforming others. Lastly, study [18] proposes a multi-stage incentive model for micro-grid project development involving multiple stakeholders.

Furthermore, research on demand-side demand timing adjustment and energy management using energy storage systems has demonstrated the effectiveness of demand-side energy management [19–22].

This study proposes a trading algorithm for electricity trading agents in a P2P electricity trading market for households with electric vehicles (EVs) and heat pump water heaters, considering the subject's preference for RE on the consumer side. The authors proposed a trading algorithm for electricity trading agents in the P2P electricity trading market in 2021 for households with EVs [23], but the mechanism to adjust the operating timing of the heat pump water heater was not considered in that algorithm. In this study, we propose a mechanism to optimize not only the charging and discharging of EVs, but also the operating timing of heat pump water heaters according to user preferences.

1.2. Related Work

Energy management on the demand side includes the use of storage batteries and changes in demand timing. Various studies have been conducted on the use of storage batteries, and consumers who own EVs can be expected to enjoy cost advantages by optimizing the recharge and discharge of their EVs, procuring more electricity from the market when electricity prices are low, storing it in batteries, and discharging it from the batteries when market electricity prices are high. This is expected to be a cost advantage. Optimizing EV recharge/discharge is also important not only from a cost perspective, but also in terms of satisfying individual users' RE preferences by inexpensively increasing the RE ratio by charging EVs when surplus cheap RE is generated. If users do not own EVs or storage batteries, their own electricity demand will be a constraint on electricity transactions, but if they own EVs or storage batteries, they can reduce costs and improve their RE ratios by recharging and discharging them at appropriate times.

Wu et al. [19] formulated a stochastic optimization problem to optimize EV charging and discharging plans with the goal of minimizing the cost of paying electricity bills while meeting household electricity demand and PEV charging requirements. Vivekananthan et al. [20] proposed an algorithm to control home appliances with the goal of reducing the total cost of residential energy consumption through real-time monitoring, stochastic scheduling, and real-time control while application combinations were considered. Langer [22] minimized costs and discomfort for market participants by mixed integer linear programming, assuming a home with a home energy management system, adjustable heat pump, and photovoltaic power generation, combined with electrical and thermal storage systems.

In addition to the energy storage facility, another example of energy management by changing demand timing would be to adjust the operating timing of the heat pump water heater. For example, excess solar power generation during the daytime can be used to operate the heat pump water heater, making effective use of surplus electricity. Studies on adjusting the operation timing of heat pump water heaters include those by Clift et al. [24] and the aforementioned Langer [22].

Studies from a preference perspective include [25–27], whereby all of them take into account the preferences of individual users while ultimately maximizing overall utility, which is a framework in which users cooperate. Our study differs significantly in that individual users move independently toward satisfying their own preferences. Reis et al. [28] assess how prosumers and consumers pursuing different goals can influence the energy self-sufficiency of a local energy community. The preferences are considered in the modeling to represent a smart community. Yet it does not take into account the orientation toward RE. Pena-Bello et al. [29] assessed P2P trading decisions of German homeowners on the basis of an online experimental study and showed that P2P energy trading based on human decision-making may lead to financial benefits for prosumers and traditional consumers, plus reduced stress for the grid. When prosumers have not only RE generation facilities but also demand-shifting facilities such as EVs and heat pumps, it is possible to reduce the load on the grid even more. Therefore, in this study, we will conduct simulations assuming that many entities own demand-shifting facilities such as EVs and heat pumps, and we will confirm the effect of such facilities.

In summary, various studies have explored energy management on the demand side, including the use of storage batteries, EVs, and adjusting the operation timing of heat pump water heaters. These studies have achieved some success in optimizing energy consumption and costs. However, there is still room for improvement in fully considering individual users' preferences and ensuring that renewable energy orientation and other user-specific priorities are adequately addressed. Further research is needed to develop more tailored energy management solutions that effectively meet the diverse preferences of users.

Therefore, the goal is to develop an operational planning model for EVs and heat pumps that takes into account the orientation toward RE in P2P transactions. It is also important to establish an actual controllable mechanism in a manner that is compatible with the standards that drive existing heat pump water heaters. In this study, the optimization of bidding contents, charging/discharging timing, and daytime shift timing is designed by keeping in mind the Application Programming Interface (API) for control and information acquisition provided by the ECHONET Lite [30] standard used by heat pump water heaters that enjoy widespread popularity in Japan.

1.3. Contribution of This Paper

In this study, we envision an electricity P2P trading market and develop an agent system that automatically performs electricity transactions on behalf of users in the market. The objective is to enable power trading based on individual user circumstances by not only bidding based on user assets such as heat pump water heaters, EVs, and solar power generation and user electricity demand, but also bidding based on user preferences for RE. The main novelty of this study is that it proposes a mechanism for energy management that takes into account individual preferences for RE, including the operation of heat pumps as well as EVs, thus achieving both economic efficiency and satisfying individual preferences for RE. In particular, this study examined how optimizing the operation of heat pumps could benefit heat pump users.

1.4. Organization of This Paper

This paper is structured as follows. First, in Section 2, we provide an overview of the P2P energy trading platform. Subsequently, in Section 3, we describe the functionalities of user agents. Then, in Section 4, we elaborate on the crucial aspects of bidding and facility operation optimization within the user agent functionalities. In Section 5, we discuss the demonstration experiments involving actual EVs and heat pumps. Since there are limitations to feasible settings in the demonstration experiments, we develop a P2P energy trading simulation in Section 6 and conduct the simulation. Finally, we present our conclusions in Section 7.

2. P2P Energy Trading Platform

Figure 1 shows an overview of the P2P power trading platform for the demonstration experiment in this study. The role of the user agent in this P2P electricity trading platform is to acquire the electricity usage and generation status of each consumer and prosumer, forecast demand and generation based on this information, plan energy procurement and usage in consideration of the facilities owned by each consumer and prosumer, such as EVs and heat pumps, and bid on the blockchain market. The system then submits bids to the blockchain market. It also has the role of issuing commands for the recharge and discharge of EVs and the daytime shift time of heat pumps based on the optimization and execution results. In this experiment, a total of five entities (HOME1 to HOME4 as demand entities and one PV as power generation entity) conduct transactions, with only HOME1 having EVs and heat pumps, and HOME2 to HOME4 having none of them. The bidding agents communicate with the Internet gateway and the Blockchain energy market to obtain information on the demand and generation of these bidding entities and to send control signals. HOME1 owns the EV and heat pump, but HOME1, EV, and heat pump have different smart meters. In the experiment, they are treated as one entity in a virtual sense. Each entity may purchase electricity from the market or from retailers. They can also sell the electricity they generate to the market or retailers. Each entity can choose whether to deal with the market or the retailer, whichever has better terms, but cannot purchase RE from the retailer. If they have EVs or storage batteries, they can purchase electricity from the market or retailers when it is cheap, store it in their batteries, and use it when the price rises.

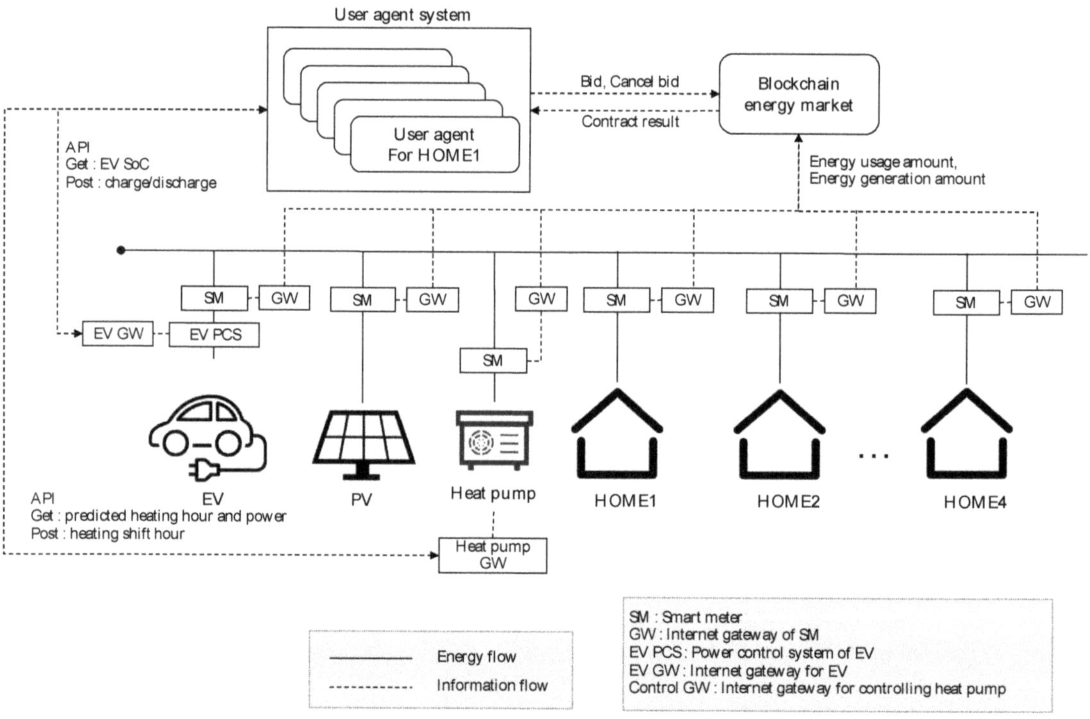

Figure 1. Diagram of demonstration experiment.

3. Functions of User Agents

Figure 2 shows the processing flow of the bidding agent. In the electricity demand forecasting function, electricity demand is forecasted based on customer demand data and weather data, and in the PV power generation function, PV power generation is forecasted based on historical power generation data and weather data. Here, the Meteorological Engineering Center's solar radiation forecasting API [31] is used to create a machine learning model using a random forest [32] to learn the relationship between actual PV power generation and forecasted solar radiation values. The inputs of this model are forecasted solar radiation, plus hour and month. In heat pump electricity demand forecasting, past electricity consumption data and schedule information on hot water demand patterns are used to forecast heat pump electricity consumption on the target day. The hot water demand patterns are determined in advance and are supposed to repeat itself on a two-week cycle. The heat pump itself has a heat demand pattern learning function and operates by predicting heat demand. The ECHONET Lite API [30] used in this project does not have a function to acquire the heat pump operation schedule, so the average value of operation during the same heat demand pattern in the past was used as a forecast. In the bid creation function, the system optimizes the power usage plan based on the transaction mode (green mode or economy mode) set by the user, demand and power generation forecast results, EV SoC, and heat pump operation forecasts, and creates a bid specifying the contents, time frame, amount of power, and price to be bid on the P2P market. In the bid execution function, the created bids are submitted to the energy market. The function for acquiring contract results gets a record of bids that have been contracted in the energy market and re-submits the results to the function for creating bids and recalculating new bids. The energy market trades power in 30-min increments, and bids can be submitted from 24 h prior to the actual power meltdown to one hour prior to the meltdown, and bidding agents change their bids for the same market every 30 min. At that time, the bid cancellation function is a function to submit a command to the market to cancel old bids made in the past. The series of processes from forecasting to bidding is repeated for each agent at 30-min intervals from the auction deadline of the previous day to the upcoming deadline (one hour before the electricity supply date and time).

The device control function is used to set charging/discharging commands to EVs and daytime boil-up shift times to heat pumps; charging/discharging commands to EVs are performed via the EV Power Control System (PCS) API based on the calculated EV charging/discharging plan once the target market is closed. The EV PCS API was developed as a mechanism that enables continuous charging and discharging at a user-defined output level for a specified duration, utilizing a REST API over HTTP. The optimal daytime shift start time for heat pumps is calculated at 9:00 p.m. on the previous day for the following day, and the results are then used to set the shift time via the ECHONET Lite API [30]. The optimal daytime shift start time is obtained by optimizing the daytime shift start time k in Bid Creation. After 21:00 on the previous day, k is fixed to the value obtained by optimization at 21:00 during the day subject to optimization.

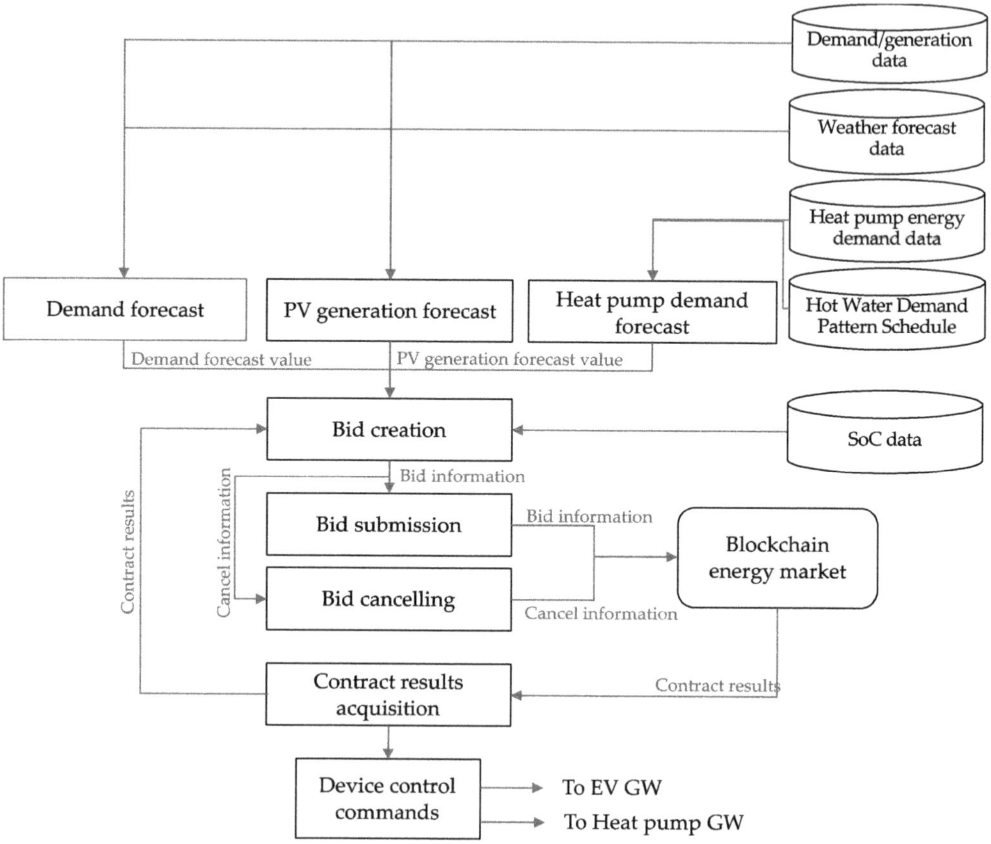

Figure 2. Calculation flow of bidding agent.

4. Bidding and Facility Operation Optimization

The bid creation function (bid creation) optimizes market bids and EV recharge/discharge based on forecasted demand, forecasted generation, SoC value, expected market contract price, and retail electricity price. The bid creation function has two modes: economy mode and green mode. In the economy mode, the optimization process aims to minimize costs by taking into account the revenues generated from electricity sales. It aims to maximize profits (minimize costs) by adjusting the timing and quantity of procurement from the market and the grid, and by controlling the shift times of its own EVs' recharge/discharge and its own heat pumps. In the green mode, the optimization process aims to minimize costs, including the revenue from electricity sales, while satisfying the target RE ratio set by the user as a constraint.

Equations (1)–(9) show the optimization equation for the economy mode. The basic equation is similar to that of Sagawa et al. [23], with the difference that a new heat pump electricity demand value $A^h_{t,k}$ is added. This is based on the value of the heat pump electricity demand forecast for no daytime shift, and the time of daytime operation shift is set to k.

Here, we provide an overview of each equation used in our model. Equation (1) represents the objective function, which aims to minimize the sum of power procurement costs and penalty amounts. The penalty term is introduced to prevent wasteful transactions that offset each other through buying and selling. For example, it serves to prevent scenarios where 100 kWh are purchased and 90 kWh are sold to procure 10 kWh.

Equations (2) and (3) impose constraints to ensure that the amount of energy traded in the market does not become negative. Equations (4) and (5) are constraints to ensure that the charging and discharging of electric vehicles (EVs) do not exceed their output limits. Equation (6) represents an energy balance constraint, ensuring that the balance between consumption, generation, selling, and purchasing is maintained. Equations (7) and (8) impose upper and lower limits on the SoC for EVs. Lastly, Equation (9) is a constraint used to determine the SoC for the next time step based on the previous time step's SoC and the charging/discharging amounts.

The expected operating hours and expected power consumption during daytime shift operation obtained from the heat pump API are taken into account to calculate the electricity consumption of the heat pump at each hour of the daytime shift operation. The following table shows the calculation of the power consumption of the heat pumps at different times of the day. The amount of electricity shifted during the daytime is newly added, and at the same time, the same amount of electricity shifted during the daytime is subtracted from the amount of electricity used during the nighttime, so that the daily electricity consumption of the heat pumps is the same with and without the daytime shift. This optimization is performed for each daytime shift time k of the heat pump, and the daytime shift time k at which the objective function is minimized and the value of the variable to be optimized at that time are adopted. This equation assumes prosumers, but mere consumers or generators can also be expressed as special cases of this equation. For example, to express consumers, it should always be $A_t^p = 0$ since they do not have PV. For example, to represent a generator, it should always be $A_t^d = 0$ since it has no demand. The penalty term C in (1) was introduced to prevent excessive transactions that would offset each other in buying and selling activities.

$$Minimize. \sum_{t=n}^{n+48\times 2} \left[P_t^m (B_t^m - S_t^m) + P_t^g B_t^g + C(B_t^m + S_t^m) \right] \tag{1}$$

$$Subject\ to.$$

$$B_t^m \geq 0 \tag{2}$$

$$S_t^m \geq 0 \tag{3}$$

$$\frac{C_{max}}{2} \geq C_t \geq 0 \tag{4}$$

$$\frac{D_{max}}{2} \geq D_t \geq 0 \tag{5}$$

$$A_t^d + A_{t,k}^h - B_t^m - \left(A_t^p - S_t^m \right) + C_t - D_t - B_t^g = 0 \tag{6}$$

$$E_t \geq E_l \tag{7}$$

$$E_t \leq E_u \tag{8}$$

$$E_{t+1} = \begin{cases} \frac{E_t E_{cap} + C_t R_c - D_t R_{dis}}{E_{cap}} \ (if\ V_t = False) \\ E_t - F_t \ (if\ V_t = True) \end{cases} \tag{9}$$

Each variable is defined as follows.

B_t^m Amount of electricity to be purchased in the market at time t [kWh] (Optimization target)

S_t^m Amount of electricity to be sold in the market at time t [kWh] (Optimization target)

B_t^g Amount of electricity to be purchased from electricity retailers at time t [yen/kWh] (Optimization target)

C_t Amount of charge to the battery at time t [kWh] (Optimization target)

D_t Amount of discharge from the battery at time t [kWh] (Optimization target)

P_t^m Expected price at time t [yen/kWh] (estimated by each agent based on expected power generation)

P_t^g Retail price of electricity at time t [yen/kWh] (defined in advance)

A_t^d Expected electricity demand except heat pump at time t [kWh] (calculated by demand forecast)

$A_{t,k}^h$ Expected heat pump electricity demand with daytime shift start time k at time t [kWh] (calculated by heat pump demand forecast)

A_t^p Expected power generation at time t [kWh] (calculated by power generation forecast)

E_t Percentage of remaining charge of the battery at time t [%]

C_{max} Maximum charging output of the battery [kW] (6.7 [kW])

D_{max} Maximum discharge output of the battery [kW] (6.0 [kW])

E_l Lower limit of SoC [%] (set to 20 percent)

E_u Upper limit of SoC [%] (set to 90%)

E_{cap} Rated capacity of battery [kWh] (40 [kWh] was set.)

R_c Battery charging efficiency [%] (set to 86.6%, so that CHARGE_RATE × DISCHARGE_RATE = 75%)

R_{dis} Discharge efficiency of the battery [%] (set to 86.6%, the same as CHARGE_RATE)

F_t Expected energy consumption by driving at time t [kWh]. This is always set to 0 because the EV is not running in this demonstration experiment.

V_t The bool value indicating whether or not the EV is running at time t. It is always set to "false" because it is not run in this verification experiment.

m A subscript indicating that it is related to the market.

g A subscript indicating that it is related to the electricity retailers.

d A subscript indicating that it is related to the electricity demand.

p A subscript indicating that it is related to the power generation.

u A subscript indicating upper limit.

l A subscript indicating lower limit.

c A subscript indicating charge.

dis A subscript indicating discharge.

cap A subscript indicating battery capacity.

C Penalty term [yen/kWh]

P_t^m is the expected market price at time t. Similar to Sagawa et al. [23], this expected price is calculated based on each agent's forecast of all the PV generation in the market on the target day based on weather information, and the expected PV generation rate p_t which is the predicted PV power generation divided by the rated maximum output. In the experiment, parameters are set to A = 3, B = 5, C = 23 and D = 5.

$$P_t^m = C * exp\left(-A * {p_t}^B\right) + D \tag{10}$$

P_t^g gives the price list for each time. We used "Hapi-e-time R" of Kansai Electric Power Co., [33] as used in Sagawa et al. [23]. This price list is shown in Table 1.

Table 1. Table of Retail price of electricity.

Hour	Price [Yen/kWh]
7:00–10:00	22.89
10:00–17:00	26.33
17:00–23:00	22.89
23:00–7:00	15.20

(Equations (11)–(20)) show the optimization equation for green mode. Again, the basic equation is the same as in Sagawa et al. [23], but a new heat pump electricity demand value, $A_{t,k}^h$, is added. If a solution does not exist, the target RE ratio will be temporarily lowered by 5% in stages until a solution is found.

In this paper, we have built upon our previous work [23] to enhance the modeling and analysis of the home energy management system with RE, EVs, and heat pump water heaters. Although the optimization problems formulated in this paper may bear some resemblance to those in [23], we have introduced a notable modification by integrating a water heater into the model. This integration affects the constraints (6) and (17) by adding a constant and, more importantly, influences the overall energy consumption patterns and optimization results. By incorporating this refinement, we contend that our current paper presents a more comprehensive contribution to the field compared to [23], addressing the intricate interplay between various components of home energy management systems and their implications for energy trading strategies.

$$Minimize. \sum_{t=n}^{n+48*2} \left[P_t^m(B_t^m - S_t^m) + P_t^g B_t^g + C(B_t^m + S_t^m)\right] \tag{11}$$

$$Subject\ to.$$

$$\sum_t \left(A_t^p - S_t^m + B_t^m\right) \geq R_{re} \sum_t \left[A_t^d + F_t + C_t(1-R_c) + D_t(1-R_d)\right] \tag{12}$$

$$B_t^m \geq 0 \tag{13}$$

$$S_t^m \geq 0 \tag{14}$$

$$\frac{C_{max}}{2} \geq C_t \geq 0 \tag{15}$$

$$\frac{D_{max}}{2} \geq D_t \geq 0 \tag{16}$$

$$A_t^d + A_{t,k}^h - B_t^m - \left(A_t^p - S_t^m\right) + C_t - D_t - B_t^g = 0 \tag{17}$$

$$E_t \geq E_{ll} \tag{18}$$

$$E_t \leq E_{hl} \tag{19}$$

$$E_{t+1} = \begin{cases} \frac{E_t E_{cap} + C_t R_c - D_t R_d}{E_{cap}} \ (if\ V_t = False) \\ E_t - F_t \ (if\ V_t = True) \end{cases} \tag{20}$$

Each variable is defined as follows.
R_{re} Target RE ratio (set by user between 0~100%)
The other items are the same as in (1–9).

5. Demonstration Experiment

5.1. Configuration of the Demonstration Experiment

Table 2 lists the entities that will conduct transactions in the demonstration experiment. The demand side consists of four entities (HOME1~HOME4) and the supply side consists of one entity (PV1). All data used in this study are based on the actual power demand and generation amounts originating from these respective facilities.

Table 2. Composition of Consumers and Generators in the Demonstration Experiment.

Entities	Description
HOME1	Laboratory A with EV and Heat pump
HOME2	Laboratory B
HOME3	Laboratory C
HOME4	Laboratory D
PV1	Solar power generation

Table 3 describes the settings for each experimental period. The period is divided into phases 1~4, each of which has a different setting. By comparing the results of these phases, we aim to show that owning both EVs and heat pumps can increase the RE ratio by having EVs absorb excess electricity generation during the daytime and by having heat pumps shift their operation to the daytime, and to confirm how this approach affects costs. The goal of the project is to determine the impact on costs.

Table 3. Experimental phase and set values.

Experimental Phase	Period	Setting
Phase 1	2022/1/26–2022/2/1	Set only HOME1 to green mode and all others to economy mode.
Phase 2	2022/2/2–2022/2/8	All users are set to economy mode.
Phase 3	2022/2/9–2022/2/15	The daytime shift of the heat pump of HOME1 is not performed and set all users to economy mode.
Phase 4	2022/2/16–2022/2/22	The daytime shift of the heat pump of HOME1 is not performed, and only HOME1 is set to Green Mode, while all other users are set to Economy Mode.

5.2. Results of the Demonstration Experiment

Table 4 shows the electricity transaction volume and price for each user for each experimental phase. For Phase 1 of the experiment, HOME1 is the only user in green mode, so HOME1 actively purchases RE even if the price is high, with 107.3 kWh of PV1's 111.6 kWh purchased by HOME1. This indicates that HOME1 purchased approximately 96% of the total contracted amount. In addition, in Phase 2 of the experiment, all users were in economy mode, and in particular, the average contract price for HOME1, which owns an EV, was lower than that of other HOME2~4. By utilizing the charging and discharging of EVs, the trading entity can purchase power on the market when prices are low and charge EVs and discharge the power from the EVs when prices are high. Experimental phase 3 is the case where all users are set to economy mode without the heat pump daytime shift in HOME 1. This phase was prepared to understand the effect of the heat pump daytime shift. It can be seen from Tables 4 and 5 that Phase 2 and Phase 3 do not differ significantly in their results. Phase 4 is the case where the daytime shift of the heat pump in HOME1 is not performed, and only HOME1 is set to green mode and the rest are set to economy mode. The average contract price of HOME1 is about 4 yen/kWh lower in Phase 4 than in Phase 1. This can be attributed to the fact that the amount of PV generation during Phase 4 was higher than in Phase 1, and more PV generation flowed into the market. In fact, total PV generation in Phase 4 was 91.11 kWh compared to 75.19 kWh in Phase 1. Since bidding prices are determined by the amount of electricity generated, it is reasonable to assume that the price difference is due to the difference in total generation rather than to the effect of the daytime shift of the heat pump.

Table 4. Amount and price of electricity traded by each user per experimental phase.

Phase	Entities	Contract Amount [kWh]	Average Contract Price [Yen/kWh]
1	PV1	111.6	30.46
	HOME1	107.3	30.68
	HOME2	2.1	24.74
	HOME3	1.3	25.78
	HOME4	0.9	24.28
2	PV1	85.4	22.10
	HOME1	22.7	16.85
	HOME2	19.6	24.01
	HOME3	14.2	24.03
	HOME4	28.9	23.97
3	PV1	94.5	22.32
	HOME1	23.7	16.31
	HOME2	22.0	24.15
	HOME3	15.9	24.48
	HOME4	32.9	24.38
4	PV1	139.9	25.52
	HOME1	100.9	26.13
	HOME2	12.1	23.97
	HOME3	9.3	24.03
	HOME4	17.6	23.91

Table 5 shows the total demand, market purchases, RE ratio, etc. for each demand-side user for each experimental phase. This study focuses on an individual's preference for RE. Therefore, we focused on the RE ratio as an indicator for individuals, rather than for the community as a whole. In Phase 1, only HOME1 is in green mode, so it can be seen that its RE ratio is higher than the others at 37.68%. In Phase 2, all users are in economy mode, but only HOME1, which owns an EV, has a smaller RE ratio. This is due to the fact that the users only execute contracts when PV, which is even cheaper than the cheap electricity purchased from nighttime retailers, is available in the market during the daytime. Phase 3 did not differ significantly from Phase 2, and it was confirmed that in the case of the economy mode, there was not much change depending on whether the customer owned a heat pump or not. In Phase 4, it can be seen that the RE ratio of HOME 2~4 has increased compared to Phase 1. As mentioned earlier, this can be attributed to the fact that PV generation was higher during Phase 4 than in Phase 1. On the other hand, the RE ratio of HOME1 has decreased. This can be attributed to the fact that the demand for the daytime operation of the heat pump in Phase 1 was no longer met by RE due to the turning off of the daytime shift of the heat pump.

Table 5. RE ratio of each demand-side user per experimental phase.

Phase	Entities	Demand	Contract Amount in the Market [kWh]	Amount Purchased from the Grid [kWh]	RE Ratio [%]
1	HOME1	177.0	107.3	110.3	37.68
	HOME2	30.4	2.1	29.0	4.61
	HOME3	33.7	1.3	32.5	3.56
	HOME4	86.6	0.9	85.8	0.92
2	HOME1	201.5	22.7	186.7	7.34
	HOME2	35.6	19.6	25.9	27.25
	HOME3	34.5	14.2	24.9	27.83
	HOME4	86.6	28.9	66.8	22.86
3	HOME1	199.0	23.7	183.8	7.64
	HOME2	32.9	22.0	20.7	37.08
	HOME3	34.1	15.9	22.2	34.90
	HOME4	86.5	32.9	61.8	28.55
4	HOME1	200.8	100.9	140.8	29.88
	HOME2	33.4	12.1	26.3	21.26
	HOME3	35.2	9.3	28.7	18.47
	HOME4	87.2	17.6	75.3	13.65

5.3. Experimental Results of Phase 1

Figure 3 shows the execution price trends for all users in Phase 1 of the experiment. The dotted line shows the retail price (Table 1). Many contracts are due to HOME1 in green mode. HOME2~4 in economy mode are always executed at or below the retail price. This is because the economy mode procures electricity from the more economically advantageous side of the market and retail.

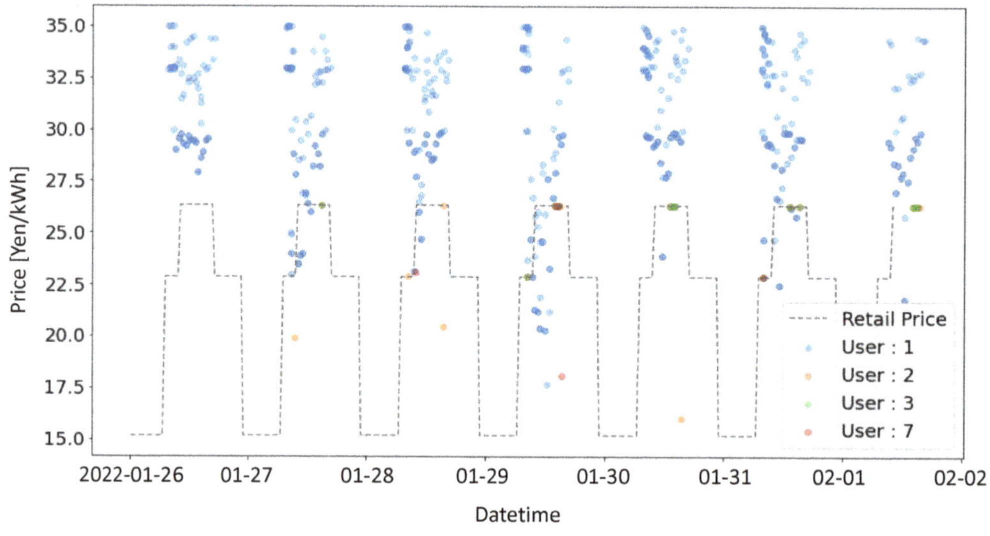

Figure 3. Trends in contract prices for all users in Phase 1 of the experiment.

Figure 4 shows the transition of electricity use by HOME1 on 28 January 2022. The blue line represents the electricity demand excluding the heat pump, the orange line shows the actual electricity demand of the heat pump, the green line indicates the predicted electricity demand of the heat pump, the red line displays the contracted amount in the market, the purple dashed line illustrates the charging amount, the brown dashed line denotes the discharging amount, and the light blue shaded area represents the SoC. HOME1 is in green mode, so it actively purchases electricity generated during the day to charge its batteries and uses the charged electricity from evening to night (increase/decrease in the light blue area). It can also be seen how the heat pump is operated during the day to provide its power with RE (rise in the orange line during the day).

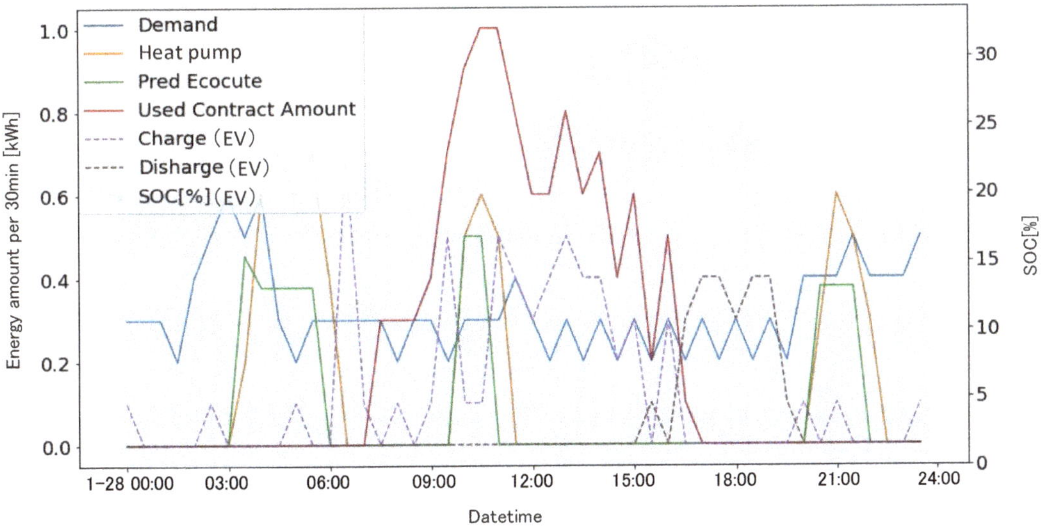

Figure 4. Electricity usage of HOME1 on one day during Phase 1 of the experiment.

5.4. Experimental Results of Phase 2

Since all users are in economy mode, they appear to be contracting at a price lower than the retail price. Figure 5 shows contract prices for all users in phase 2. In particular, since HOME1 owns an EV, it can take the strategy of storing inexpensive late-night power and using it during the daytime, so the contract price is not only lower than the retail price at the same time but is optimized so that the contract is only executed at a price that is economically advantageous, including charging and discharging inexpensive late-night power, so the contract price is particularly low. This is especially true for low contract prices.

In Figure 6, HOME1 is in economy mode, which means that the strategy is to store inexpensive electricity at night in the battery and use it during the day, and the optimization does not appear to result in a daytime shift of the heat pump.

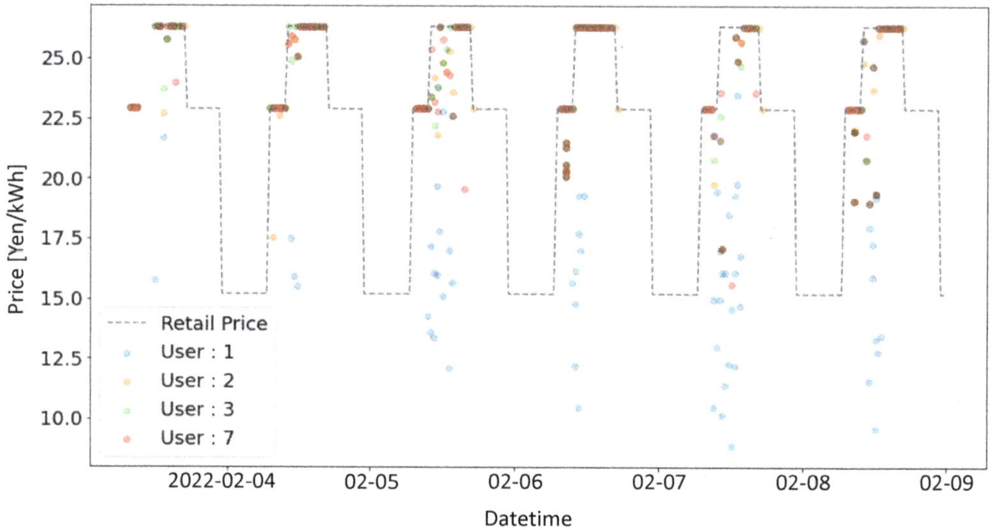

Figure 5. Trends in contract prices for all users in Phase 2 of the experiment.

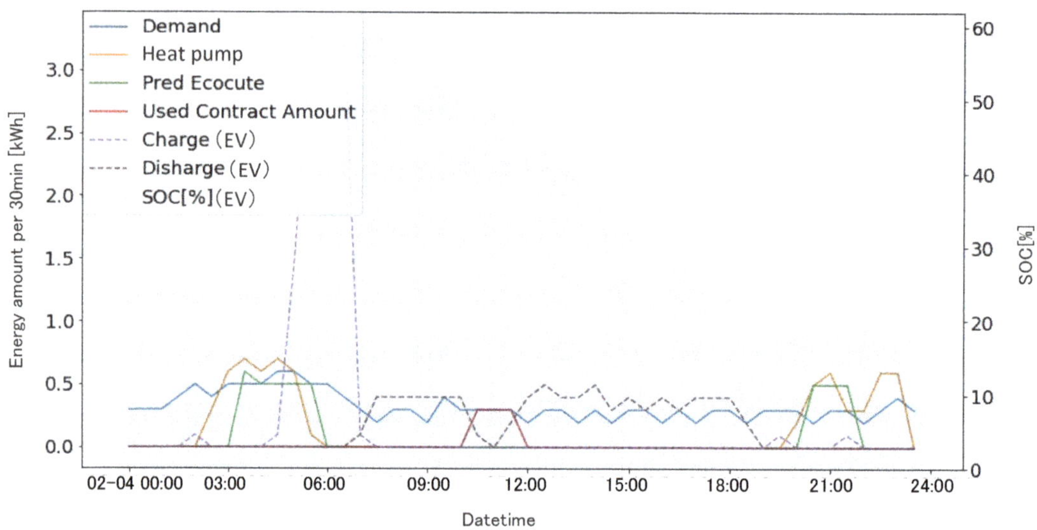

Figure 6. Electricity usage of HOME1 on one day during Phase 2 of the experiment.

5.5. Experimental Results of Phase 3

Figure 7 shows the price transition. Since all entities are in economy mode, it can be seen that they are moving in a similar manner to Phase 2.

In this phase, all entities are in economy mode, so inexpensive power at night is stored in batteries and used during the day. Therefore, it can be seen from Figure 8 that the situation is basically the same as in Phase 2. In addition, since there was a lot of PV generation on this day and inexpensive electricity was available in the market during the day, it can be confirmed that the electricity is stored in batteries during the day as well. In Phase 3, the heat pump is set to not shift during the daytime, but in Phase 2, the heat pump did not shift during the daytime, so the timing of the heat pump operation is also unchanged.

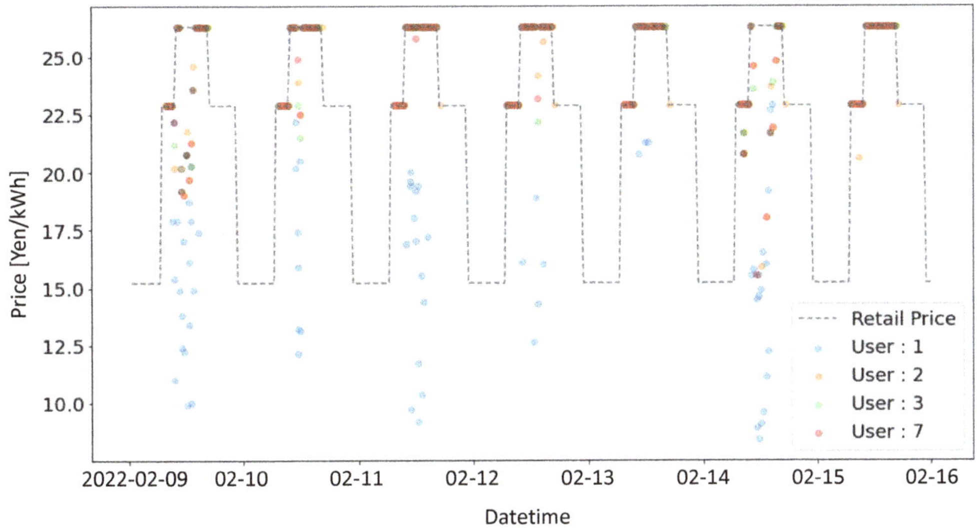

Figure 7. Trends in contract prices for all users in Phase 3 of the experiment.

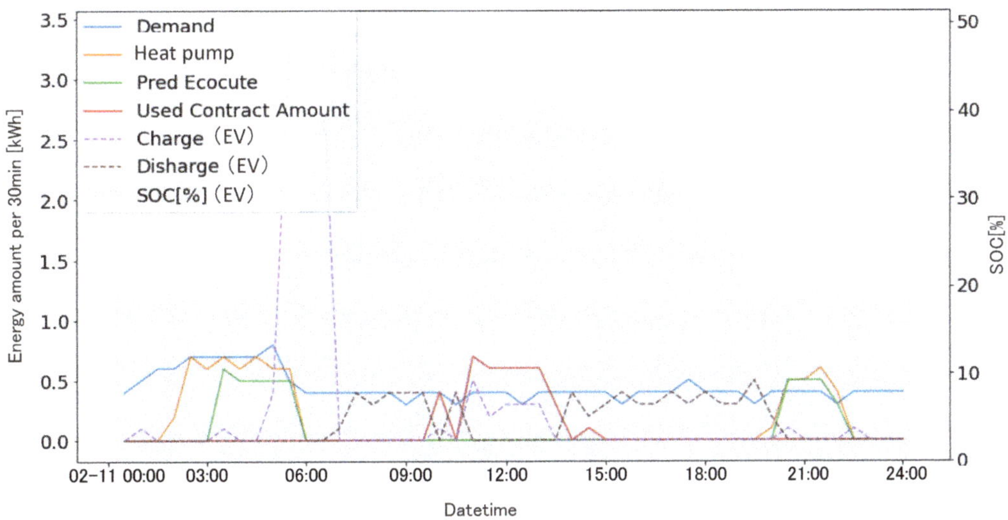

Figure 8. Electricity usage of HOME1 on one day during Phase 3 of the experiment.

5.6. Experimental Results of Phase 4

It is similar to Phase 1 in that the heat pump is set to not shift during the day and is in green mode, so it actively purchases RE during the day. The movement of contract price is also similar in Figures 3 and 9 of Phase 1. The difference between Phase 4 and Phase 1 is that the heat pump is not shifted during the daytime in Phase 4.

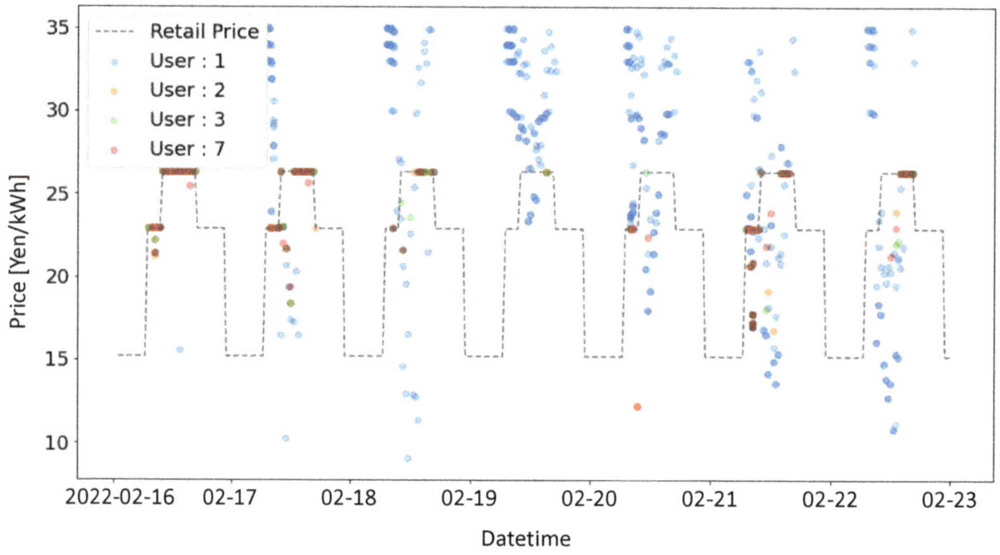

Figure 9. Trends in contract prices for all users in Phase 4 of the experiment.

Figure 10 shows the electricity usage of HOME1, and it can be seen that the heat pump is not running during the daytime. This is thought to have caused the RE ratio to be lower than in Phase 1. It can be said that in Phase 1, the RE ratio could be improved by shifting the heat pump operation to daytime compared to Phase 3.

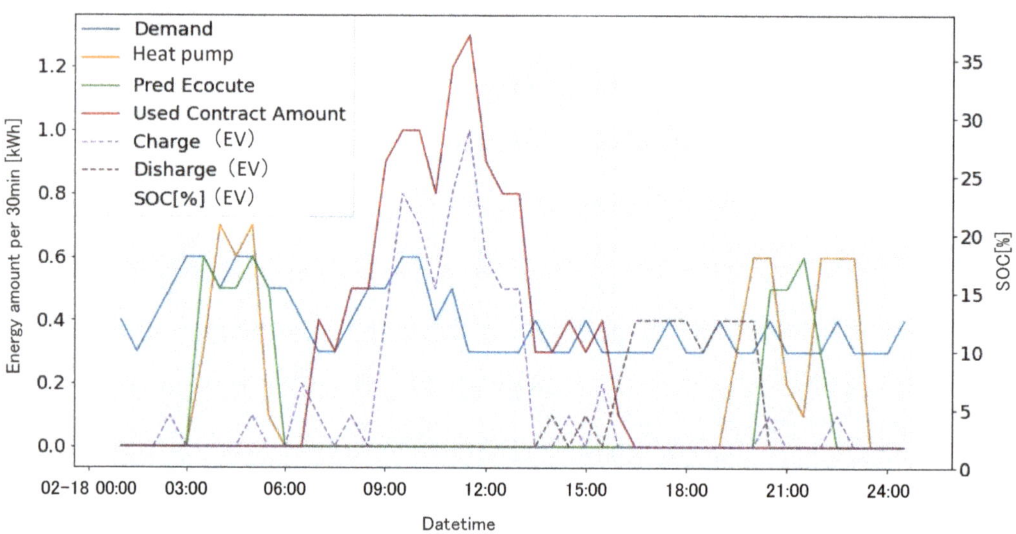

Figure 10. Electricity usage of HOME1 on one day during Phase 4 of the experiment.

5.7. Discussions

P2P trading allows consumers to make better use of surplus energy by proactively controlling their equipment while taking into account their own RE orientation. This study aimed to enable participants to automatically and optimally control their own devices in P2P transactions to reflect their own preferences.

In this experimental setup, the comparison of Phase 1 and Phase 4 showed that the RE ratio can be increased during green mode by optimizing the operating timing of the heat pumps. In the economy mode of Phase 2 and Phase 3, the benefit of optimizing the heat pump operation timing was not confirmed, but this was due to the low amount of PV generation in Phase 2 (the contract amount of PV in Phase 2 in Table 4 is small.) This may be because the amount of PV generation in Phase 2 was small (the contract amount of PV in Phase 2 in Table 4 is small), and there were not many times when the market price was low during the day, so the heat pumps could not be operated at those times. It should be added that in a demonstration experiment such as this, there is a limit to the comparison between phases because the weather conditions cannot be matched between phases. The emphasis of this experiment was on proof of concept, including the actual operation of the equipment.

6. P2P Energy Trading Simulation

Since it was practically difficult to conduct the demonstration experiment with various configurations in terms of the number and attributes of participants and distributed resources owned by each participant due to constraints such as the number of participants and procurement of resource equipment, a simulator was developed to verify various scenarios. In addition, the heat pump operation strategy in Demonstration Experiment 2 was limited to selecting from a limited number of options concerning what time to start heating during the daytime according to the operation plan created by the heat pump equipment based on its own internal logic. Therefore, it was not a truly efficient operation strategy. To solve this problem, we constructed a simulator as a mechanism to create operation plans on a zero-based basis according to the demand for hot water, and to evaluate potentials that are not affected by manufacturers' proprietary algorithms. Using such a simulator, we conducted experiments with various participant and resource configurations to simulate P2P electricity trading assuming a virtual community and evaluated differences in results depending on the configuration of participants' preferences and differences in the resources they possess. The agent and market system constructed in the demonstration experiment was reconstructed as a simulator. The bid creation function was developed based on Equations (21)–(34).

Equation (21) represents the objective function, which, similar to the one previously introduced, aims to minimize the energy procurement cost and penalty amount. Equation (22) is a constraint related to the energy balance of consumption, generation, purchase, and sale. Equation (23) is a constraint regarding the amount of available renewable energy for sale. Equation (24) is a constraint that represents the amount of energy that can be purchased when all the power in the market is derived from solar energy, preventing plans to purchase solar-generated power during non-generation periods. Equation (25) is a constraint related to the trading of grid-derived power. Equation (26) is a constraint used to calculate the next time step's state of charge (SoC) based on the previous time step's SoC and charging/discharging amounts. Equation (27) is a constraint used to determine the next time step's hot water storage amount based on the previous time step's hot water storage amount, additional hot water amount, and used hot water amount. Equations (28) and (29) are constraints related to the trading amount in the market. Equation (30) sets the upper and lower limits for the charging/discharging amount per 30 min. Equations (31) and (32) set the upper and lower limits for the SoC. Equation (33) sets the lower limit for hot water storage amount. Equation (34) is valid only during the RE mode and serves as a constraint to ensure that the RE ratio exceeds the desired value.

$$Minimize. \sum_{t=n}^{n+48*2} \begin{bmatrix} P_t^{mr}(B_t^{mr} - S_t^{mr} + C_t^{mr}) + P_t^{mb}\left(B_t^{mb} - S_t^{mb} + C_t^{mb}\right) + \\ P_{buy,t}^{g} B_t^{g} - P_{sell,t}^{g} S_t^{g} + C\left(B_t^{mr} + S_t^{mr} + B_t^{mb} + S_t^{mb}\right) \end{bmatrix} \tag{21}$$

$$Subject\ to.$$

$$A_t^d + A_t^h - B_t^{mr} - B_t^{mb} - A_t^p + S_t^{mr} + S_t^{mb} + C_t - C_t^{mr} - C_t^{mb} - B_t^g + S_t^g = 0 \tag{22}$$

$$S_t^{mr} - C_t^{mr} \leq A_t^p \tag{23}$$

$$B_t^{mr} \leq kR_t \tag{24}$$

$$S_t^{mb} - C_t^{mb} - B_t^{mb} \leq E_{cap}(E_t - E_{ll}) \tag{25}$$

$$E_{t+1} = \begin{cases} \frac{E_t E_{cap} + C_t}{E_{cap}} (if\ V_t = False) \\ \frac{E_t E_{cap} - F_t}{E_{cap}} (if\ V_t = True) \end{cases} \tag{26}$$

$$W_{t+1}^h = W_t^h + \frac{A_t^h * K_{COP} * K_{kWhtoJ}}{K_s * (T^h - T^g)} - \frac{W_t^d \left(T^d - T^g\right)}{T^h - T^g} \tag{27}$$

$$B_t^m \geq 0 \tag{28}$$

$$S_t^m \geq 0 \tag{29}$$

$$\frac{C_{max}}{2} \geq C_t \geq -\frac{C_{max}}{2} \tag{30}$$

$$E_t \geq E_{ll} \tag{31}$$

$$E_t \leq E_{hl} \tag{32}$$

$$W_t^h \geq W_{ll} \tag{33}$$

$$if\ RE\ mode$$

$$\sum_{t=n}^{n+48*2} \left(A_t^p - S_t^{mr} + B_t^{mr} + C_t^{mr} - S_t^{mb}\right) \geq R_{re} \sum_{t=n}^{n+48*2} \left(A_t^d + A_t^h\right) \tag{34}$$

B_t^{mr} Amount of electricity of RE origin purchased in the market at time t [kWh] (Optimization target).

B_t^{mb} Grid-derived electricity purchased in the market at time t [kWh] (Optimization target).

S_t^{mr} Electricity of RE origin sold in the market at time t [kWh] (Optimization target).

S_t^{mb} Amount of grid-derived electricity sold in the market at time t [kWh] (Optimization target).

B_t^g Amount of electricity purchased from electricity retailers at time t [yen/kWh] (Optimization target).

S_t^g Amount of electricity sold to electricity retailers at time t [yen/kWh] (Optimization target).

C_t Amount of charge/discharge to battery at time t [kWh] +: charge −: discharge (Optimization target).

A_t^h Electricity consumption of heat pump at time t [kWh] (Optimization target).

A_t^d Estimated demand at time t [kWh] (calculated by demand forecast).

A_t^p Estimated electricity generation at time t [kWh] (calculated by electricity generation forecast).

C_t^{mr} Amount of electricity of RE origin purchased and sold in the market at time t that has already been contracted through past bids [kWh] +: amount purchased, −: amount sold.

C_t^{mb} Amount of grid-derived electricity purchased and sold in the market at time t that has already been contracted through past tenders [kWh] +: Purchased, −: Sold.

R_t Predicted solar radiation [MJ/m^2].

k Constant that allows the purchase of electricity derived from RE from the market only when solar radiation is available.

E_{cap} Rated capacity of batteries [kWh].

E_{ll} Lower limit of SoC [%].

E_{hl} Upper limit of SoC [%].

E_t Remaining charge of the battery at time t (SoC).

F_t Expected energy consumption by driving at time t [kWh]. In this simulation, the value is always set to 0 because the EV is not driving.

V_t The bool value indicating whether or not the EV is running at time t. In this demonstration experiment, the value is always set to "false" because the EV is not driving.

W_t^h Volume of hot water stored in the heat pump thermal storage tank at T^h °C at time t [L].

T^h Temperature of stored hot water in the heat pump storage tank [°C].

W_t^d Hot water demand at time t [L].

T^d Hot water temperature when using hot water supply [°C].

T^g Temperature of tap water [°C].

K_s Specific heat of water [J/L-K] 4186.

K_{COP} COP of heat pump (=heat quantity [J]/energy consumption [J]).

K_{kWhtoJ} Constant for converting kWh to J. 3.6×10^6.

P_t^{mr} Expected price of RE-derived electricity in the market at time t [yen/kWh] (estimated by each agent based on expected solar radiation).

P_t^{mb} Expected market price of grid-derived electricity at time t [yen/kWh].

P_t^g Retail price of electricity at time t [yen/kWh].

C_{max} Maximum battery charge/discharge output [kW].

C Penalty term [yen/kWh].

In this study, we have formulated two optimization problems to model the automatic bidding strategy for home energy management systems with RE, EVs, and heat pump water heaters. It is important to note that both of these optimization problems are linear programs. Linear programming is a mathematical optimization technique that is widely used to find the best possible solution to problems that can be represented by linear relationships. Linear programs are relatively easier to solve and have efficient algorithms, such as the simplex method, available for finding optimal solutions. We have employed the PuLP library, a Python-based linear programming library, to solve the linear programs formulated for the automatic bidding strategy. For our analysis, we used the default PuLP solver, which is the open-source COIN-OR Linear Programming (CLP) solver. The CLP solver is a robust and high-performance linear programming solver, making it suitable for tackling the optimization problems in our proposed approach.

Bid Creation of the Simulator

We assume a case in which a large number of houses have photovoltaic power generation facilities. This is in consideration of the fact that the Tokyo Metropolitan Government will require new buildings with a total floor area of less than 2000 m^2 to be equipped with RE facilities such as solar power generation starting in April 2025, targeting house builders and other businesses with a certain annual total floor area of supply in Tokyo. The purpose of this study is to confirm the effectiveness of these distributed resources in utilizing RE in the community under P2P transactions by varying the ownership rates of EVs and heat pump water heaters of individual entities. Table 6 shows the resource retention scenarios for which the resource retention rates of the entities were set. In all scenarios, the ownership rate is set to 75%, assuming a situation where PV is widely deployed. Resource Ownership Scenario One assumes a 20% EV/heat pump ownership rate. The current ownership rate of EVs is 1.51%, heat pumps 19% (owner-occupied in Tokyo) and 2% (renters in Tokyo), so a 20% ownership rate is a large value, but it is the closest to the future assumed in this scenario. Resource Ownership Scenario Two assumes a future in which the EV/heat pump ownership rate is further increased to 50%. Resource Ownership Scenario Three is a future scenario in which the EV/heat pump ownership rate is increased to 80%. These resource ownership scenarios are applied to a community of 20 households. The demand data utilized in this study was obtained from actual household consumption.

Table 6. Equipment ownership scenario.

Equipment Ownership Scenario	Percentage of PV Owned	Percentage of EV Owned	Percentage of HP Owned
1	75%	20%	20%
2	75%	50%	50%
3	75%	80%	80%

The agent system, which represents the individual entities, will be able to make decisions on resource operation, taking into account the overall situation. This is expected to enable the effective use of RE generated by the community as a whole. In this section, we evaluate the difference in energy use by the community as a whole depending on the presence or absence of the coordination mechanism.

Figure 11 shows the difference in the degree of load on the grid for the entire community depending on the resource ownership scenario with and without P2P trading. The two indicators of the community's overall load on the grid are the rate of inflow of RE generation into the grid and the percentage of grid-derived electricity consumed, which are plotted on the vertical and horizontal axes, respectively. The inflow rate of renewable electricity into the grid is defined as the amount of PV electricity sold to the grid/the total amount of PV electricity generated, and the consumption rate of grid-derived electricity is defined as the amount of electricity purchased from the grid/the total amount of electricity consumed. In other words, the inflow rate of RE into the grid is the proportion of RE generated by the community as a whole that is not fully consumed by the community and is taken back by the power grid. Currently, the handling of RE surpluses is an issue, as RE output is curtailed to stabilize the power system, and a small value is desirable from the perspective of stabilizing the power system. The consumption ratio of grid-derived electricity is also a desirable indicator from the perspective of effectively using RE in the community. In the figure, • represents the case with a cooperative mechanism and × represents the case without a cooperative mechanism, and the color indicates the resource ownership scenario. In both resource ownership scenarios, the introduction of the cooperative mechanism reduces two indicators: the inflow rate of RE into the grid and grid-derived electricity. In other words, the cooperative mechanism allows the community to consume more of the RE generated within the community and to reduce the amount of electricity procured from external sources. In addition, the two indices are smaller in the scenarios with higher resource ownership, indicating that the proportion of resources owned by the community has a significant impact on the differences among the ownership scenarios. The difference between the scenarios with and without a cooperative mechanism within the ownership scenarios shows that the difference between the scenarios with and without a cooperative mechanism is smaller in the scenarios with higher ownership ratios. This indicates that the higher ownership of demand-shifting resources such as storage batteries, EVs, and heat pumps enables the individual entities to consume RE to some extent even without a cooperative mechanism, but the introduction of a cooperative mechanism still reduces the two indicators more. In addition, especially in Resource Ownership Scenario Three, both the EV ownership rate and the heat pump ownership rate are extremely high at 80%, and it is considered that it will take a considerable period of time to reach this point. In addition to the installation of solar power generation facilities, the introduction of demand-shifting equipment such as storage batteries, EVs, and heat pumps and their appropriate control systems is required.

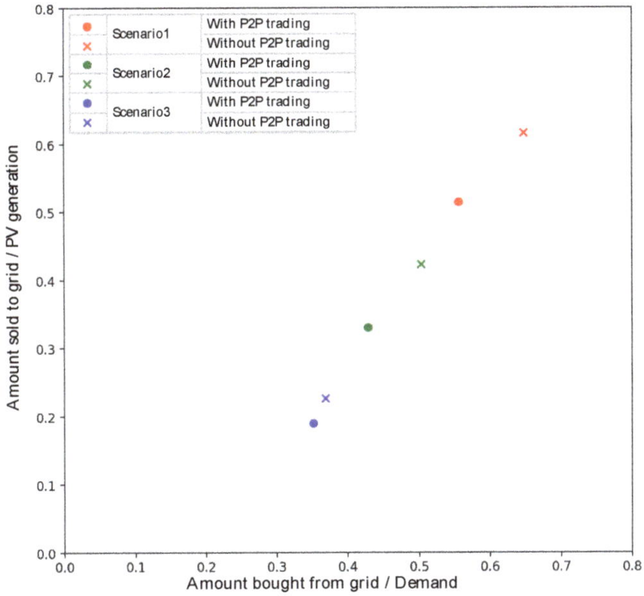

Figure 11. Differences in the degree of load on the electricity system for the entire community with and without P2P trading based on resource ownership scenarios.

Table 7 shows the community-wide demand and supply for the six patterns in Figure 11. In particular, in Scenario 3 with P2P, it can be seen that only 480 kWh of the 2517 kWh of electricity generated by the PV goes to the grid, and the rest can be consumed within the community. It can also be seen that the amount of electricity purchased from the grid is thereby reduced.

Table 7. The supply and demand of electricity throughout the community.

Scenario	with or without P2P	PV Generation [kWh]	Demand [kWh]	Amount Sold to Grid [kWh]	Amount Bought from Grid [kWh]	Amount Sold to Grid ÷ PV Generation [%]	Amount Bought from Grid ÷ Demand [%]
Scenario1	With	2517	2755	1295	1533	51.5%	55.6%
	Without	2517	2743	1551	1777	61.6%	64.8%
Scenario2	With	2517	2954	832	1269	33.0%	42.9%
	Without	2517	2924	1065	1472	42.3%	50.3%
Scenario3	With	2517	3143	480	1105	19.1%	35.2%
	Without	2517	3086	571	1140	22.7%	36.9%

The fact that the consumption rate of RE is increased and the amount of electricity procured from the power grid can be reduced through this cooperative mechanism is due to the fact that consumers without RE generation equipment can use RE through the flexibility of surplus generation, and that the community as a whole can coordinate their use of RE. This effect is due to the fact that when surpluses occur, consumers actively utilize demand-shifting equipment to absorb the excess power. Figures 12 and 13 show the results of electricity use by the entire community without the coordination mechanism and with the coordination mechanism, respectively. The green line indicates PV power generation, the yellow area indicates power consumption by heat pumps, the light blue area indicates power consumption by non-heat pumps, the brown area indicates discharge from EVs, and the purple area indicates charge to EVs. The timing of the heat pump operation and the EV recharge/discharge timing change depending on the presence or absence of the coordination mechanism. In Figure 12, the individual entities are planning the operation of their resources in a closed manner, so the heat pumps are often operating and the EVs are storing electricity even when there is no surplus electricity in the community, while in Figure 13, the individual entities are planning to procure surplus electricity through the market. In Figure 13, we can see that the demand shift reflects the whole community situation, such as running heat pumps and charging EVs when there is a surplus of electricity in the community, because the individual entities plan the operation of resources considering the procurement of surplus electricity through the market. The system is flexible in its charging plan according to the situation.

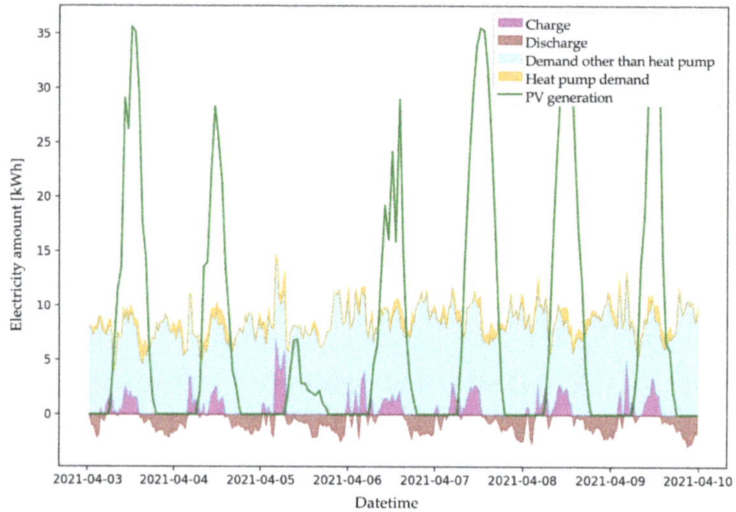

Figure 12. Demand changes for one week without P2P transactions (Scenario 1).

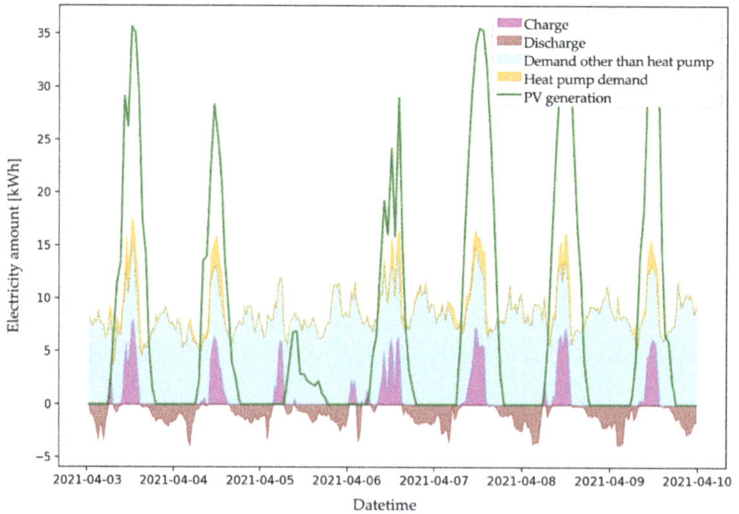

Figure 13. Demand changes for one week with P2P transactions (Scenario 1).

7. Conclusions and Future Work

7.1. Conclutions

In this study, we developed an agent system that automatically performs electricity transactions on behalf of users, assuming an electricity P2P trading market. We developed a bidding agent system that not only performs bidding based on the user's assets such as heat pump water heaters, EVs, and solar power generation plus the user's electricity demand, but also performs bidding based on the user's preferences for RE, enabling power transactions tailored to individual users. The main novelty of this study is that, in contrast to the automatic bidding agent system proposed in the existing study [23] that takes into account the charging and discharging of EVs and users' preferences for RE, we proposed a mechanism to optimize the system by including the operating timing of heat pumps as part of controlling demand itself.

We also conducted a demonstration experiment using the proposed model. The following results were obtained in the demonstration experiment.

- The demonstration experiment confirmed that users in economy mode who own an EV can lower the average contract price.
- In the case of economy mode, there was no significant difference depending on whether the user owned a heat pump or not. This is thought to be because the economic benefits of the daytime shift did not arise due to the low power generation in Phase 2.
- In the case of green mode, it was confirmed that optimizing the daytime shift of a heat pump can improve the RE ratio while reducing the unit cost of electricity compared to the case where a heat pump is not shifted during the daytime.

Finally, we constructed a P2P trading simulator based on the agent algorithm developed in the demonstration experiment, and conducted a P2P transaction simulation assuming a future in which photovoltaic power generation facilities are widely used and showed that a cooperative mechanism of P2P transactions can improve the utilization ratio of RE for the entire community. The results show that the P2P trading cooperative mechanism can improve the ratio of RE use in the community as a whole. In particular, we showed that it is important to increase the share of demand-shiftable resources such as EVs and heat pumps in order to improve the utilization ratio of RE.

7.2. Future Work

One of the potential areas for future research is to address the uncertainty in RE generation, which is not explicitly considered in our current model. Incorporating stochastic optimization techniques could be an effective way to handle such uncertainties. By formulating the problem as a stochastic optimization, we could account for the variability and unpredictability of RE generation, leading to more robust and adaptive bidding strategies. Additionally, investigating the impact of various forecasting methods on the proposed approach would provide valuable insights into the performance and reliability of the automatic bidding strategy under uncertain conditions. We believe that addressing these aspects would significantly enhance the applicability of our work in real-world settings with uncertain RE generation.

Although the present study demonstrates the effectiveness of our proposed algorithm in the context of peer-to-peer energy trading with heat pumps, we have not yet assessed its scalability. In future work, we plan to investigate the performance of our algorithm when applied to larger-scale energy systems, involving a greater number of participants and more diverse energy sources. This analysis will help us understand how our approach performs in more complex energy networks and identify any potential challenges or limitations. By addressing scalability, we aim to contribute further to the development of efficient and sustainable energy management solutions that can be widely adopted in various energy market scenarios.

Author Contributions: Conceptualization, D.S., K.T. and F.I.; methodology, D.S.; writing—original draft preparation, D.S.; project administration; K.T., F.I., H.S., N.T. and K.S. All authors have read and agreed to the published version of the manuscript.

Funding: Grant-in-Aid for Scientific Research (A) 20H00285 from the Japan Society for the Promotion of Science (JSPS).

Data Availability Statement: Restrictions apply to the availability of these data. Data was obtained from The Kansai Electric Power Co., Inc. and are available from the authors with the permission of The Kansai Electric Power Co., Inc.

Acknowledgments: The second and the third authors are supported by Grant-in-Aid for Scientific Research (A) 20H00285 from the Japan Society for the Promotion of Science (JSPS).

Conflicts of Interest: The authors declare no conflict of interest.

References

1. Pörtner, H.-O.; Roberts, D.C.; Poloczanska, E.S.; Mintenbeck, K.; Tignor, M.; Alegría, A.; Craig, M.; Langsdorf, S.; Löschke, S.; Möller, V.; et al. (Eds.) IPCC, 2022: Summary for Policymakers. In *Climate Change 2022: Impacts, Adaptation, and Vulnerability. Contribution of Working Group II to the Sixth Assessment Report of the Intergovernmental Panel on Climate Change*; Cambridge University Press: Cambridge, UK, 2022; *in press*.
2. Liu, W.; Qi, D.; Wen, F. Intraday Residential Demand Response Scheme Based on Peer-to-Peer Energy Trading. *IEEE Trans. Ind. Inform.* **2020**, *16*, 1823–1835. [CrossRef]
3. AlSkaif, T.; Crespo-Vazquez, J.L.; Sekuloski, M.; van Leeuwen, G.; Catalão, J.P.S. Blockchain-Based Fully Peer-to-Peer Energy Trading Strategies for Residential Energy Systems. *IEEE Trans. Ind. Inform.* **2022**, *18*, 231–241. [CrossRef]
4. Tanaka, K.; Abe, R.; Nguyen-Van, T.; Yamazaki, Y.; Kamitamari, T.; Sako, K.; Koide, T. A Proposal on an Electricity Trading Platform Using Blockchain. In *Transdisciplinary Engineering Methods for Social Innovation of Industry 4.0*; IOS Press: Amsterdam, The Netherland, 2018; pp. 976–983.
5. Green, J.; Newman, P. Citizen utilities: The emerging power paradigm. *Energy Policy* **2017**, *105*, 283–293. [CrossRef]
6. Dang, C.; Zhang, J.; Kwong, C.-P.; Li, L. Demand Side Load Management for Big Industrial Energy Users Under Blockchain-Based Peer-to-Peer Electricity Market. *IEEE Trans. Smart Grid* **2019**, *10*, 6426–6435. [CrossRef]
7. Sikorski, J.J.; Haughton, J.; Kraft, M. Blockchain technology in the chemical industry: Machine-to-machine electricity market. *Appl. Energy* **2017**, *195*, 234–246. [CrossRef]
8. Shukla, S.; Thakur, S.; Hussain, S.; Breslin, J.G. A Blockchain-Enabled Fog Computing Model for Peer-To-Peer Energy Trading in Smart Grid. In *Blockchain and Applications*; Prieto, J., Partida, A., Leitão, P., Pinto, A., Eds.; Springer International Publishing: Cham, Switzerland, 2021; pp. 14–23.
9. Matsuda, Y.; Yamazaki, Y.; Oki, H.; Takeda, Y.; Sagawa, D.; Tanaka, K. Demonstration of Blockchain Based Peer to Peer En-ergy Trading System with Real-Life Used PHEV and HEMS Charge Control. *Energies* **2021**, *14*, 7484. [CrossRef]
10. Mengelkamp, E.; Gärttner, J.; Rock, K.; Kessler, S.; Orsini, L.; Weinhardt, C. Designing microgrid energy markets: A case study: The Brooklyn Microgrid. *Appl. Energy* **2018**, *210*, 870–880. [CrossRef]
11. Cali, U.; Çakir, O. Energy policy instruments for distributed ledger technology empowered peer-to-peer local energy markets. *IEEE Access* **2019**, *7*, 82888–82900. [CrossRef]
12. Takeda, Y.; Nakai, Y.; Senoo, T.; Tanaka, K. Designing a User-Centric P2P Energy Trading Platform: A Case Study—Higashi-Fuji Demonstration. *Energies* **2021**, *14*, 7289. [CrossRef]
13. Kontani, R.; Tanaka, K.; Yamada, Y. Feasibility Conditions for Demonstrative Peer-to-Peer Energy Market. *Energies* **2021**, *14*, 7418. [CrossRef]
14. Ahl, A.; Yarime, M.; Goto, M.; Chopra, S.S.; Kumar, N.M.; Tanaka, K.; Sagawa, D. Exploring blockchain for the energy transition: Opportunities and challenges based on a case study in Japan. *Renew. Sustain. Energy Rev.* **2020**, *117*, 109488. [CrossRef]
15. Ahl, A.; Yarime, M.; Tanaka, K.; Sagawa, D. Review of blockchain-based distributed energy: Implications for institutional development. Renew. Sustain. *Energy Rev.* **2019**, *107*, 200–211. [CrossRef]
16. Hanna, R.; Ghonima, M.; Kleissl, J.; Tynan, G.; Victor, D.G. Evaluating business models for microgrids: Interactions of technology and policy. *Energy Policy* **2017**, *103*, 47–61. [CrossRef]
17. Zhou, Y.; Wu, J.; Long, C. Evaluation of peer-to-peer energy sharing mechanisms based on a multiagent simulation framework. *Appl. Energy* **2018**, *222*, 993–1022. [CrossRef]
18. Long, Y.; Wang, Y.; Pan, C. Incentive mechanism of micro-grid project development. *Sustainability* **2018**, *10*, 163. [CrossRef]
19. Wu, X.; Hu, X.; Yin, X.; Moura, S.J. Stochastic Optimal Energy Management of Smart Home with PEV Energy Storage. *IEEE Trans. Smart Grid* **2018**, *9*, 2065–2075. [CrossRef]
20. Vivekananthan, C.; Mishra, Y.; Li, F. Real-Time Price Based Home Energy Management Scheduler. *IEEE Trans. Power Syst.* **2015**, *30*, 2149–2159. [CrossRef]
21. Englberger, S.; Gamra, K.A.; Tepe, B.; Schreiber, M.; Jossen, A.; Hesse, H. Electric vehicle multi-use: Optimizing multiple value streams using mobile storage systems in a vehicle-to-grid context. *Appl. Energy* **2021**, *304*, 117862. [CrossRef]

22. Langer, L. An Optimal Peer-to-Peer Market Considering Modulating Heat Pumps and Photovoltaic Systems under the German Levy Regime. *Energies* **2020**, *13*, 5348. [CrossRef]
23. Sagawa, D.; Tanaka, K.; Ishida, F.; Saito, H.; Takenaga, N.; Nakamura, S.; Aoki, N.; Nameki, M.; Saegusa, K. Bidding Agents for PV and Electric Vehicle-Owning Users in the Electricity P2P Trading Market. *Energies* **2021**, *14*, 8309. [CrossRef]
24. Dean Holland Clift, Harry Suehrcke, Control optimization of PV powered electric storage and heat pump water heaters. *Sol. Energy* **2021**, *226*, 489–500. [CrossRef]
25. Morstyn, T.; McCulloch, M.D. Multiclass Energy Management for Peer-to-Peer Energy Trading Driven by Prosumer Preferences. *IEEE Trans. Power Syst.* **2019**, *34*, 4005–4014. [CrossRef]
26. Park, D.-H.; Park, J.-B.; Lee, K.Y.; Son, S.-Y.; Roh, J.H. A Bidding-Based Peer-to-Peer Energy Transaction Model Considering the Green Energy Preference in Virtual Energy Community. *IEEE Access* **2021**, *9*, 87410–87419. [CrossRef]
27. Park, D.-H.; Park, Y.-G.; Roh, J.-H.; Lee, K.Y.; Park, J.-B. A Hierarchical Peer-to-Peer Energy Transaction Model Considering Prosumer's Green Energy Preference. *Int. J. Control. Autom. Syst.* **2021**, *19*, 311–317. [CrossRef]
28. Reis, I.F.G.; Gonçalves, I.; Lopes, M.A.R.; Antunes, C.H. Assessing the Influence of Different Goals in Energy Communities' Self-Sufficiency—An Optimized Multiagent Approach. *Energies* **2021**, *14*, 989. [CrossRef]
29. Pena-Bello, A.; Parra, D.; Herberz, M.; Tiefenbeck, V.; Patel, M.K.; Hahnel, U.J. Integration of prosumer peer-to-peer trading decisions into energy community modelling. *Nat. Energy* **2022**, *7*, 74–82. [CrossRef]
30. ECHONET Lite, "ECHONET Lite API". Available online: https://echonet.jp/web_api_appendix_v100/ (accessed on 1 December 2022).
31. Meteorological Engineering Center, Inc. Apollon: Solar Power Generation Forecast System. Available online: https://www.meci.jp/apollon.html (accessed on 1 December 2022).
32. Breiman, L. Random Forests. *Mach. Learn.* **2001**, *45*, 5–32. [CrossRef]
33. The Kansai Electric Power Co., Inc. Hapi-E-Time R. Available online: https://kepco.jp/ryokin/menu/hapie_r/ (accessed on 29 November 2021).

energies

MDPI

Article

Construction of Mixed Derivatives Strategy for Wind Power Producers

Yuji Yamada [1,*] and Takuji Matsumoto [2,*]

1 Faculty of Business Sciences, University of Tsukuba, Tokyo 112-0012, Japan
2 Faculty of Transdisciplinary Sciences for Innovation, Kanazawa University, Ishikawa 920-1192, Japan
* Correspondence: yuji@gssm.otsuka.tsukuba.ac.jp (Y.Y.); mtakuji@staff.kanazawa-u.ac.jp (T.M.)

Abstract: Due to the inherent uncertainty of wind conditions as well as the price unpredictability in the competitive electricity market, wind power producers are exposed to the risk of concurrent fluctuations in both price and volume. Therefore, it is imperative to develop strategies to effectively stabilize their revenues, or cash flows, when trading wind power output in the electricity market. In light of this context, we present a novel endeavor to construct multivariate derivatives for mitigating the risk of fluctuating cash flows that are associated with trading wind power generation in electricity markets. Our approach involves leveraging nonparametric techniques to identify optimal payoff structures or compute the positions of derivatives with fine granularity, utilizing multiple underlying indexes including spot electricity price, area-wide wind power production index, and local wind conditions. These derivatives, referred to as mixed derivatives, offer advantages in terms of hedge effectiveness and contracting efficiency. Notably, we develop a methodology to enhance the hedge effects by modeling multivariate functions of wind speed and wind direction, incorporating periodicity constraints on wind direction via tensor product spline functions. By conducting an empirical analysis using data from Japan, we elucidate the extent to which the hedge effectiveness is improved by constructing mixed derivatives from various perspectives. Furthermore, we compare the hedge performance between high-granular (hourly) and low-granular (daily) formulations, revealing the advantages of utilizing a high-granular hedging approach.

Keywords: mixed derivatives; wind power producers; hedging; nonparametric regressions

1. Introduction

Wind power is considered to be sustainable and useful for green energy generation, and the installed capacity of wind power has rapidly increased in many countries and regions. Such an escalation in the installed capacity of wind power and the variability in wind power generation may give rise to considerable uncertainty for power industries operating in competitive electricity markets, as wind power output is volatile and largely contingent upon weather conditions. Owing to the uncertainty of the future and current wind conditions, wind power producers may suffer from the risk of volume fluctuations as well as price movement according to supply and demand in a short period (e.g., every half an hour). Consequently, price and volume simultaneously fluctuate, resulting in extreme volatility for the revenue or income cash flow of a wind power producer. To expand and support wind power generation businesses, even in a competitive environment, it is key to know how to control revenue and stabilize cash flow fluctuations effectively.

In this study, we construct a quantitative strategy for reducing the cash flow fluctuations of wind power producers in the electricity trading market using adequate financial instruments that are related to electricity prices, wind power generation, and weather indexes. Such financial instruments are called electricity derivatives [1] when the underlying asset is the price of electricity or the amount of electricity generated, and they are referred to as weather derivatives [2] when the underlying asset is an index composed of observed weather values, etc. Various instruments have been developed and proposed, as well as a number of optimal hedging techniques using derivatives in energy markets [3]. The use of derivatives and hedging transactions are widely practiced in today's electricity market and have become an integral part of risk management for electric utilities [4,5]. In particular, weather derivatives are extremely useful for hedging energy market-related risks, and in recent years, there have been a large number of applications in both practical and research contexts [6]. Moreover, pricing methods and specific applications of weather derivatives are presented in several pieces of literature [7–10], where previous studies have been mainly related to weather derivatives on a single underlying index such as temperature [11–16], but also solar radiation/irradiation [17,18] and rainfall [19]. In the context of wind power businesses, the recent increase in hedging needs, coupled with the increased demand for wind power to achieve decarbonization goals, has led to recent studies on wind derivatives. There is a wide variety of derivatives, including the European put-type quanto options [20], barrier options [21,22], collar options [23], and wind call options [24], as well as the wind power futures (WPFs) [25–27], which are traded. Additionally, in a more recent study [28], the authors note that WPFs with area-wide generation as the underlying asset are not suitable for the effective hedging of individual generators, and factor futures with some principal components of individual generation as the underlying asset were proposed.

However, most of these studies have focused on the "pricing" of existing/standardized derivatives with single underlying indexes (Note that of the above studies, refs. [23,25,28] deal with hedging effects; however, it is still true that pricing challenges exist there). This fact is not surprising as the need for pricing, or the practical challenge, is inherent in those financial instruments themselves, i.e., standard-type derivatives. In other words, in the trading of standard-type derivatives, because the form of the derivative's payoff function is determined in advance as a univariate function, the procedure whereby the buyer pays the seller an amount equivalent to the expected value of that payoff at the contract time (obtained from the "pricing") is inevitable (assuming transactions between risk-neutral players and ignoring interest rates here). Moreover, the payoff functions of existing derivatives of the put or call types are not necessarily ideal ones for individual hedgers. For example, some studies [29–31] have proposed the use of a large number of put/call electricity derivatives along with the hedge volume risk in the electricity businesses. However, if the optimal derivatives can be designed and are tradable, they allow for more effective hedging while at the same time consolidating derivative contracts. Therefore, from the perspective of enhancing the hedge effect for individual hedgers, we adopt the so-called "flexible approach" that optimally synthesizes the payoff function (possibly a multivariate nonlinear function) to reflect individual hedge requests instead of existing standard-type derivatives.

The flexible approach has another advantage of solving the pricing issues of derivatives as illustrated in [32,33]. That is, it is demonstrated that any (European type) derivative can be designed by optimizing the payoff function with an expected value constraint of zero, which does not require payment procedures at the contract time. In other words, considering the "net payoff" of a derivative contract is the payoff at the maturity time minus the (fair) price at the contract time, and the essential issue of effective derivative design is to optimize this net payoff under fair trading conditions (i.e., expected net payoff equals zero). This derivative design was first applied in [32] for designing temperature derivatives in electric utilities and has been developed into various forms of derivatives transactions for prediction error losses [33], simultaneous hedging of price and volume [34,35], and mediated hedge trading model among different electric utility players [36]. A common feature of these studies is to use nonparametric regression techniques and to seek derivatives with arbitrarily nonlinear payoffs (nonparametric derivatives) for optimal hedging. Note that another approach to hedging that replicates arbitrarily nonlinear payoffs has recently been proposed in [37] based on the standardized derivatives, which were shown to be superior to nonparametric derivatives in terms of market transactions at the expense of hedging effect.

Thus, nonparametric derivatives have proven to be flexible and advantageous in terms of both hedge effect and contracting procedures efficiency. However, it is an open research question whether they can be applied to hedging the fluctuation risk of the product of price and volume in wind power businesses or whether (and how much) their effectiveness can be improved when using nonparametric derivatives with multiple underlying indexes. In other words, there exists a research gap in the absence of a modeling methodology or empirical analysis of this issue. Additionally, while recognizing the limitations of area-wide wind derivatives, as in [28], it may be possible to solve the problem of improving hedge effects by combining derivatives on weather variables as well as wind power generation; therefore, this study also addresses this interesting hypothesis.

In light of the aforementioned background, the present study proposes the utilization of nonparametric techniques to find optimal payoff structures or positions for mixed derivatives with fine granularity on the underlying indexes given by spot electricity price, area-wide wind power production index, and local wind speed and direction. Moreover, we develop a methodology for constructing mixed derivatives by modeling multivariate functions (related to payoff functions for multivariate derivatives) of wind speed and wind direction, where tensor product spline functions are applied to incorporate periodicity constraints on the wind direction. In the end, we demonstrate the enhancement of the hedge effect for mixed derivatives from various aspects based on empirical data. We also compare the hedge performance between high-granular (hourly) and low-granular (daily) formulations and reveal the advantages of utilizing a high-granular hedging formulation.

Note that this paper provides a new perspective on the simultaneous price and volume hedging problem by constructing derivative portfolios of multiple underlying assets and by elaborately modeling the nonlinear relationships among variables, which is distinct from our previous studies that have only dealt with volume risk [33,37]. Other related studies dealing with simultaneous price and volume hedging problems have been conducted for retailers [29–31,38], solar power producers [39], wind power producers [40], etc., but all of them deal with derivatives having a single underlying asset (e.g., electricity prices or temperatures). While some of our previous studies have also constructed multiple derivative portfolios for solar power producers and retailers [34,36], this study has a different objective in that it provides useful insights for wind power risk management practices and modeling challenges based on actual wind power generation data of wind farms (WFs) and other observation data for the electricity market and weather. In the context of a simultaneous hedging of price and volume, the quanto option, which has both of them as underlying assets, could be utilized (see, e.g., [41,42]). However, as previously noted, such conventional derivatives encounter challenges in terms of pricing and hedge effectiveness (thus, this approach was not employed). Furthermore, this study is not only new in constructing effective portfolios for the simultaneous price and volume hedging problem in the wind power business but also makes an unprecedented effort to examine in detail the modeling techniques that enhance the hedge effect. For example, the viewpoint that hedge effects can be improved by combining (original) local wind speed derivatives in addition to (commonly traded) area-wide wind generation derivatives would provide a useful perspective for both power producers and financial product designers. In addition, we demonstrate other multifaceted empirical analyses that have not been focused on in previous studies, such as the use of lag variables in the design of underlying assets, testing the robustness by varying in-sample and out-of-sample periods in multiple scenarios, and understanding hedge effects using different indicators.

The remainder of this paper is organized as follows. In Section 2, we start with motivational arguments on the policy mechanisms related to renewable power trading and formulate the optimal hedging problem with mixed derivatives. We construct a concrete hedging model, detail the underlying assets considered and the method of modeling nonlinearity, and provide data description and performance measures. Section 3 presents the results of the empirical analysis from various perspectives. In particular, we clarify the hedge effects across multiple hedging models and the choice of underlying indexes with a detailed discussion of effective modeling approaches and their implications. In Section 4, we compare the hedge performance between high-granular (hourly) and low-granular (daily) formulations and reveal the advantages of utilizing a high-granular hedging formulation. Finally, we provide conclusions in Section 5.

2. Optimal Hedging with Mixed Derivatives

2.1. Motivation

We begin with motivational arguments on policy mechanisms related to renewable power trading. Among several policy mechanisms for enhancing renewable energy development, the feed-in tariff (FIT) is probably the most commonly introduced method in many countries [43]. Under the FIT mechanism, wind power producers are generally protected from price risk; that is, the sales price of the power output is guaranteed for a certain long-term period. In this case, it is sufficient for wind power producers to hedge volume risk using derivatives with suitable underlying indexes related to wind power variability. A potential candidate for such an underlying index is the wind power production (WPP) index, which denotes an area-wide production of wind power in a country, such as the German wind power index listed in Nasdaq [44], to hedge the production of wind power in Germany. Such a WPP may be effective for groups of WFs covering a wide area but may be challenging to use for individual WFs because of the strong local effect of weather. In addition, the WPP index must be standardized and managed under a solid rule to guarantee the fairness of data. Another candidate for a wind-related index is to use publicly available wind weather condition (WWC) data, such as wind speed and direction observed by a meteorological weather station in the local area. The advantage of such a WWC index is that weather data are objective and observed at many different places with the same quality and that WFs may incorporate local wind conditions near their locations.

There is an updated policy mechanism called the feed-in premium (FIP; see, e.g., [45]), under which wind power producers sell the power output at a market price and receive an additional payment (premium). Both FIT and FIP mechanisms provide an incentive for business owners to develop wind power generation projects in the corresponding country, but an essential difference is that, under FIP, wind power producers need to be concerned with the risk of price as well as volume fluctuations. The introduction of FIP is aimed to integrate the renewable energy development business into a market-oriented environment where the sales price changes according to the competitive market price. This situation is the focus of this study, under which wind power producers may suffer from the risk of simultaneous fluctuations in price and volume, and it is the key to know how to stabilize their revenues (or cash flows) for trading wind power output effectively in the electricity market.

To manage the risk of cash flow fluctuations determined by the price and volume, wind power producers may use electricity derivatives whose underlying index is given by the spot electricity price, in addition to wind derivatives on WPP and WWC indexes. However, it would be more desirable for wind power producers to use mixed derivatives of electricity prices and wind-related indexes using a multivariate payoff function if there exists a suitable methodology. Our study is based on this point of view, i.e., we would like to develop a methodology to optimally construct mixed derivatives on wind indexes and electricity prices for wind power producers (i.e., WFs) and control their cash flows against the simultaneous fluctuation of electricity price and wind power generation.

2.2. Minimum Variance Hedging Problem

The goal of hedging is to replicate the target cash flow using another cash flow that is defined by the payoff of a portfolio of derivative contracts with effective underlying indexes. In this context, we provided a methodology to construct a derivative portfolio of weather and energy derivatives using nonparametric regression techniques for power retailers and photovoltaic (PV) generators [34]. This idea has been extended to the case where power retailers and generators (PV or thermal generators) trade through the central exchange market, and an insurance company handles weather and electricity derivatives or forward contracts between them [36]. In this study, we extend our previous work to the hedging problem of wind power producers and formulate mixed derivatives by specifying various types of payoff functions.

Let V_t [kWh] be the power output of a WF that can be sold at the spot electricity price, denoted by S_t, in the corresponding area for delivering electricity in the (hourly) time interval $[t, t+1)$, where we assume that the basic period is an hour and all variables are observed at least hourly. Note that V_t changes with wind conditions, whereas S_t is volatile depending on the supply and demand at the moment; hence, the revenue (or the cash flow) from electricity sales, defined by V_tS_t, may be extremely volatile.

To reduce the risk of cash flow fluctuations, we introduce the following minimum variance hedging problem using mixed derivatives on spot electricity price S_t and WPP and WCC indexes WPP_t and WWC_t:

$$\begin{aligned} &\min_{f} \mathrm{Var}[V_tS_t - f(S_t, WPP_t, WWC_t)] \\ &\text{s.t.} \overline{f(S_t, WPP_t, WCC_t)} = 0 \end{aligned} \tag{1}$$

where Var[·] represents the variance and f denotes a payoff function of mixed derivatives, which is optimized subject to the zero-mean condition $\overline{f(S_t, WPP_t, WWC_t)} = 0$. In (1), WWC_t may contain the wind speed W_t and direction D_t observed by a meteorological weather station near the WF; WPP_t is the total wind power generation in a country or region, covering all locations of the WFs. We use the total wind power generation data in the (hourly) time interval $[t, t+1)$, denoted by P_t, in the electricity network observed by the independent system operator (ISO). Note that P_t is an area-wide index and that only one representative variable is observed. On the other hand, the dimensions of W_t and D_t depend on the number of meteorological weather stations near the wind farm. If there are multiple observation points, each variable of W_t or D_t becomes multidimensional to incorporate multiple wind speeds or directions observed at different stations.

In our approach, we formulate European-type derivatives by optimizing payoff functions, subject to the expected value constraint of zero, without involving any upfront payment obligations at the contract initiation, thereby leading to derivative contracts with an initial value of zero. In this context, the crux of effective derivative design lies in the optimization of the "net payoff" under fair trading conditions wherein the expected net payoff is set at zero; the "net payoff" of a derivative contract is defined as the payoff at maturity minus the (fair) price at the time of contract initiation (which is set as zero in our approach).

Furthermore, it is noteworthy that wind power generation can be technically formulated by a cubic equation for wind speed, which accounts for Betz's law [46,47]. However, in practical application, there is also a certain correlation between the wind direction and the wind power generation, as wind speed is measured at a specific point. As a result, wind direction is often incorporated as an explanatory variable in contemporary wind power forecasting models [48]. Thus, the incorporation of local wind speed and wind direction information in addition to area-wide power generation data is anticipated to provide a more precise depiction of individual wind power generation patterns and potentially enhance the effectiveness of hedge effects with mixed derivatives.

2.3. Construction of Hedging Models

In this study, we apply three types of regression models, corresponding to linear hedging, spline hedging, and tensor hedging, respectively, for solving (1), and evaluate their hedge effects.

2.3.1. Linear Hedging

At first, we demonstrate the following linear regression of V_tS_t on covariates (W_t, P_t, S_t) as a benchmark, namely "linear hedging", which provides the simplest hedging strategy for solving (1):

$$V_tS_t = \beta[W_t, P_t, S_t, W_tS_t, P_tS_t, 1]^\top + \epsilon_t, \quad (2)$$

where β is a coefficient vector to be estimated and ϵ_t is the residual term in the linear regression. In Equation (2), the regression coefficients provide optimal investment units of forward contracts on the underlying indexes $(W_t, P_t, W_tS_t, P_tS_t)$ for controlling the cash flow deviation from $\overline{V_tS_t}$ by assuming that the physical probability measure provides a risk-neutral probability measure. This is because forward prices are given by their mean values under the risk-neutral probability measure, and the following condition holds with the optimal regression coefficient vector β^* and residual ϵ_t^*:

$$\epsilon_t^* = V_tS_t - \overline{V_tS_t} - \beta^*\left[W_t - \overline{W_t}, P_t - \overline{P_t}, W_tS_t - \overline{W_tS_t}, P_tS_t - \overline{P_tS_t}\right]^\top.$$

(In this paper, the overline notation, e.g., $\overline{V_tS_t}$, is used to denote the mean value. Note that the notation $a \equiv b$ below is used to denote that a is defined to be b).

Then, we see that minimizing $\mathrm{Var}[\epsilon_t^*]$ provides the optimal value of the objective function by letting:

$$f(S_t, WPP_t, WCC_t) \equiv \beta^*\left[W_t - \overline{W_t}, P_t - \overline{P_t}, W_tS_t - \overline{W_tS_t}, P_tS_t - \overline{P_tS_t}\right]^\top.$$

In other words, the residual ϵ_t^* corresponds to the optimally hedged cash flow using forward contracts on $(W_t, P_t, W_tS_t, P_tS_t)$; therefore, we may conclude that solving the minimum variance hedging problem (1) with forward contracts can be achieved through linear regression of (2).

2.3.2. Spline Hedging

Next, we introduce a nonparametric regression,

$$V_tS_t = \varphi_W(W_t)S_t + \varphi_P(P_t)S_t + \beta_0 + \epsilon_t, \quad (3)$$

characterized by smoothing spline functions φ_W and φ_P being multiplied by S_t (called a by variable; see [49]), where β_0 is the intercept and ϵ_t is the residual. The regression problem (3) may be solved as a generalized additive model (GAM [50,51]) using univariate smoothing spline functions s_W and s_P to minimize the residual sum of squares with a penalty on smoothness. In this study, we refer to the hedging strategy obtained from (3) as "spline hedging."

A common structure of each regression function in (3) is to have S_t as a by variable of smooth terms [49] that is introduced precisely against the price risk, and its interaction with wind power generation is incorporated. That is, the sum of spline functions, $\varphi_W(W_t) + \varphi_P(P_t)$, is multiplied by S_t to hedge the simultaneous fluctuation of price and wind power generation.

The regression Equation (3) provides minimum variance hedging using derivatives on (W_t, P_t, S_t). Even though derivative prices are not explicitly incorporated, it can be shown that the resulting residuals from the regressions provide hedged cash flows in the objective function of (1) under the same assumptions as those for (2). Then, the optimal payoff function may be obtained using the regression functions in (3) as

$$f(S_t, WPP_t, WCC_t) \equiv \varphi_W(W_t)S_t - \overline{\varphi_W(W_t)S_t} + \varphi_P(P_t)S_t - \overline{\varphi_P(P_t)S_t}.$$

2.3.3. Tensor Hedging

Third, we further incorporate the wind direction in the first term of (3) and consider the following regression:

$$V_t S_t = te(D_t, W_t)S_t + \varphi_P(P_t)S_t + \beta_0 + \epsilon_t. \quad (4)$$

Compared with (3), we see that the first term in (4) is extended to incorporate the nonlinearities of wind speed W_t regarding power generation and the cyclic feature of wind direction D_t. Note that the bivariate smooth function te is represented as a tensor product spline function of (D_t, W_t), which is a multidimensional smooth function that is expressed by a tensor product. Because separate basis functions may be specified in tensor product spline functions, we set a cyclic smoothing basis for the D_t-direction and a cubic spline basis for the W_t-direction to solve the regression problem in (4), where the cyclic smoothing basis enables us to model a periodic function whose endpoint smoothly connects with the initial point (see [51] for the definitions of tensor product and cyclic spline functions). Then, the regression problem (4) may be solved as GAM and the optimal payoff function may be obtained using the regression functions in (4) as

$$f(S_t, WPP_t, WCC_t) \equiv te(D_t, W_t)S_t - \overline{te(D_t, W_t)S_t} + \varphi_P(P_t)S_t - \overline{\varphi_P(P_t)S_t}.$$

2.3.4. Incorporation of Lagged and forward Variables

Moreover, we assume that each variable related to the WPP and WCC indexes may be multidimensional when observation data for multiple meteorological weather stations and/or lagged variables of (D_t, W_t) and P_t are used. In such a case, additional spline functions or linear terms are added to (2)–(4). For example, when W_t and D_t are both two-dimensional with $W_t = \left[W_t^{(1)}, W_t^{(2)}\right]^\top$ and $D_t = \left[D_t^{(1)}, D_t^{(2)}\right]^\top$, (2)–(4) are modified as

$$\begin{aligned}
V_t S_t &= \beta\left[W_t^{(1)}, W_t^{(2)}, P_t, S_t, \left(W_t^{(1)}, W_t^{(2)}\right)S_t, P_t S_t, 1\right]^\top + \epsilon_t, \\
V_t S_t &= \varphi_W^{(1)}\left(W_t^{(1)}\right)S_t + \varphi_W^{(2)}\left(W_t^{(2)}\right)S_t + \varphi_P(P_t)S_t + \beta_0 + \epsilon_t, \\
V_t S_t &= te^{(1)}\left(D_t^{(1)}, W_t^{(1)}\right)S_t + te^{(2)}\left(D_t^{(2)}, W_t^{(2)}\right)S_t + \varphi_P(P_t)S_t + \beta_0 + \epsilon_t,
\end{aligned}$$

where $\varphi_W^{(i)}$ and $te^{(i)}$, $i = 1, 2$ are univariate smoothing spline and bivariate tensor spline functions, respectively, and β is a coefficient vector with appropriate dimensions.

Another noteworthy issue is the enhancement of covariates by incorporating lagged and forward variables. The effect of wind conditions may not be direct, and lagged or forward variables may contribute to the improvement in hedge performance because the meteorological weather stations chosen in our experiment are not neighbors (approximately 10 km or more from the wind farm). A similar consideration may be applied to the relationship with the WPP index. In this study, we examine the effect of lagged and forward variables in covariates, (D_t, W_t, P_t) on the hedge performance (see Section 3.2).

2.4. *Performance Evaluation Measures*

We estimate the optimal regression functions and other required parameters in (2)–(4) using the data in the learning period and subsequently evaluate the hedge performance using the out-of-sample data in the test period. More specifically, we execute out-of-sample hedge simulations by substituting the observed data in the test period into the estimated regression Equations of (2)–(4) to compute the predicted values of the target variable $V_t S_t$. In this case, the deviation of $V_t S_t$ from the expected cash flow $\overline{V_t S_t}$ is hedged using the payoff of derivatives $f(S_t, WPP_t, WWC_t)$, where $V_t S_t - \overline{V_t S_t}$ may be considered as an unhedged cash flow for the wind power producer. Note that the payoff functions are extracted from the estimated regression Equations in (2)–(4), and the difference between the predicted and realized values for $V_t S_t$ provides out-of-sample hedge errors. Like our previous study in [32,36], we use the following variance reduction rates (VRRs) and normalized mean absolute errors (NMAEs) to evaluate the hedge performance in the out-of-sample case:

$$VRR^{out} := \frac{\mathrm{Var}\left[\epsilon_t^{out}\right]}{\mathrm{Var}[V_t S_t]}, \; NMAE^{out} := \frac{\overline{\left|\epsilon_t^{out}\right|}}{\overline{\left|V_t S_t - \overline{V_t S_t}\right|}}, \tag{5}$$

where ϵ_t^{out} denotes the out-of-sample hedge errors in the test period.

To test the robustness of the proposed model, we also consider the following hypothetical in-sample case. That is, instead of using the estimation result of the learning period to compute out-of-sample hedge errors, we can estimate the regression functions and compute hedge errors using the same data as the test period. In this case, the test period becomes in-sample, and we refer to the resulting hedge errors as in-sample hedge errors. The in-sample hedge errors provide a measure of the "best hedge" based on the same model and data, and the comparison of out-of-sample hedge errors with in-sample hedge errors allows us to understand the performance degradation from the best hedge. For in-sample hedge errors ϵ_t^{in}, we define VRR and NMAE in the in-sample case as

$$VRR^{in} := \frac{\mathrm{Var}\left[\epsilon_t^{in}\right]}{\mathrm{Var}[V_t S_t]}, \; NMAE^{in} := \frac{\overline{\left|\epsilon_t^{in}\right|}}{\overline{\left|V_t S_t - \overline{V_t S_t}\right|}}. \tag{6}$$

2.5. Data Description

We used the power generation data V_t [kWh] from two individual WFs located in the Kyushu and Chubu areas, Japan, which have the data periods of 1 October 2019 to 30 September 2022 for the WF in Kyushu and 1 October 2019 to 31 March 2022 for the WF in Chubu. The data were observed every 30 min, but we used hourly data to adjust the data frequency to those of the WPP and WCC index data. For the WPP index, we used area-wide wind power generation data, P_t [MWh], released by the general power transmission and distribution business operator, whose function corresponds to the ISO in Japan and is obligated to guarantee final power transmission and distribution in addition to balancing the entire electricity network in the area. For the WCC index, we used wind speed W_t [m/s] and cardinal direction categorized into 16 directions. We assigned $D_t = -16, \ldots, -1$ to the cardinal direction starting from the north clockwise direction. Finally, we used the JEPX spot price, which is the price of spot electricity to deliver electricity during a predetermined half-hour slot and which is executed under the single price auction system in the day-ahead market (see http://www.jepx.org/english/outline/index.html, accessed on 10 April 2023). Here, we take the hourly average to obtain hourly data, denoted by S_t [JPY/kWh].

For the wind farm located in the Kyushu area, we chose two weather observation stations whose distances from the wind farm were approximately 10 and 12 km, respectively. For the wind farm in the Chubu area, we chose four weather observation stations whose distances from the wind farm were approximately 13, 16, 18, and 25 km, respectively. The JEPX spot prices, WPP indexes for Kyushu and Chubu, and weather data were downloaded from the following websites, respectively:

- http://www.jepx.org/market/index.html (JEPX Kyushu area price and Chubu area price, accessed on 27 March 2023).
- https://www.kyuden.co.jp/td_service_wheeling_rule-document_disclosure (WPP index, Kyushu, accessed on 27 March 2023).
- https://powergrid.chuden.co.jp/denkiyoho/ (WPP index, Chubu, accessed on 27 March 2023).
- https://www.data.jma.go.jp/gmd/risk/obsdl/ (WWC Index, Japan Meteorological Agency, accessed on 27 March 2023. Note that missing values were imputed using linear interpolation.).

Panels (a)–(c) in Figure 1 respectively demonstrate the historical data of the Kyushu area spot electricity price, WF power generation with the WPP index, and wind speed data observed at two meteorological weather stations for two months in the data period (i.e., 1 August–30 September 2021). Note that the WF power generation and WPP index are normalized by the mean and standard deviation for comparison based on the same scale.

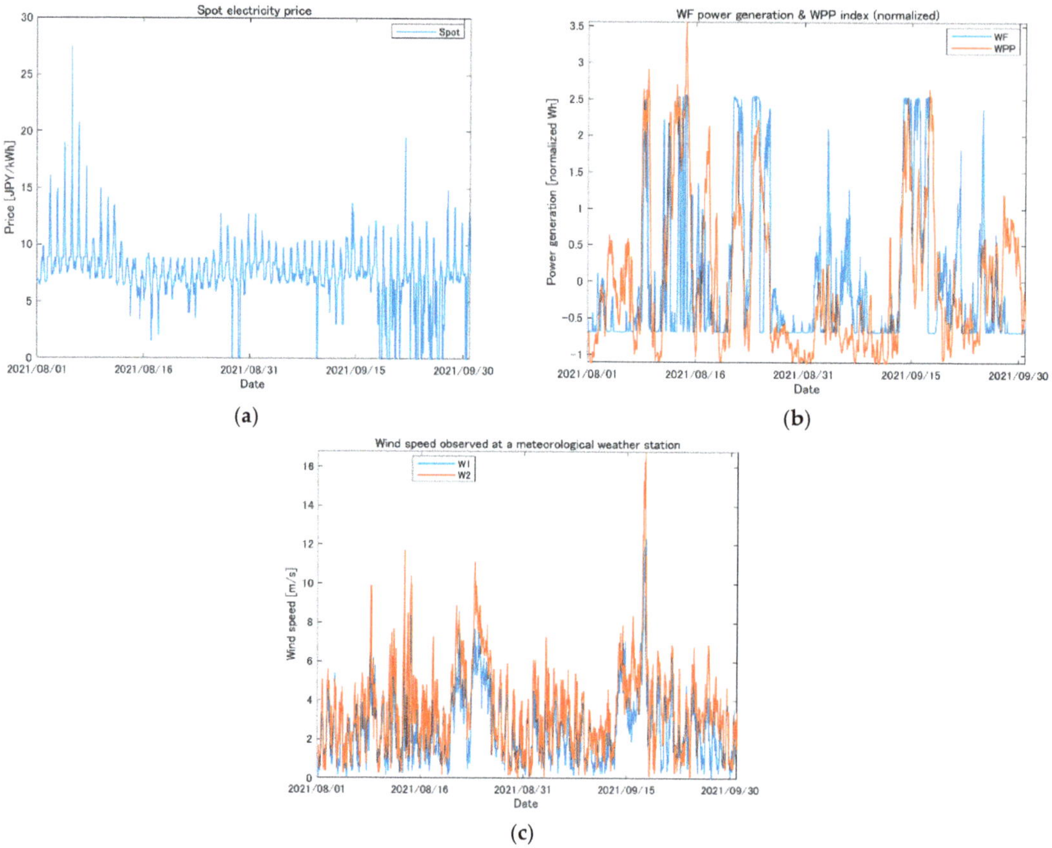

Figure 1. Historical data of the Kyushu area spot electricity price (Panel (**a**)), WF power generation with the WPP index (Panel (**b**)), and wind speed data observed at two meteorological weather stations (Panel (**c**)) during the two months of the data period (i.e., 1 August–30 September 2021).

In this study, all regression models were estimated using R 4.2.2 software (https://www.R-project.org/, accessed on 19 March 2023) using the mgcv package [49] for applying GAM, wherein the smoothing parameter is calculated using the generalized cross-validation criterion. Some figures were plotted using MATLAB 2022a (MathWorks, Inc., Natick, MA, USA).

3. Empirical Performance of Hourly Hedges

We estimate the optimal regression functions and other required parameters in (2)–(4) using the data in the learning period and subsequently evaluate the hedge performance using the out-of-sample data in the test period. Here, the learning and test data periods were set as follows (see Appendix A for descriptive statistics of variables):

- Learning period: 1 October 2019 to 30 September 2020. (Hourly data, number of data observations: 8784)
- Test period: 1 October 2020 to 30 September 2021. (Hourly data, number of data observations: 8760)

In this section, the two-year data period was split in half as shown above, but a further investigation into the relation of the lengths between learning and test periods would be interesting and should be considered as a future research topic.

3.1. Estimated Smooth Functions

Here, we demonstrate estimated smooth functions when regression Equation (4) is applied for the learning period of the data to hedge the cash flow fluctuation of WF in Kyushu. Note that such smooth functions provide a payoff function of mixed derivatives as:

$$f(S_t, WPP_t, WCC_t) = te(D_t, W_t)S_t - \overline{te(D_t, W_t)S_t} + \varphi_P(P_t)S_t - \overline{\varphi_P(P_t)S_t}.$$

Panels (a) and (b) in Figure 2 illustrate the estimated univariate and tensor spline functions of $\varphi_P(P_t)$ and $te(D_t, W_t)$, respectively, where two tensor spline functions (Panels (a) and (b)) and one univariate spline function (Panel (c)) were estimated as there were two observation points of WCC indexes in the case of the WF in Kyushu. We observe that the estimated univariate spline function reflects the positive effect of the WPP index on cash flow and is nonlinear for large values of the WPP index. On the other hand, the estimated tensor spline function depicted in Panels (a) and (b) represents the interaction between wind speed and direction concerning the target cash flow and can model the cyclic property of the wind direction.

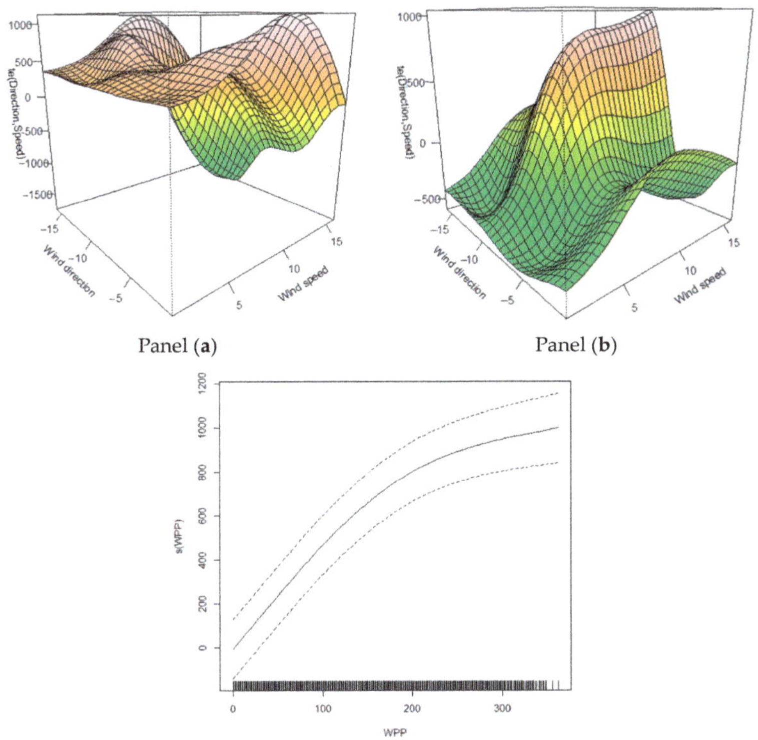

Panel (**a**) Panel (**b**)

Panel (**c**)

Figure 2. Estimated tensor spline functions (Panels (**a**,**b**)) and univariate spline function (Panel (**c**)) when regression Equation (4) is applied for the learning period of the data to hedge the cash flow fluctuation of the WF in Kyushu. Note that because there are two observation points of WCC indexes in the case of the WF in Kyushu, two tensor spline functions (Panels (**a**,**b**)) and one univariate spline function (Panel (**c**)) were estimated.

Based on the estimated tensor spline functions presented in Panels (a) and (b), it becomes evident that an intricate interplay exists between the variables of wind direction and wind speed. Furthermore, this interaction exhibits dynamic changes that are contingent upon the location of the observation point relative to the installation site of WF. It is noteworthy that the configuration of the payoff function is prone to undergo significant changes in the presence of high wind speeds, owing to the limited number of samples available for such cases (For example, outlier data may arise in the context of catastrophic typhoons, leading to the shutdown of wind turbines. There may be potential for exploring additional insurance mechanisms, such as compensation for generator emergency shutdowns that are infrequent in occurrence. However, it should be noted that such considerations are beyond the scope of this study).

It should be noted that one of the advantages of using nonparametric regression models such as GAM is not only the incorporation of the nonlinear effects of covariates but also the ease of interpretation by visualizing the estimated functions [52]. Such a visualization helps us understand the structure of payoff functions for derivatives, as multivariate derivatives may have complex payoffs that need to be specified before executing derivative contracts. Moreover, the ease of interpretation provides an important factor in building consensus and ensuring the reliability of the model as well as accuracy in electric utility practice [53]. In addition, the use of GAM allows us to obtain statistical interpretation as well, similar to other statistical models, although we have omitted the details for brevity.

3.2. *Effect of Lagged and Forward Variables*

Now, we demonstrate the out-of-sample results for our hedge simulations. As aforementioned, the out-of-sample hedge simulations were executed by substituting the observed data in the test period into the regression equations and computing the hedge errors given the estimated smooth functions and coefficients of (2)–(4) using the learning period of the data. Here we examine the hedge effect using VRR and NMAE mainly based on the bar graphs shown in Figures 3 and 4. Note that the numerical results are listed in table formats in Appendix B.

The left and right three sets of bins in Panels (a) and (b) in Figure 3 compare the VRR in the out-of-sample case with those of the in-sample case when different hedges (i.e., linear hedging, spline hedging, and tensor hedging) are applied for the cash flows of the wind farm in Kyushu and Chubu, respectively. The red bins represent the cases in which the target cash flow is regressed concerning covariates with the same observation period only. On the other hand, one period of lagged variables are added for the green bins and one period of forward variables are further added for the blue bins.

When we focus on the same choice of covariates, we see that the hedge performance measured by VRR improves by taking nonlinear effects into account in the regression equations, even for the out-of-sample case. Although the hedge model using tensor product splines (i.e., tensor hedging) can reflect the effect of wind direction and provide the best hedge performance in Figure 3 among all the cases, it sometimes becomes worse compared with spline hedging. However, tensor hedging achieves the smallest VRR among all out-of-sample simulations when lagged and forward variables are added, and the incorporation of lagged and forward variables almost always improves the out-of-sample VRR. In these simulations, the smallest VRR for the out-of-sample case was 11.5% for Kyushu and 18.2% for Chubu, indicating that the variance of the target cash flow was reduced by 88.5% or 81.8% using derivatives with tensor product spline functions.

Note that the in-sample VRR is also plotted as the three sets of bins on the right. In the case of in-sample simulations, we see the improvement of the VRR concerning the choice of methods and effect of adding variables more clearly. In addition, although the out-of-sample VRR for Chubu using tensor hedging achieves a reasonably good value of 18.2%, the performance was degraded compared with the best (in-sample) VRR of 8.30%. This indicates that there is some room for improvement for out-of-sample hedges, and further investigation to reduce the gap may be effective.

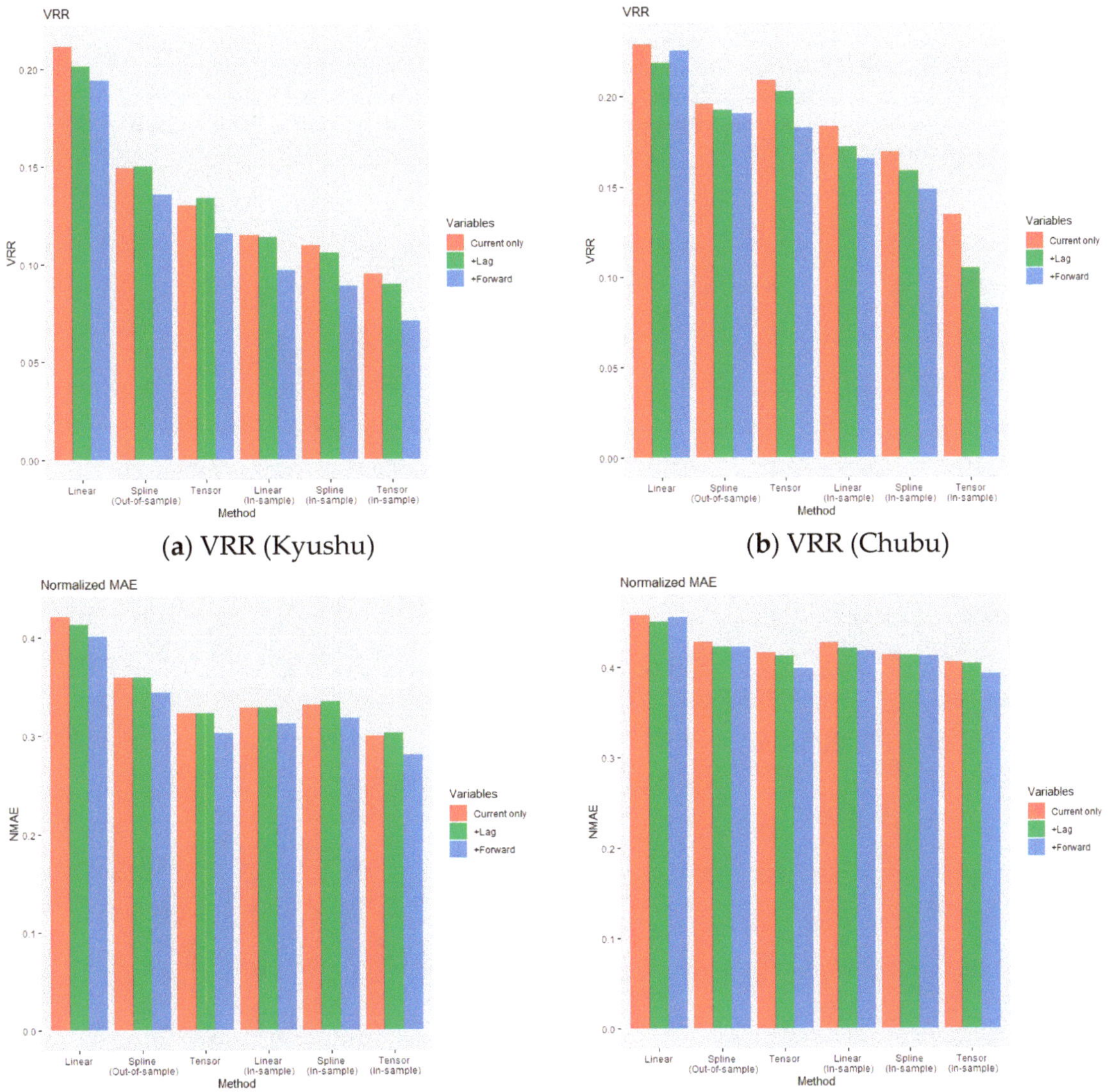

(a) VRR (Kyushu) (b) VRR (Chubu)

(c) NMAE (Kyushu) (d) NMAE (Chubu)

Figure 3. Out-of-sample VRR and NMAE vs. in-sample cases with the comparison for the effect of lagged and forward variables (see Tables A5 and A6 in Appendix B for the numerical results). Panel (**a**)—VRR for the WF in Kyushu. Panel (**b**)—VRR for the WF in Chubu. Panel (**c**)—NMAE for the WF in Kyushu. Panel (**d**)—NMAE for the WF in Chubu.

A similar tendency was observed when NMAE is plotted instead of VRR as shown in Panels (c) and (d) in Figure 3 where NMAE, which is defined in Section 2.4, provides a measure of improvement for the hedges in terms of mean absolute errors relative to unhedged cash flows. In the two simulations for Kyushu and Chubu, the in-sample NMAE and out-of-sample NMAE are not significantly different, indicating that the performance degradation is not very noticeable. Note that the residual sum of squares minimized in the regression problems is closely related to variance minimization; in this sense, VRR may be more suitable as a performance measure for hedges in our approach. However, an evaluation using mean absolute errors or mean squared errors is also worthwhile because they have often been used to evaluate the out-of-sample performance of recent machine learning approaches. In the case of out-of-sample simulations using tensor hedging, we see that the sizes of deviations for the cash flows were reduced by 70% and 61% for Kyushu and Chubu, respectively.

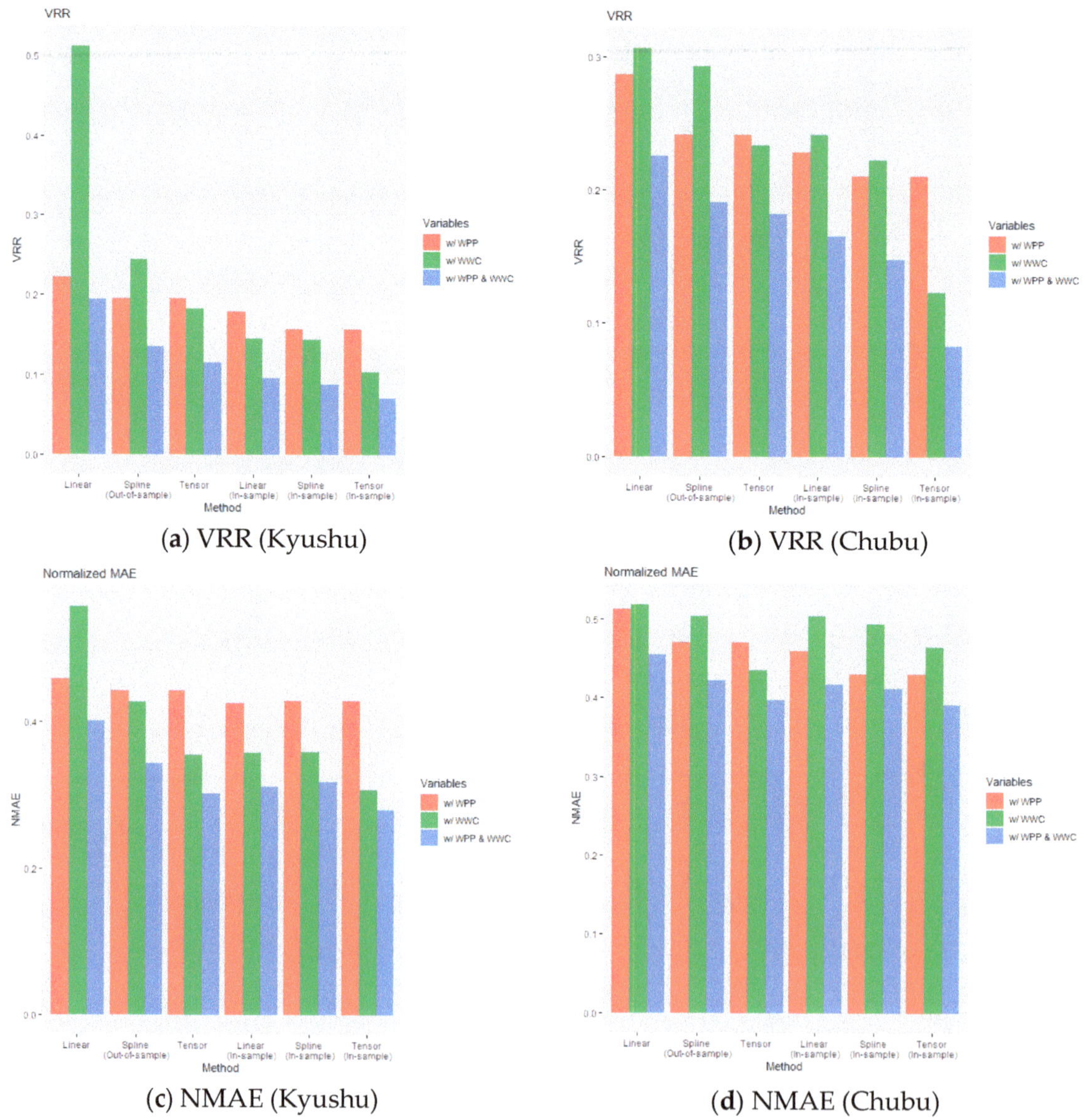

Figure 4. Out-of-sample VRR and NMAE vs. in-sample cases with the comparison for the effect of the incorporation of the WPP or WWC index (see Tables A7 and A8 in Appendix B for the numerical results). Panel (**a**)—VRR for the WF in Kyushu. Panel (**b**)—VRR for the WF in Chubu. Panel (**c**)—NMAE for the WF in Kyushu. Panel (**d**)—NMAE for the WF in Chubu.

3.3. Effect of Individual Wind Power Related Covariates

Next, we compare the hedge effects for different combinations of underlying indexes. Given that the focus of this study pertains to the hedging problem associated with the combined risks of price and volume, it is imperative to utilize electricity futures, with the spot electricity price serving as the underlying index, as a prerequisite for mitigating price risk. Subsequently, it is indispensable to design mixed derivatives that are aimed at addressing the "volume risk" stemming from the uncertainties in wind power generation based on a judicious selection of volume-related underlying indexes. In this context, we examine the effect of volume-related index selection on hedge enhancement and compared the three distinct cases of incorporating WPP, WCC, or both in mixed derivatives.

Panels (a)–(d) in Figure 4 illustrate the relationship between the hedge effects measured by VRR and NMAE, respectively, for the WFs in Kyushu (Panels (a) and (c)) and Chubu (Panels (b) and (d)), where the incorporation of the WPP and/or WCC indexes are compared. Similar to Figure 3, the three sets of bins on the left denote the out-of-sample hedge effects with different choices of covariates, whereas the three sets of bins on the right compare the in-sample hedge effects, in which both lagged and forward variables of (D_t, W_t, P_t) are included along with the current observations. The red bins represent cases where the target cash flow is regressed with the WPP index only, as well as their lagged and forward variables. On the other hand, the green bins are cases with the WCC index, and both WPP and WCC indexes are incorporated for the blue bins. Note that the lengths of the blue bins are the same as those in Figure 3. From these figures, we see that the hedge effects including WPP vs. WWC indexes are almost even or depend on the hedging methods, but the incorporation of both WPP and WWC indexes always provides a better VRR or NMAE for both in-sample and out-of-sample cases. In this sense, we may conclude that derivatives with WWC indexes largely contribute to the improvement of hedge performance.

Given the estimated smooth functions and coefficients of (2)–(4), which were obtained using the learning period of data, we can compute the predicted values of the target cash flows in the out-of-sample period by substituting the observed data into the regression equations. Note that such values do not necessarily provide forecasts from the past period because they contain explanatory variables that are measured at the same time as when the target value V_tS_t is observed. However, plotting the predicted values with the realized values of V_tS_t may help us intuitively grasp how hedging is performed in the out-of-sample period.

Panels (a) and (b) in Figure 5 illustrate the realized versus predicted values of V_tS_t for the WFs in Kyushu and Chubu, respectively, during the out-of-sample period. The red lines denote the predicted values of V_tS_t, computed by (4) using lagged and forward variables, whereas the blue lines denote the realized values of V_tS_t. Note that the difference between the red and blue lines provides the hedge errors used to compute the out-of-sample VRR and NMAE in the previous figures. In both cases, the predicted values seem to replicate the realized values in the upward direction but sometimes overshoot or stay away from zero, even though the realized values are zero for some periods. This is because power generation sometimes becomes zero even though the predicted values stay away from zero, which may be a situation where power generation is controlled to be zero to avoid an accident under strong wind. In addition, the target values become zero when the spot electricity price is zero, and this usually occurs when solar generation is large. Both situations are difficult to predict in advance, but it is important to incorporate the effect of unexpected zero values of target cash flows because they may lead to a large loss for wind power generators. This is an interesting topic to explore further in a future work.

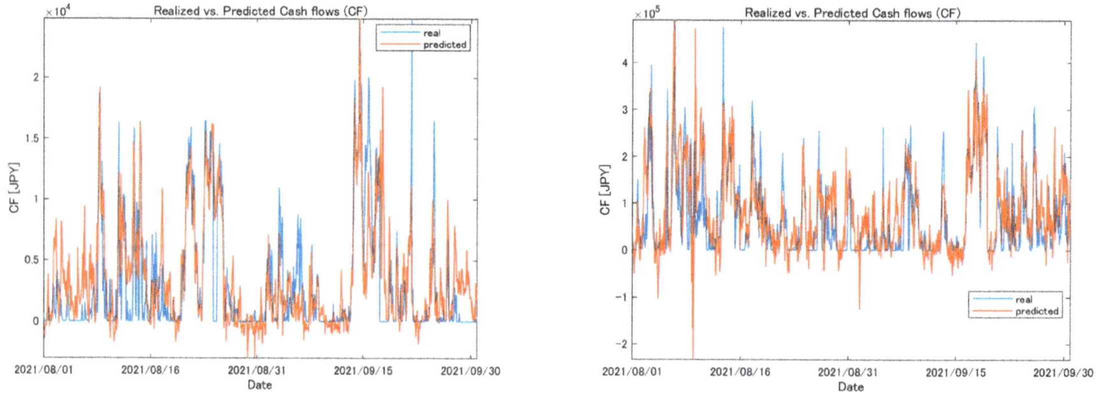

(**a**) Kyushu (**b**) Chubu

Figure 5. Realized vs. predicted values of $V_t S_t$ in the out-of-sample period.

4. Hedging with Longer Horizons

In the previous sections, we have assumed to apply hedges at every hour using derivatives with the hourly settlement. The advantage of considering hedging with high frequency is that it enables the hedger to have the flexibility of selecting the time interval and adjusting the hedge ratio when executing derivative contracts. On the other hand, the hedger may consider applying hedges for a longer time horizon or less frequently, e.g., every day or so. In this section, we introduce a hedging problem with a longer horizon (i.e., low-granular hedging) to reduce the fluctuation of daily revenues. Then, we make a comparison with the high-granular hedging developed in the previous sections.

4.1. Daily Hedging Models

Let $\hat{R}_m$ be a revenue at day m given by the sum of hourly revenues as

$$\hat{R}_m \equiv \sum_{t \in \mathcal{T}_m} V_t S_t, \tag{7}$$

where $\mathcal{T}_m$ stands for a set of 24-hour indexes at day m (e.g., $\mathcal{T}_1 = \{1, \ldots, 24\}$, $\mathcal{T}_2 = \{25, \ldots, 48\}, \ldots$). Similarly, we define the underlying daily indexes related to the wind speed and area-wide wind power production, W_t and P_t, as

$$\hat{W}_m \equiv \frac{1}{24} \sum_{t \in \mathcal{T}_m} W_t, \ \hat{P}_m \equiv \frac{1}{24} \sum_{t \in \mathcal{T}_m} P_t. \tag{8}$$

For the daily spot electricity price index, we take the average of 24-hour prices at day m, i.e.,

$$\hat{S}_m \equiv \frac{1}{24} \sum_{t \in \mathcal{T}_m} S_t. \tag{9}$$

With this notation, we introduce the following hedging problem with respect to the daily revenue as

$$\begin{aligned} & \min_{\hat{f}} \mathrm{Var}\left[\hat{R}_m - \hat{f}(\hat{S}_m, \hat{P}_m, \hat{W}_m)\right] \\ & \text{s.t.} \hat{f}(\hat{S}_m, \hat{P}_m, \hat{W}_m) = 0 \end{aligned} \tag{10}$$

where $\hat{f}$ is a payoff function of the derivative contract and is optimized under the zero-mean condition. Based on a similar argument to that of solving (1), the optimization problem in (10) may be solved by constructing suitable regression equations.

In this study, we construct the following two regression equations, corresponding to linear hedging and spline hedging, respectively:

$$\hat{R}_m = \hat{\beta}\left[\hat{W}_m, \hat{P}_m, \hat{S}_m, \hat{W}_m\hat{S}_m, \hat{P}_m\hat{S}_m, 1\right]^{\top} + \hat{\epsilon}_m \tag{11}$$

$$\hat{R}_m = \hat{\varphi}_W(\hat{W}_m)\hat{S}_m + \hat{\varphi}_P(\hat{P}_m)\hat{S}_m + \hat{\varphi}_{Seasonal}(Seasonal_m)\hat{S}_m + \hat{\beta}_0 + \hat{\epsilon}_m \tag{12}$$

where $Seasonal_m$ denotes a yearly repetitive dummy variable to reflect the seasonality of wind speed in a year, $\hat{\varphi}_W$ and $\hat{\varphi}_P$ are cubic spline functions, $\hat{\varphi}_{Seasonal}$ is a cyclic spline function, and $\hat{\epsilon}_m$ is a residual. The first equation is a linear regression equation, whereas the second may be solved as a GAM to minimize the residual sum of squares with a penalty on smoothness. Then, we obtain linear hedging and spline hedging strategies to solve the optimal hedging problem for the daily setting defined in (10).

As stated in Section 2.3, the cyclic smoothing basis used in the cyclic spline function $\hat{\varphi}_{Seasonal}$ in (12) enables us to model a periodic function whose endpoint smoothly connects with the initial point (see [51] for the definitions of tensor product and cyclic spline functions). Note that the consideration of wind direction is limited to hourly hedging in this study as it is observed on an hourly basis and poses challenges in defining a representative daily value. Nevertheless, incorporating the influence of wind direction in daily hedging could be a potential avenue for future research by exploring its frequency or other relevant parameters.

The coefficient vector $\hat{\beta}$ in (11) is estimated as a linear regression problem, whereas spline functions $\hat{\varphi}_W$, $\hat{\varphi}_P$, and $\hat{\varphi}_{Seasonal}$ (as well as the intercept $\hat{\beta}_0$) in (12) are estimated using the GAM given the observed data in the learning period. Note that once the regression problems are solved, the optimal linear payoff function and nonlinear payoff function are given by estimated regression equations in (11) and (12), and the residuals provide (daily) hedging errors. Because there is no wind direction term for the daily hedges, we only have two hedging models, but a seasonal trend is incorporated in (12) in the case of daily hedges.

Another way for deriving daily hedging is to apply hourly hedging consecutively using regression equations in (2)–(4). In this case, the daily hedging errors, denoted by $\hat{\epsilon}_m^{hourly}$, may be computed by the sum of hourly hedging errors, i.e.,

$$\hat{\epsilon}_m^{hourly} \equiv \sum\nolimits_{t\in\mathcal{T}_m} \epsilon_t. \tag{13}$$

where ϵ_t is an hourly hedging error when the optimal hedging is performed hourly.

To estimate the daily hedge performance in the out-of-sample case, we compute the VRR and NMAE given as follows, similar to that of the hourly hedges defined in Section 2.4:

$$V\hat{R}R^{out} := \frac{Var\left[\hat{\epsilon}_m^{out}\right]}{Var\left[\hat{R}_m\right]}, \; NM\hat{A}E^{out} := \frac{\overline{\left|\hat{\epsilon}_m^{out}\right|}}{\overline{\left|\hat{R}_m - \overline{\hat{R}_m}\right|}}, \tag{14}$$

where $\hat{\epsilon}_m^{out}$ denotes the daily hedge errors in the test period and is replaced by $\hat{\epsilon}_m^{hourly}$ when hourly hedging is applied consecutively to estimate the daily hedge errors.

4.2. Empirical Performance of Daily Hedges

We estimate the optimal regression functions and other required parameters in (11) and (12) using the data in the learning period and apply daily hedging for the out-of-sample data in the test period to evaluate the out-of-sample hedge performance. Similarly, we apply hourly hedging based on the regression Equations (2)–(4) and measure daily hedging errors in the out-of-sample test period. We use the same set of data described in Section 2.5 but consider extending the learning period by sliding the one-year window for the test period as follows (see Appendix A for descriptive statistics of variables):

Period (1) Learning period: 1 October 2019 to 30 September 2020 (Number of daily observations: 366). Test period: 1 October 2020 to 30 September 2021 (Number of daily observations: 365). Kyushu and Chubu.

Period (2) Learning period: 1 October 2019 to 31 March 2021. (Number of daily observations: 548). Test period: 1 April 2021 to 31 March 2022 (Number of daily observations: 365). Kyushu and Chubu.

Period (3) Learning period: 1 October 2019 to 30 September 2021. (Number of daily observations: 731). Test period: 1 October 2021 to 30 September 2022 (Number of daily observations: 365). Kyushu only.

Noting that the data period of the wind farm in Chubu is only until 31 March 2022 (because a portion of the wind farm was separated from the balancing group after 31 March 2022), we performed the empirical test using data in Periods (1)–(3) for the wind farm in Kyushu and in Periods (1) and (2) for the wind farm in Chubu. We examine the hedge effect using VRR and NMAE mainly based on the bar graphs shown in Figures 6 and 7 in this section. Note that the numerical results are listed in table format in Appendix B.

Panels (a), (c), and (e) in Figure 6 illustrate the relationship between the hedge effects estimated by VRR for the out-of-sample periods in Periods (1), (2), and (3), respectively, whereas Panels (b), (d), and (f) illustrate those by NMAE in the corresponding periods. The cyan and magenta bins are estimated by applying (11) and (12), respectively, for the daily observation data, which are labeled as "Linear (Daily)" and "Spline (Daily)". On the other hand, the rest of the bins are estimated using hourly hedges based on (2)–(4) and are labeled as "Linear (Hourly)", "Spline (Hourly)", and "Tensor (Hourly)" on the x-axis, where the three sets of bins compare the VRR and NMAE while incorporating lagged (and forward) variables in addition to current observations as in Figure 3.

At first, we see that nonlinear hedging using the GAM (12) always provided a better hedge performance in terms of both VRR and NMAE than linear hedging using linear regression (11) for all periods. Second, when the rest of the three hedging models are compared, both VRR and NMAE obtained via tensor hedging, i.e., the ones labeled as "Tensor (Hourly)", demonstrated a tendency to outperform those obtained through the utilization of linear and spline hedging techniques, labeled as "Linear (Hourly)" and "Spline (Hourly)", respectively; this agrees with the aforementioned results in Section 3 for the hourly hedges. In addition, the effect of lagged and forward variables was not significant for the improvement of VRR and NMAE. Third, if the linear hedging using daily data is compared with the linear hedging using hourly data, the hedge performance tended to be improved when using hourly data or provided no significant difference between the two. A similar tendency was observed for nonlinear hedging, where VRR and NMAE obtained from hedging with hourly data were at least as good as those from hedging with daily data or improved when the hedges with tensor spline functions, i.e., "Tensor (Hourly)", were applied.

Panels (a)–(d) in Figure 7 provide the results for the wind farm in Chubu when using the data in Periods (1) and (2). Note that there exist certain cases wherein nonlinear hedging does not necessarily exhibit superior performance to linear hedging in Period (2). However, hedging with hourly data showed a better performance at least for "Spline (Hourly)" and "Tensor (Hourly)" compared with both linear and nonlinear hedging using daily data.

In conclusion, our empirical analysis in this section reveals that hedging with the high-granular (hourly) formulation generally provides a better hedge effect than that with the low-granular (daily) formulation and has the advantage of improving the hedge effect.

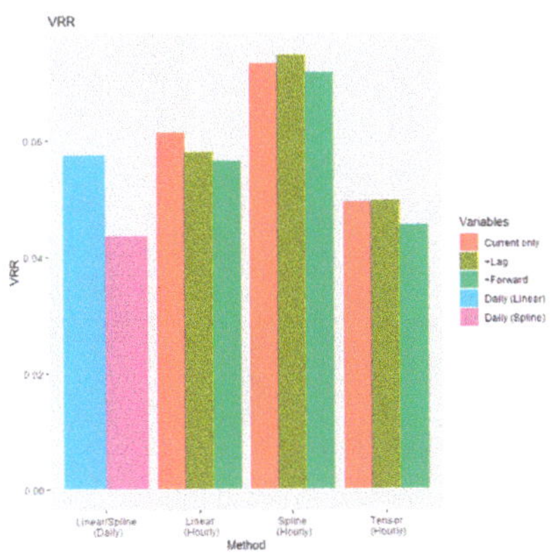

(a) VRR (1 October 2020–30 September 2021)

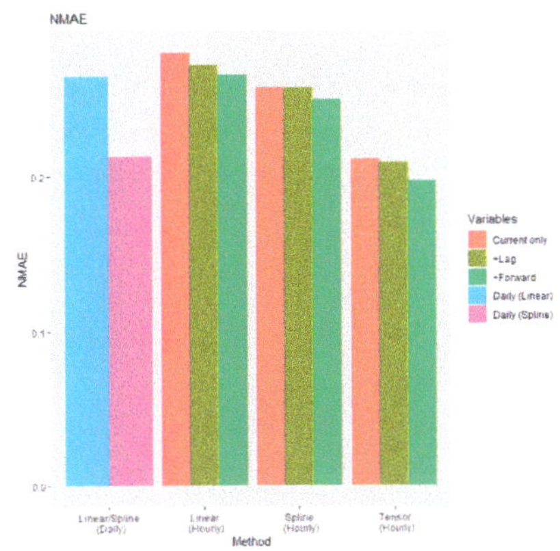

(b) NMAE (1 October 2020–30 September 2021)

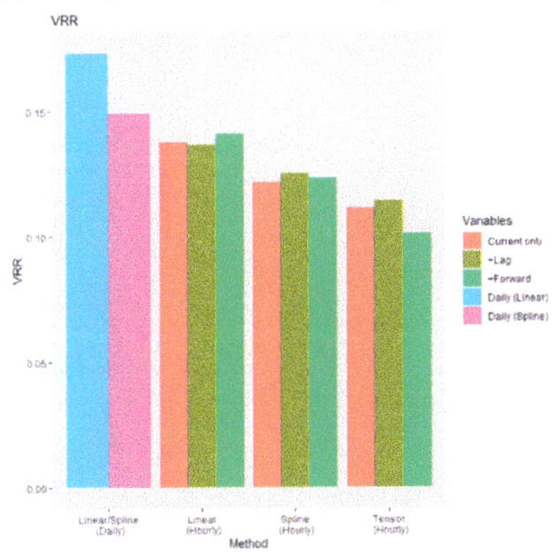

(c) VRR (1 April 2021–31 March 2022)

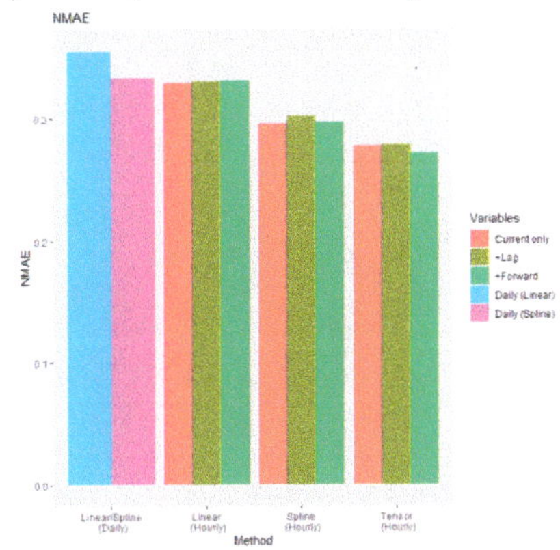

(d) NMAE (1 April 2021–31 March 2022)

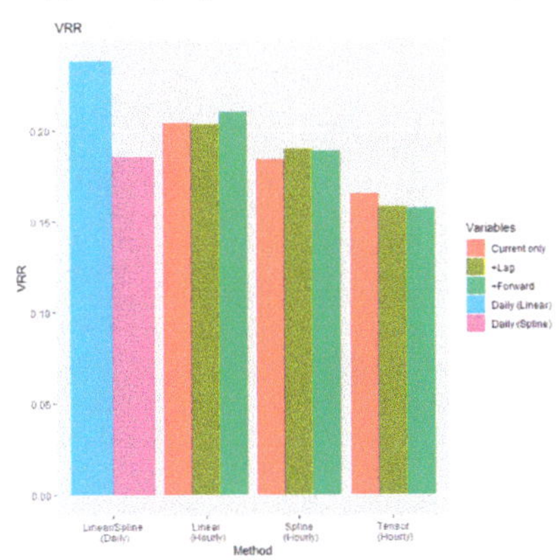

(e) VRR (1 October 2021–30 September 2022)

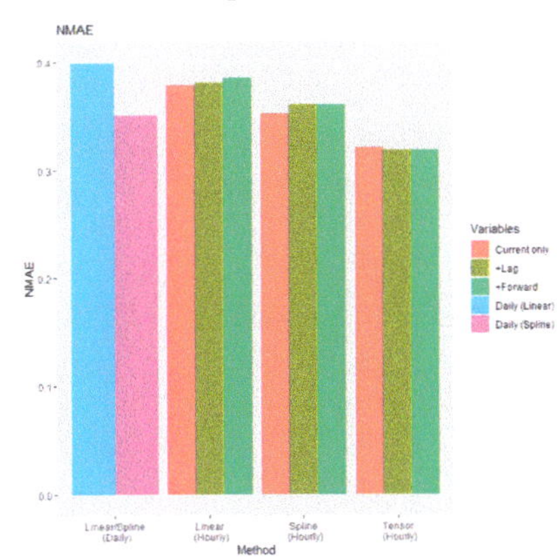

(f) NMAE (1 October 2021–30 September 2022)

Figure 6. Comparison of daily hedges with hourly hedges in Kyushu (see Table A9 in Appendix B for the numerical results). Panels (**a**,**b**): VRR and NMAE for the out-of-sample period of 1 October 2020–30 September 2021. Panels (**c**,**d**): VRR and NMAE for the out-of-sample period of 1 April 2021–31 March 2022. Panels (**e**,**f**): VRR and NMAE for the out-of-sample period of 1 October 2021–30 September 2022.

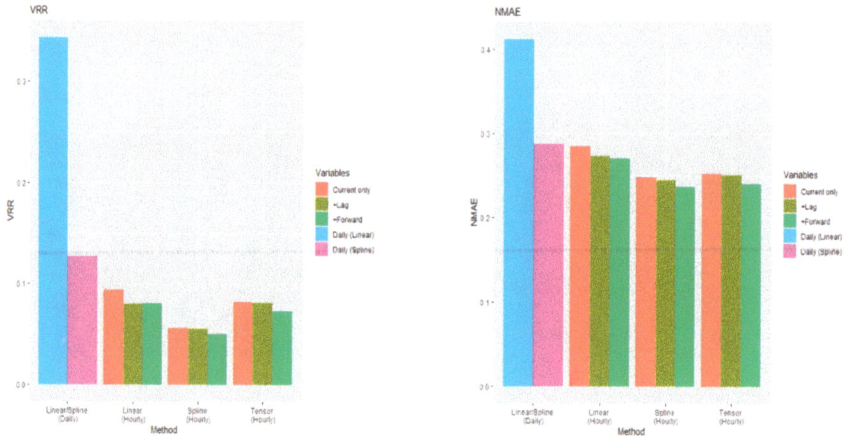

(**a**) VRR (1 October 2020–30 September 2021) (**b**) NMAE (1 October 2020–30 September 2021)

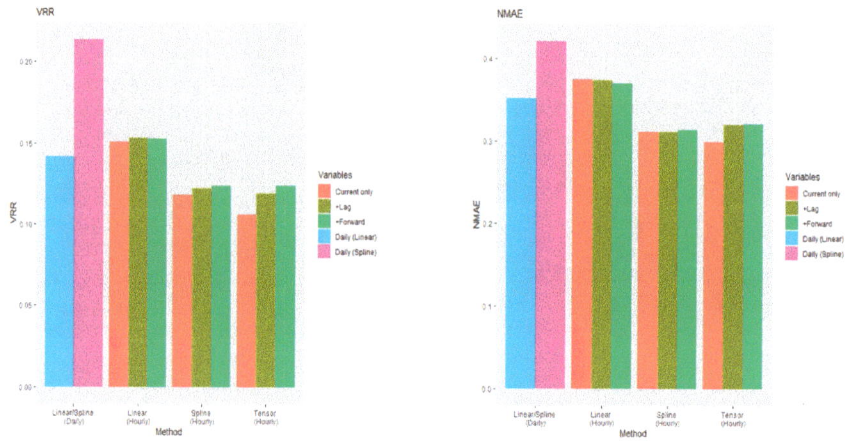

(**c**) VRR (1 April 2021–31 March 2022) (**d**) NMAE (1 April 2021–31 March 2022)

Figure 7. Comparison of daily hedges with hourly hedges in Chubu (see Table A10 in Appendix B for the numerical results). Panels (**a**,**b**): VRR and NMAE for the out-of-sample period of 1 October 2020–30 September 2021. Panels (**c**,**d**): VRR and NMAE for the out-of-sample period of 1 April 2021–31 March 2022.

5. Discussion and Concluding Remarks

In this study, we developed a quantitative strategy for controlling cash flow fluctuations for wind power producers using mixed derivatives on wind indexes and electricity prices. We adopted a flexible approach that optimally synthesizes the payoff function by applying nonparametric regression techniques to reflect the individual wind power generation with fine granularity. An empirical back test was conducted to illustrate our proposed hedging strategy. The out-of-sample simulation results demonstrated that the incorporation of both wind power production and wind weather condition indexes always improved VRR and NMAE, enhancing the value of our high-resolution weather derivatives and the effectiveness of the hedging strategy with mixed derivatives. We also compared the hedge performance between high-granular (hourly) and low-granular (daily) formulations and showed the advantages of utilizing a high-granular hedging formulation for improving the hedge effect.

In addition, the new modeling method using a bivariate tensor product spline function consisting of a wind direction trend with periodicity constraints and a wind speed trend provides a novel perspective in the context of wind power forecasting models, which has attracted much attention in recent years. In wind power forecasting, controlling noise effects and performing a robust model estimation of nonlinear components has been challenging, and various methods have been proposed [54]. Our approach provides a suggestion for this challenge as well.

Finally, we provide supplementary deliberations, encompassing prospective observations.

In the design and trading of derivatives that exhibit high granularity in both temporal and spatial dimensions, it is noteworthy to acknowledge that transaction costs must be taken into account when implementing them in practice. However, the derivatives proposed in this study do not entail payment procedures at the time of contracting and can be calculated relatively easily based solely on historical generation, prices, and weather data, thereby mitigating potential concerns regarding transaction costs. In fact, with the advent of digitalized platforms that support automatic calculations and contracts, the transaction costs are expected to be minimal. While the practical application of such a trading scheme may necessitate further consideration, the rapid advancement of cutting-edge information technology in recent years is anticipated to facilitate the efficient trading of derivatives.

In the policy context of transitioning from FIT to FIP, the significance of price risk hedging is expected to be further accentuated. Furthermore, considering the evolutionary trajectory of the market development, characterized by the advancement of various financial instruments and the changing landscape of decentralized wind power deployment across diverse regions, the issues addressed in this study, namely the need for a sophisticated approach to concurrently manage wind power price and volume risk, as well as the development of such enabling trading environments, are projected to become progressively indispensable in practice. Additionally, with the growing prevalence of variable renewable energy sources, the demand for diversified risk management methods in the electricity market is anticipated to escalate.

As power generators transition towards decentralization, such as local wind and solar power generation, employing finely grained weather indexes may prove to be a more effective strategy for hedging against risks through the utilization of derivative instruments. Considering this situation, the fundamental concepts presented in this study regarding the integration of mixed derivatives that account for both localized and regional indexes, along with the associated technical modeling approaches, are expected to be further developed and investigated, including their applicability to other domains within the power sector.

Author Contributions: Conceptualization, Y.Y. and T.M.; data curation, Y.Y.; formal analysis, Y.Y.; funding acquisition, Y.Y. and T.M.; investigation, Y.Y. and T.M.; methodology, Y.Y. and T.M.; project administration, Y.Y.; resources, Y.Y.; software, Y.Y.; supervision, Y.Y.; validation, Y.Y.; visualization, Y.Y.; writing—original draft, Y.Y.; writing—review and editing, Y.Y. and T.M. All authors have read and agreed to the published version of the manuscript.

Funding: Grant-in-Aid for Scientific Research (A) 20H00285, Grant-in-Aid for Challenging Research (Exploratory) 19K22024, and Grant-in-Aid for Young Scientists 21K14374 from the Japan Society for the Promotion of Science (JSPS).

Data Availability Statement: The JEPX spot prices, WPP indexes for Kyushu and Chubu, and weather data used in this study were obtained from the following websites: https://www.kyuden.co.jp/td_service_wheeling_rule-document_disclosure (accessed on 19 March 2023); https://powergrid.chuden.co.jp/denkiyoho/ (accessed on 19 March 2023); and https://www.data.jma.go.jp/gmd/risk/obsdl/ (accessed on 19 March 2023). The power generation data of the WFs located in the Kyushu and Chubu areas, Japan, were provided by UPDATER Co., Ltd. (Tokyo, Japan) and are available from the first author upon reasonable request.

Acknowledgments: This work was supported by Grant-in-Aid for Scientific Research (A) 20H00285, Grant-in-Aid for Challenging Research (Exploratory) 19K22024, and Grant-in-Aid for Young Scientists 21K14374 from the Japan Society for the Promotion of Science (JSPS). We express our sincere gratitude to UPDATER Co., Ltd. (Tokyo, Japan) for providing the power generation data for the wind farms.

Conflicts of Interest: The authors declare no conflict of interest.

Nomenclature

V	The power output of a WF
S	Spot electricity price
W	Wind speed
D	Wind direction
P	Area-wide wind power generation
WPP	Wind power generation indexes
WCC	Wind weather condition indexes
ϵ	Hedging errors
$\varphi(\cdot)$	Cubic spline function
$te(\cdot)$	Bivariate tensor product spline function
β	Coefficient vector
VRR	Variance reduction rates
NMAE	Normalized mean absolute errors

Note: The subscripts t and m are omitted in the above list, where t and m stand for hour and day, respectively. Moreover, variables, functions, and indexes for date granularity data are denoted with a hat in the text.

Appendix A Descriptive Statistics of Variables for Hourly Hedging

In this appendix, we demonstrate the descriptive statistics of variables used for the hourly and daily hedging performed in Sections 3 and 4. Tables A1 and A2 show the descriptive statistics of the hourly data in Kyushu and Chubu, respectively, for the learning and test periods, whereas Tables A3 and A4 displays those of the daily data in Kyushu and Chubu, respectively. Note that, for daily data, only the case of Period (1) is shown.

Table A1. Descriptive statistics of hourly data in Kyushu for learning and test periods.

	WF Generation [kWh]	Spot Price [JPY/kWh]	Cash Flow [JPY]	WPP Index [MWh]	W1 [m/s]	W2 [m/s]
Learning period						
Mean	544	5.63	3.21×10^3	78.2	2.93	3.94
SD	626	3.81	4.62×10^3	74.3	1.88	2.43
Min	0.00	0.01	0.00	0.00	0.00	0.00
Max	1.89×10^3	50.2	8.84×10^4	361	19.7	28.8
Test period						
Mean	650	22.7	2.06×10^4	80.8	1.81	2.33
SD	567	11.8	7.42×10^3	90.9	2.87	4.00
Min	0.00	0.01	0	0	0	0
Max	1.89×10^3	235	2.90×10^5	370	12.5	16.8

Learning period: 1 October 2019 to 30 September 2020; test period: 1 October 2020 to 30 September 2021. WF generation: power generation of WF in Kyushu [kWh]; spot price: area price in Kyushu (hourly average) [JPY]; cash flow: cash flow of WF [JPY]; WPP index: area-wide wind power generation data in Kyushu [MWh]; W1,W2: wind speeds observed at two meteorological weather stations near the WF [m/s].

Table A2. Descriptive statistics of hourly data in Chubu for learning and test periods.

	WF Generation [kWh]	Spot Price [JPY/kWh]	Cash Flow [JPY]	WPP Index [MWh]	W1 [m/s]	W2 [m/s]	W3 [m/s]	W4 [m/s]
Learning period								
Mean	1.46×10^4	6.03	9.28×10^4	59.3	1.42	2.58	1.53	2.27
SD	1.51×10^4	3.82	1.20×10^5	52.8	1.15	1.70	1.20	1.78
Min	0.00	0.01	0.00	0.00	0.00	0.00	0.00	0.00
Max	4.59×10^4	50.2	1.79×10^6	237	11.3	12.6	9.90	12.3
Test period								
Mean	1.33×10^4	12.3	1.92×10^5	63.3	1.32	2.54	1.49	2.13
SD	1.31×10^4	22.4	5.58×10^5	57.4	1.06	1.72	1.20	1.69
Min	0.00	0.01	0.00	0.00	0.00	0.00	0.00	0.00
Max	4.31×10^4	235	8.55×10^6	239	7.80	11.2	9.30	9.90

Learning period: 1 October 2019 to 30 September 2020; test period: 1 October 2020 to 30 September 2021. WF generation: power generation of WF in Chubu [kWh]; spot price: area price in Kyushu (hourly average) [JPY]; cash flow: cash flow of WF [JPY]; WPP index: area-wide wind power generation data in Kyushu [MWh]; W1–W4: wind speeds observed at four meteorological weather stations near the WF [m/s].

Table A3. Descriptive statistics of daily data in Kyushu for learning and test periods.

	WF Generation [kWh]	Spot Price [JPY/kWh]	Cash Flow [JPY]	WPP Index [MWh]	W1 [m/s]	W2 [m/s]
Learning period						
Mean	1.31×10^4	5.63	7.71×10^4	78.2	2.93	3.94
SD	1.18×10^4	2.18	8.05×10^4	62.1	1.33	1.76
Min	0.00	2.10	0.00	0.792	1.07	1.42
Max	4.46×10^4	17.2	5.65×10^5	302	11.2	12.3
Test period						
Mean	1.36×10^4	11.8	1.78×10^5	90.9	2.87	4.00
SD	1.28×10^4	19.8	4.02×10^5	70.0	1.31	1.77
Min	0.00	3.06	0.00	1.04	0.992	1.13
Max	4.52×10^4	140	3.51×10^6	305	7.55	11.0

Learning period: 1 October 2019 to 30 September 2020; test period: 1 October 2020 to 30 September 2021. WF generation: total daily power generation of WF in Kyushu [kWh]; spot price: area price in Kyushu (daily average) [JPY]; cash flow: total daily cash flow of WF [JPY]; WPP index: area-wide wind power generation data in Kyushu (daily average) [MWh]; W1,W2: wind speeds observed at two meteorological weather stations near the WF (daily average) [m/s].

Table A4. Descriptive statistics of daily data in Chubu for learning and test periods.

	WF Generation [kWh]	Spot Price [JPY/kWh]	Cash Flow [JPY]	WPP Index [MWh]	W1 [m/s]	W2 [m/s]	W3 [m/s]	W4 [m/s]
Learning period								
Mean	3.50×10^5	6.03	2.23×10^6	59.3	1.42	2.58	1.53	2.27
SD	2.79×10^5	2.36	2.15×10^6	44.6	0.810	1.15	0.679	1.18
Min	0.00	2.10	0.00	2.29	0.367	0.971	0.429	0.671
Max	1.07×10^6	17.4	1.29×10^7	196	5.35	8.01	4.11	6.43
Test period								
Mean	3.19×10^5	12.3	4.61×10^6	63.3	1.32	2.54	1.49	2.13
SD	2.38×10^5	19.3	1.03×10^7	50.3	0.705	1.15	0.698	1.08
Min	4.90×10^3	3.07	4.03×10^4	1.04	0.279	1.01	0.458	0.504
Max	9.97×10^5	132	9.83×10^7	206	3.85	7.30	4.19	6.23

Learning period: 1 October 2019 to 30 September 2020; test period: 1 October 2020 to 30 September 2021. WF generation: total daily power generation of the WF in Chubu [kWh]; spot price: area price Chubu (daily average) [JPY]; cash flow: total daily cash flow of WF [JPY]; WPP index: area-wide wind power generation data in Kyushu (daily average) [MWh]; W1–W4: wind speeds observed at four meteorological weather stations near WF (daily average) [m/s].

Appendix B Numerical Results for VRR and NMAE

In Sections 3 and 4, we conducted a comprehensive analysis of the hedge effect of our proposed methodologies, employing VRR and NMAE as the primary evaluation metrics, predominantly based on the analysis of bar graphs from Figures 3, 4, 6 and 7. In this appendix, we arrange the numerical results in tabular formats to present their actual values in Tables A5–A10 below.

Table A5. Out-of-sample VRR and NMAE vs. in-sample cases with the comparison for the effect of lagged and forward variables (Kyushu; see bar graphs in Figure 3).

	VRR (Current only)	VRR (+Lag)	VRR (+Forward)	NMAE (Current only)	NMAE (+Lag)	NMAE (+Forward)
In-sample						
Linear	11.5%	11.4%	9.7%	32.8%	32.8%	31.2%
Spline	10.9%	10.6%	8.9%	33.1%	33.4%	31.8%
Tensor	9.4%	8.9%	7.0%	29.9%	30.3%	27.9%
Out-of-sample						
Linear	21.2%	20.1%	19.4%	42.1%	41.3%	40.1%
Spline	14.9%	15.0%	13.6%	35.9%	35.9%	34.4%
Tensor	12.5%	13.1%	11.4%	32.1%	32.1%	30.2%

Learning period: 1 October 2019 to 30 September 2020; test period: 1 October 2020 to 30 September 2021.

Table A6. Out-of-sample VRR and NMAE vs. in-sample cases with the comparison for the effect of lagged and forward variables (Chubu; see bar graphs in Figure 3).

	VRR (Current only)	VRR (+Lag)	VRR (+Forward)	NMAE (Current only)	NMAE (+Lag)	NMAE (+Forward)
In-sample						
Linear	18.3%	17.2%	16.6%	42.7%	42.1%	41.7%
Spline	16.9%	15.9%	14.8%	41.3%	41.3%	41.2%
Tensor	13.4%	10.7%	8.3%	40.6%	40.4%	39.0%
Out-of-sample						
Linear	22.9%	21.9%	22.6%	45.8%	45.0%	45.5%
Spline	19.6%	19.3%	19.1%	42.7%	42.2%	42.2%
Tensor	22.0%	20.8%	19.3%	42.5%	42.0%	40.9%

Learning period: 1 October 2019 to 30 September 2020; test period: 1 October 2020 to 30 September 2021.

Table A7. Out-of-sample VRR and NMAE vs. in-sample cases with the comparison for the effect of the incorporation of the WPP or WWC index (Kyushu; see bar graphs in Figure 4).

	VRR (w/WPP)	VRR (w/WWC)	VRR (w/WPP and WWC)	NMAE (w/WPP)	NMAE (w/WWC)	NMAE (w/WPP and WWC)
In-sample						
Linear	17.9%	14.5%	9.7%	42.6%	35.8%	31.2%
Spline	15.7%	14.5%	8.9%	42.9%	35.9%	31.8%
Tensor	15.7%	10.4%	7.1%	42.9%	30.8%	28.0%
Out-of-sample						
Linear	22.2%	51.2%	19.4%	45.9%	55.9%	40.1%
Spline	19.5%	24.4%	13.6%	44.3%	42.8%	34.4%
Tensor	19.5%	18.3%	11.5%	44.3%	35.5%	30.3%

Learning period: 1 October 2019 to 30 September 2020; test period: 1 October 2020 to 30 September 2021.

Table A8. Out-of-sample VRR and NMAE vs. in-sample cases with the comparison for the effect of the incorporation of the WPP or WWC index (Chubu; see bar graphs Figure 4).

	VRR (w/WPP)	VRR (w/WWC)	VRR (w/WPP and WWC)	NMAE (w/WPP)	NMAE (w/WWC)	NMAE (w/WPP and WWC)
In-sample						
Linear	22.8%	24.2%	16.6%	46.0%	50.4%	41.7%
Spline	21.1%	22.3%	14.8%	43.0%	49.4%	41.2%
Tensor	21.1%	12.3%	8.3%	43.0%	46.5%	39.2%
Out-of-sample						
Linear	28.7%	30.6%	22.6%	51.3%	51.9%	45.5%
Spline	24.2%	29.3%	19.1%	47.1%	50.4%	42.2%
Tensor	24.2%	23.3%	18.2%	47.1%	43.5%	39.8%

Learning period: 1 October 2019 to 30 September 2020; test period: 1 October 2020 to 30 September 2021.

Table A9. Comparison of daily hedges with hourly hedges (Kyushu; see bar graphs in Figure 6).

	VRR (Current only)	VRR (+Lag)	VRR (+Forward)	NMAE (Current only)	NMAE (+Lag)	NMAE (+Forward)
Period (1)						
Linear (Hourly)	6.1%	5.8%	5.6%	28.1%	27.3%	26.7%
Spline (Hourly)	7.3%	7.5%	7.2%	25.8%	25.8%	25.0%
Tensor (Hourly)	4.9%	5.0%	4.5%	21.1%	20.9%	19.7%
Period (2)						
Linear (Hourly)	13.7%	13.6%	14.1%	32.9%	33.1%	33.1%
Spline (Hourly)	12.2%	12.5%	12.3%	29.6%	30.2%	29.7%
Tensor (Hourly)	11.1%	11.4%	10.1%	27.7%	27.8%	27.2%
Period (3)						
Linear (Hourly)	20.4%	20.3%	21.0%	37.8%	38.1%	38.5%
Spline (Hourly)	18.4%	19.0%	18.8%	35.2%	36.0%	36.0%
Tensor (Hourly)	16.5%	15.8%	15.7%	32.0%	31.8%	31.8%
	VRR Period (1)	**NMAE Period (1)**	**VRR Period (2)**	**NMAE Period (2)**	**VRR Period (3)**	**NMAE Period (3)**
Linear (Daily)	5.7%	26.5%	17.3%	35.5%	23.8%	39.9%
Spline (Daily)	4.4%	21.3%	14.9%	33.3%	18.5%	35.0%

Table A10. Comparison of daily hedges with hourly hedges (Chubu; see bar graphs in Figure 7).

	VRR (Current only)	VRR (+Lag)	VRR (+Forward)	NMAE (Current only)	NMAE (+Lag)	NMAE (+Forward)
Period (1)						
Linear (Hourly)	9.4%	8.0%	8.1%	28.5%	27.4%	27.1%
Spline (Hourly)	5.6%	5.5%	5.0%	24.8%	24.5%	23.7%
Tensor (Hourly)	8.2%	8.1%	7.3%	25.2%	25.1%	24.0%
Period (2)						
Linear (Hourly)	15.1%	15.3%	15.3%	37.5%	37.4%	37.0%
Spline (Hourly)	11.8%	12.2%	12.4%	31.1%	31.2%	31.3%
Tensor (Hourly)	10.6%	11.9%	12.4%	30.0%	32.0%	32.1%
	VRR Period (1)	**NMAE Period (1)**	**VRR Period (2)**	**NMAE Period (2)**		
Linear (Daily)	34.4%	41.2%	14.1%	35.2%		
Spline (Daily)	12.7%	28.7%	21.3%	42.1%		

References

1. Deng, S.J.; Oren, S.S. Electricity derivatives and risk management. *Energy* **2006**, *31*, 940–953. [CrossRef]
2. Brockett, P.L.; Wang, M.; Yang, C. Weather derivatives and weather risk management. *Risk Manag. Insur. Rev.* **2005**, *8*, 127–140. [CrossRef]
3. Halkos, G.E.; Tsirivis, A.S. Energy commodities: A review of optimal hedging strategies. *Energies* **2019**, *12*, 3979. [CrossRef]
4. Eydeland, A.; Wolyniec, K. *Energy and Power risk Management: New Developments in Modeling, Pricing, and Hedging*; John Wiley & Sons, Ltd.: Hoboken, NJ, USA, 2002; Volume 97.
5. Clewlow, L.; Strickland, C. *Energy Derivatives: Pricing and Risk Management*; Lacima: Sydney, NS, Australia, 2000.
6. Wieczorek-Kosmala, M. Weather risk management in energy sector: The Polish case. *Energies* **2020**, *13*, 945. [CrossRef]
7. Alexandridis, A.K.; Zapranis, A.D. *Weather Derivatives, Modeling and Pricing Weather-Related Risk*; Springer: New York, NY, USA, 2013; ISBN 978-1-4614-6070-1.
8. Burger, M.; Klar, B.; Muller, A.; Schindlmayr, G.A. spot market model for pricing derivatives in electricity markets. *Quant. Financ.* **2004**, *4*, 109–122. [CrossRef]
9. Jewson, S.; Brix, A. *Weather Derivative Valuation: The Meteorological, Statistical, Financial and Mathematical Foundations*; Cambridge University Press: Cambridge, UK, 2005; p. 392.
10. Roncoroni, A.; Fusai, G.; Cummins, M. *Handbook of Multi-Commodity Markets and Products: Structuring, Trading and Risk Management*; John Wiley & Sons, Ltd.: Hoboken, NJ, USA, 2015; pp. 255–277.
11. Berhane, T.; Shibabaw, A.; Awgichew, G.; Walelgn, A. Pricing of weather derivatives based on temperature by obtaining market risk factor from historical data. *Model. Earth Syst. Environ.* **2021**, *7*, 871–884. [CrossRef]
12. Elias, R.S.; Wahab, M.I.M.; Fang, L. A comparison of regime-switching temperature modeling approaches for applications in weather derivatives. *Eur. J. Oper. Res.* **2014**, *232*, 549–560. [CrossRef]
13. Zapranis, A.; Alexandridis, A. Modelling the temperature time-dependent speed of mean reversion in the context of weather derivatives pricing. *Appl. Math. Financ.* **2008**, *15*, 355–386. [CrossRef]
14. Benth, F.E.; Benth, J.Š. The volatility of temperature and pricing of weather derivatives. *Quant. Financ.* **2007**, *7*, 553–561. [CrossRef]
15. Gyamerah, S.A.; Ngare, P.; Ikpe, D. Hedging Crop Yields Against Weather Uncertainties—A Weather Derivative Perspective. *Math. Comput. Appl.* **2019**, *24*, 71. [CrossRef]
16. Alaton, P.; Djehiche, B.; Stillberger, D. On modelling and pricing weather derivatives. *Appl. Math. Financ.* **2002**, *9*, 1–20. [CrossRef]
17. Boyle, C.F.; Haas, J.; Kern, J.D. Development of an irradiance-based weather derivative to hedge cloud risk for solar energy systems. *Renew. Energy* **2021**, *164*, 1230–1243. [CrossRef]
18. Mosquera-López, S.; Uribe, J.M. Pricing the risk due to weather conditions in small variable renewable energy projects. *Appl. Energy* **2022**, *322*, 119476. [CrossRef]
19. Leobacher, G.; Ngare, P. On modelling and pricing rainfall derivatives with seasonality. *Appl. Math. Financ.* **2011**, *18*, 71–91. [CrossRef]
20. Benth, F.E.; Di Persio, L.; Lavagnini, S. Stochastic modeling of wind derivatives in energy markets. *Risks* **2018**, *6*, 56. [CrossRef]
21. Xiao, Y.; Wang, X.; Wang, X.; Wu, Z. Trading wind power with barrier option. *Appl. Energy* **2016**, *182*, 232–242. [CrossRef]
22. Rodríguez, Y.E.; Pérez-Uribe, M.A.; Contreras, J. Wind put barrier options pricing based on the Nordix index. *Energies* **2021**, *14*, 1177. [CrossRef]
23. Masala, G.; Micocci, M.; Rizk, A. Hedging Wind Power Risk Exposure through Weather Derivatives. *Energies* **2022**, *15*, 1343. [CrossRef]
24. Kanamura, T.; Homann, L.; Prokopczuk, M. Pricing analysis of wind power derivatives for renewable energy risk management. *Appl. Energy* **2021**, *304*, 117827. [CrossRef]
25. Christensen, T.S.; Pircalabu, A. On the spatial hedging effectiveness of German wind power futures for wind power generators. *J. Energy Mark.* **2018**, *11*, 71–96. [CrossRef]
26. Benth, F.E.; Pircalabu, A. A non-Gaussian Ornstein–Uhlenbeck model for pricing wind power futures. *Appl. Math. Financ.* **2018**, *25*, 36–65. [CrossRef]
27. Gersema, G.; Wozabal, D. An equilibrium pricing model for wind power futures. *Energy Econ.* **2017**, *65*, 64–74. [CrossRef]
28. Thomaidis, N.S.; Christodoulou, T.; Santos-Alamillos, F.J. Handling the risk dimensions of wind energy generation. *Appl. Energy* **2023**, *339*, 120925. [CrossRef]
29. Oum, Y.; Oren, S.; Deng, S. Hedging quantity risks with standard power options in a competitive wholesale electricity market. *Nav. Res. Logist.* **2006**, *53*, 697–712. [CrossRef]
30. Oum, Y.; Oren, S. VaR constrained hedging of fixed Price load-following obligations in competitive electricity markets. *J. Risk Decision Anal.* **2009**, *1*, 43–56. [CrossRef]
31. Oum, Y.; Oren, S. Optimal static hedging of volumetric risk in a competitive wholesale electricity market. *Decis. Anal.* **2010**, *7*, 107–122. [CrossRef]
32. Yamada, Y. Valuation and hedging of weather derivatives on monthly average temperature. *J. Risk* **2007**, *10*, 101–125. [CrossRef]
33. Yamada, Y. Optimal Hedging of Prediction Errors Using Prediction Errors. *Asia Pac. Financ. Mark.* **2008**, *15*, 67–95. [CrossRef]
34. Matsumoto, T.; Yamada, Y. Simultaneous hedging strategy for price and volume risks in electricity businesses using energy and weather derivatives. *Energy Econ.* **2021**, *95*, 105101. [CrossRef]
35. Matsumoto, T.; Yamada, Y. Customized yet Standardized temperature derivatives: A non-parametric approach with suitable basis selection for ensuring robustness. *Energies* **2021**, *14*, 3351. [CrossRef]

36. Yamada, Y.; Matsumoto, T. Going for Derivatives or Forwards? Minimizing Cashflow Fluctuations of Electricity Transactions on Power Markets. *Energies* **2021**, *14*, 7311. [CrossRef]
37. Matsumoto, T.; Yamada, Y. Improving the Efficiency of Hedge Trading Using High-er-Order Standardized Weather Derivatives for Wind Power. *Energies* **2023**, *16*, 3112. [CrossRef]
38. Coulon, M.; Powell, W.B.; Sircar, R. A model for hedging load and price risk in the Texas electricity market. *Energy Econ.* **2013**, *40*, 976–988. [CrossRef]
39. Bhattacharya, S.; Gupta, A.; Kar, K.; Owusu, A. Risk management of renewable power producers from co-dependencies in cashflows. *Eur. J. Oper. Res.* **2020**, *283*, 1081–1093. [CrossRef]
40. Kaufmann, J.; Kienscherf, P.A.; Ketter, W. Modeling and managing joint price and volumetric risk for volatile electricity portfolios. *Energies* **2020**, *13*, 3578. [CrossRef]
41. Benth, F.E.; Lange, N.; Myklebust, T.A. Pricing and hedging quanto options in energy markets. *J. Energy Mark.* **2015**, *8*, 1–35. [CrossRef]
42. Caporin, M.; Preś, J.; Torro, H. Model based Monte Carlo pricing of energy and temperature quanto options. *Energy Econ.* **2012**, *34*, 1700–1712. [CrossRef]
43. Ministry of Economy, Trade and Industry; METI. Present Status and Promotion Measures for the Introduction of Renewable Energy in Japan. 2011. Available online: https://www.meti.go.jp/english/policy/energy_environment/renewable/index.html (accessed on 17 March 2023).
44. Nasdaq. Wind Power Futures Based on the German Wind Index NAREX WIDE. 2023. Available online: https://www.nasdaq.com/solutions/wind-power-futuresl (accessed on 17 March 2023).
45. Japan Electric Power Information Center, Inc.; JEPIC. The Electric Power Industry in Japan. 2022. Available online: https://en.wikipedia.org/wiki/Feed-in_premium (accessed on 17 March 2023).
46. Betz, A. *Wind-Energie und ihre Ausnutzung durch Windmühlen*; Vandenhoek and Ruprecht: Göttingen, Germany, 1926; 64p.
47. Villanueva, D.; Feijóo, A. Wind power distributions: A review of their applications. *Renew. Sustain. Energy Rev.* **2010**, *14*, 1490–1495. [CrossRef]
48. Hanifi, S.; Liu, X.; Lin, Z.; Lotfian, S. A critical review of wind power forecasting methods—Past, present and future. *Energies* **2020**, *13*, 3764. [CrossRef]
49. Wood, S.N. Package 'mgcv' v. 1.8–42. 2023. Available online: https://cran.r-project.org/web/packages/mgcv/mgcv.pdf (accessed on 25 March 2023).
50. Hastie, T.; Tibshirani, R. *Generalized Additive Models*; Chapman & Hall: Boca Raton, FL, USA, 1990.
51. Wood, S.N. *Generalized Additive Models: An Introduction with R*, 2nd ed.; Chapman and Hall: New York, NY, USA, 2017.
52. Wood, S.N.; Goude, Y.; Fasiolo, M. Interpretability in Generalized Additive Models. In *Interpretability for Industry 4.0: Statistical and Machine Learning Approaches*; Springer: Cham, Switzerland, 2022.
53. Matsumoto, T.; Yamada, Y. Comprehensive and Comparative Analysis of GAM-Based PV Power Forecasting Models Using Multidimensional Tensor Product Splines against Machine Learning Techniques. *Energies* **2021**, *14*, 7146. [CrossRef]
54. Liu, H.; Chen, C. Data processing strategies in wind energy forecasting models and applications: A comprehensive review. *Appl. Energy* **2019**, *249*, 392–408.

energies

Article

Capacity Market and Investments in Power Generations: Risk-Averse Decision-Making of Power Producer

Naoki Makimoto [1] and Ryuta Takashima [2,*]

1 Institute of Business Sciences, University of Tsukuba, Bunkyo-ku, Tokyo 112-0012, Japan; makimoto@gssm.gsbs.tsukuba.ac.jp
2 Department of Industrial and Systems Engineering, Tokyo University of Science, Chiba, Noda-shi 278-8510, Japan
* Correspondence: takashima@rs.tus.ac.jp

Abstract: The penetration of power generations from renewable energy sources into the power market has a significant impact on the capacity factor of existing power generations. This is because power producers cannot recover a capital cost of power generations with high operating cost possibly due to underinvestment. One solution to address this problem includes a capacity mechanism; that is, the capacities of the power generations can be sold through a market or a bilateral contract. Many schemes of the capacity mechanism have been used worldwide. In this study, we examine an investment in a power plant in both the electricity and capacity markets. The effect of investment opportunity on uncertainty and risk aversion is analyzed by a real options approach that is one of analytical methods for investment decisions under uncertainty. The investment timing for the standard energy-only market is compared with that for the capacity market. When the risk averse for the power producer is relatively small, the income in the energy-only market is obtained whereas, when the risk averse is relatively high, the income is gained in both the electricity and capacity markets for the sake of enough profit.

Keywords: capacity market; electricity market; investment in power generation; risk-averse decision making; mean-variance utility; threshold price

1. Introduction

1.1. Background

Recently, policymakers in the power sector have implemented policies and regulations for solving various problems such as mitigating greenhouse gas emissions and promoting renewable energy. As electricity cannot be economically stored, power producers obtain their profit only when it is delivered to consumers immediately after it is produced via power markets and/or bilateral contracts. Therefore, if producers cannot operate power generations due to the market, policy and regulations, and/or physical conditions for power generations, they might not obtain sufficient profit from selling electricity. Particularly, a capacity factor of power generations with relatively high fuel cost becomes low, and then these are closed due to the operating management depending on the relationship between operating costs and market prices. Meanwhile, the penetration of power generations from renewable energy sources into the power market has a significant impact on the capacity factor of existing power generations. This is because operations of renewable energy sources has become a priority due to their relatively low operating cost, that is, the merit-order effect. Consequently, power producers cannot recover a capital cost of power generations with high operating costs possibly due to underinvestment. This is called the "missing money" problem [1]. European utilities such as RWE, Vattenfall, and E.ON closed or mothballed 21 GW_e of gas-fired generation in 2012–2013 although 11 GW_e of this generation was less than 10-years-old because of the rise of renewable energy [2].

The penetration of renewable energy into the power market leads to low market prices. Reference [3] indicates that, in the California power market, the price increases slow down due to the high penetration of renewable energy, although the shutdown of nuclear power plants leads to increasing prices.

One of the policies for mitigating this problem includes a capacity mechanism, indicating that power producers can sell their generating capacities kW_e through a market of the capacity or via a bilateral contract. Many schemes of the capacity mechanism (e.g., strategic reserve, capacity market, capacity obligation, reliability options, and capacity payments) have been used in many countries. Although the capacity mechanism has a beneficial aspect for the capacity adequacy, some operational capacity adequacy occurs. Even if the revenue is priced adequately through the capacity mechanism, risks such as missing money, cannot be efficiently addressed when it is not perceived by producers. This is the so-called "missing market" problem [4]. Thus, producers should recognize how much effect of risk hedge occurs in the capacity mechanism.

In this study, we consider an investment problem of capacity expansion considering the scheme of capacity mechanisms. We analyze an investment timing for different ratios of selling the capacity. We also compare the investment timing for the standard energy-only market with that for the capacity market.

1.2. Literature Review

There is a growing body of literature that addresses the impact of capacity mechanisms on power market prices and investment decisions for power producers and mitigates the missing money problem by using analytical models [5,6]. Reference [7] evaluates a stability of capacity mechanisms in the presence of uncertainty of the demand growth rate by using a system dynamics model and a Markov Chain Monte Carlo approach. They show that all capacity mechanisms such as capacity payments, operating reserve pricing, and capacity obligation, can contribute to less volatility of power prices and mitigate the shortage. Among these mechanisms, capacity obligation is the most stable both in terms of investment and prices and reduces the market power and offers power producers efficient operating incentives. Reference [8] examines an effect of capacity payment on investment in gas-fired generations under uncertainties of electricity and fuel prices in a real option framework. In a circumstance of this model, they show that capacity payment increases investment opportunity. In cases of capacity payments of 30,000–50,000 EUR/MW p.a., the generations can have positive values even if their capacity factors are low as 5–10%. Thus, they indicate that capacity payment is an important policy for gas-fired generation of a low-capacity factor when shares of renewable energy is high. Reference [9] evaluates a market-wide centralized capacity market and targeted strategic reserves by formulating a game-theoretic market equilibrium model. They find that strategic reserves have significant drawbacks such as an additional investment by replacing the lost flexibility of a contracted capacity. They also show that a centralized capacity market provides a beneficial outcome from both system's and generator's perspective. This is because the scheme obtains a suitable reserve margin that avoids energy non-served at the lowest cost and meets the renewable energy target at a lower cost. Reference [10] investigates the robustness and total cost of power generation when the capacity market is adopted, rather than the energy-only market. They analyze a case of the Great Britain market in the framework of a stochastic dynamic model for the capacity investment. They find that the capacity market has a stabilizing effect on unplanned possibilities as no-load of conventional power generations forms the aspects of the cost and price volatility. The introduction of the capacity market leads to an increase in the total cost via payments for the capacity. In the parameter setting, they find that there is an effective region as trade-offs between robustness and costs in the range of reserve margins between 0 and 15%. Reference [11] addresses the dynamics of capacity adequacy and market efficiency with capacity remuneration policy by means of a stochastic simulation model based on the framework of real options. They show that policymakers replace supply security for market prices with payment for consumers when the capacity remuneration mechanism is implemented. They find that when uncertainty affects the growth of electricity demand and power capacity, the capacity market has a beneficial effect in increases in supply security.

The capacity mechanisms have a characteristic of hedging the profit from the energy-only market as forward or option contracts. Thus, the effect of these mechanisms on risk-averse decision-making might become prominent. Reference [12] explores the effect of contracts for the difference, reliability options, and forward capacity market on welfare and investment by using stochastic equilibrium models of risk-averse agents. They find that these contracts are quite effective complements to the energy-only market. However, adding the forward capacity market to the energy-only market can have quite different impacts depending on the extent to which the demand for capacity is calibrated. The contract for the difference and reliability options might also require trade volumes that have never been encountered in power markets. Reference [13] analyzes how the capacity mechanism can address the supply security in power markets by using a system dynamics model that integrates investment and closure decisions. They discuss the energy-only market with a price cap, with and without the capacity mechanisms compared to scarcity pricing in investment decision scenarios with and without risk aversion. In risk-neutral decision-making, the energy-only market with scarcity pricing and the capacity mechanism are two efficient schemes to reach similar levels of load loss. Meanwhile, under risk aversion the capacity mechanism is more effective than scarcity pricing. Reference [14] analyzes the risk aversion of power producers facing an uncertain peak load in the capacity mechanism by using a simulation model based on system dynamics. Particularly, they investigate the impact of the capacity mechanism on the ability to limit shortage and total generation cost, compared to the energy-only market, capacity market, and strategic reserve mechanism. The analytical results indicate that the scheme of the capacity market is the least affected by the degree of risk aversion both with respect to reliability and cost. Therefore, the capacity market is a robust scheme for reliability and cost even if the producers in the market have various forms of risk aversion.

Our work contributes to the literature by exploring how risk-averse decision-making affects investment in power generation and capacity selling in a capacity market. We use the dynamic mean-variance portfolio of [15] by incorporating the real option framework of [16]. Some works focus on investment decisions for risk-averse firms in the real option framework. Reference [17] extends a general model of [16,18] to introduce risk aversion for a constant relative risk aversion utility function in the case where firms face an incomplete market. They show that risk aversion decreases the investment opportunity and detracts the value of investment projects. Reference [19] examines the extent to which operational flexibility in the form of suspension and resumption options with no costs mitigates the effect of risk aversion in the same framework as in [17]. They show how operational flexibility becomes more valuable as risk aversion increases and volatility becomes large. Reference [20] develops a utility-based, regime-switching model to evaluate different technology-adoption strategies under price and technological uncertainty. They find that, when market regimes are asymmetric, larger risk aversion has a non-monotonic impact on the decision to abandon an old market regime and may either increase or decrease the abandonment threshold in contrast with [19].

1.3. Significance and Contribution

This paper contributes to a growing literature on the relationship between the capacity mechanism and producers' decision-making under uncertainty. Our main results are as follows.

First, we show that the use of the capacity market depends on risk aversion. When the risk averse for the power producer is relatively small, the income in the energy-only market is obtained. In contrast, when the risk averse is relatively high, the income is gained in both the electricity and capacity markets for the sake of enough profit. Thus, the risk averse power producer might use the capacity market for the risk hedge.

Second, unlike previous papers such as [17,19], an effect of risk aversion on investment decisions is non-monotonic by the existence of the capacity market. When the risk aversion is relatively low, the risk averse power producer increases the investment opportunity in the energy-only market, whereas when the risk aversion is relatively high, they obtain the selling income in both the electricity and capacity market.

The third result is the finding that the relationship between the investment opportunity and the profit rate of the capacity market is ambiguous. When the profit rate is relatively low, that is, a low capacity price compared to electricity (Since the unit of capacity price is different from that of electricity, in this paper suppose that the unit of capacity is converted to the same as electricity), the investment threshold increases with the profit rate. This implies that while the power producers have an incentive to sell the capacity in the market, the investment opportunity decreases because the profit from the electricity market decreases. However meanwhile, when the profit rate is relatively high, the power producers have a large incentive to invest because enough profit from the capacity market can be obtained.

The remainder of this paper is organized as follows. Section 2 develops the model after considering disinvestment and operating flexibility. In Section 3, we derive the optimal policy of the investment in power plants, and examine the effects of parameters on the optimal policy. We use the model to analyze the interaction between equilibrium investments and a renewable energy policy. Figure 1 shows a flowchart of the analysis and discussion in this study. Finally, Section 4 summarizes and concludes the study.

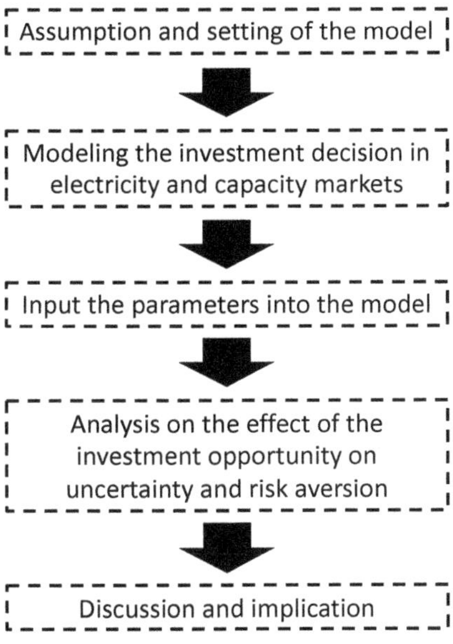

Figure 1. Flowchart of the model analysis.

2. Methodology

2.1. Investment Problem to the Power Generator without the Capacity Market

We first consider the investment decision problem to the power generator in the energy-only market (i.e., without the capacity market). As used in previous studies for analytical convenience [21–23], we assume that the electricity price follows a geometric Brownian motion (GBM); that is

$$\frac{dP_t}{P_t} = \mu dt + \sigma dW_t \quad (1)$$

where μ is the instantaneous expected growth rate, σ is the instantaneous volatilities, and W_t is a standard Brownian motion (The uncertainty of the electricity price is represented by the GBM, but the model considers no periodicity as a peak hour. If the periodicity is embedded into the model, another stochastic process needs to be used).

Suppose that a single price-taking power producer with a capacity has the perpetual option to invest in a power plant. This study assumes that the power producer sells the total electricity with a capacity factor of 100% that the power producer generates in one power plant, and that there exists no surplus electricity in the power plant.

After the investment at time s, a plant can produce electricity immediately and forever. From this plant, the power producer can sell electricity at exogenous price P_t at time $t > s$. Therefore, the power producer can obtain a discounted accumulated income for selling electricity after the investment as follows:

$$\int_s^\infty e^{-\rho(t-s)} P_t dt = P_s R \tag{2}$$

where

$$R = \int_0^\infty e^{-\rho t} \exp\left[\left(\mu - \frac{\sigma^2}{2}\right)t + \sigma W_t\right] dt \tag{3}$$

and ρ is the discount rate. (3) implies that R fluctuates corresponding to electricity price after the investment time, s. Suppose that the discounted profit from selling electricity is given by

$$S_1 = BP_s R - C - J, \tag{4}$$

where B is a product of a capacity of the power plant and a profit rate, C is the accumulated operating costs including fuel and O&M, and J is an investment cost that is not dependent on investment time. In this study, suppose that the power producer is risk averse, and his/her utility for the discounted profit is the mean-variance type as follows (Although the power utility function has been used in real options analysis [17,19,20], the mean-variance utility is used in this paper because the payoff function is difficult to obtain analytically when the optimal ratio between the energy and capacity markets is derived. Although previous papers of real options have used the power utility in the investment problem, the electricity procurement problem faced by retailers utilizes utility functions with the risk measure as the variance and the value-at-risk as in [24]):

$$f_1(p) = \mathrm{E}(S_1 \mid P_s = p) - \frac{\lambda}{2}\mathrm{V}(S_1 \mid P_s = p) \tag{5}$$

where λ is a risk-averse coefficient. Since P_t follows the GBM, that is, a Markov process, the expected value and the variance of $P_s R$ can be obtained as follows (see Appendix A.1):

$$\mathrm{E}(P_s R \mid P_s = p) = p\mathrm{E}(R) = \frac{p}{\rho - \mu}, \tag{6}$$

$$\mathrm{V}(P_s R \mid P_s = p) = p^2\mathrm{V}(R) = \frac{p^2\sigma^2}{2(\rho-\mu)(\rho-\mu-\frac{\sigma^2}{2})}. \tag{7}$$

Using (6) and (7), the mean-variance utility, $f_1(p)$, is rewritten as

$$f_1(p) = -\frac{\gamma}{2}p^2 + B_0 p - I \tag{8}$$

where

$$B_0 = \frac{B}{\rho - \mu}, \tag{9}$$
$$I = J + C, \tag{10}$$
$$\gamma = \frac{\lambda B_0^2}{2\eta}, \tag{11}$$
$$\eta = \frac{\rho - \mu}{\sigma^2} - \frac{1}{2}. \tag{12}$$

Utility function $f_1(p)$ is a convex quadratic function and has an maximum value at $p = p^\flat \equiv \frac{B_0}{\gamma}$. When the electricity price, p, is high, $f_1(p)$ increases with the income of electricity selling, whereas when the electricity price considerably increases, $f_1(p)$ decreases due to the large risk for price fluctuation. If $f_1(p^\flat) \leq 0$, the optimal policy is not to invest at all. Therefore, we assume that $f_1(p^\flat) > 0$. This condition is rewritten as

$$\lambda I < \eta. \tag{13}$$

Under (13), it is assured that $\mathrm{E}(R) < \infty$ and $\mathrm{V}(R) < \infty$.

When the present time is $t = 0$, the optimal investment problem for the expected utility is formulated by

$$V_1(p) = \sup_{\tau \in \mathcal{T}} \mathrm{E}\big(e^{-\rho\tau} f_1(P_\tau) \,\big|\, P_0 = p\big). \tag{14}$$

$\mathcal{T}$ is a set of all stopping times that are adapted to the filtration generated by a Brownian motion $\{W_t\}$.

2.2. Investment Problem to the Power Generator with the Capacity Market

In this section, the optimal investment problem in the presence of the capacity market is formulated. The power producer holds the plant by a ratio of $r \in [0,1]$ and sells electricity in the energy market, whereas the capacity of the remaining $1 - r$ is sold at the investment time in the capacity market. When the electricity price at the investment time, s, is P_s, suppose that the profit in the case of selling the full capacity in the market is $AP_s - C$ (Although there exists no correlation between electricity and capacity prices in any realistic case, we assume the correlation for analytical convenience). Unlike in the case of income for electricity selling, the power producer is not effected by the price fluctuation in the future due to selling the capacity in the market. Since S_1 of (4) is the profit from electricity selling when the power producer holds the power plant, the weighted average profit based on the holding ratio is given by

$$S_2 = rS_1 + (1-r)(AP_s - C - J) = (BR - A)P_s r + AP_s - I. \tag{15}$$

Similarly, in Section 2.1, the mean-variance utility for S_2 is calculated as follows:

$$\mathrm{E}(S_2 \,|\, P_s = p) - \frac{\lambda}{2}\mathrm{V}(S_2 \,|\, P_s = p) = -\frac{\gamma}{2}p^2r^2 + \alpha pr + Ap - I \tag{16}$$

where

$$\alpha = B_0 - A. \tag{17}$$

In this study, we assume that $\alpha > 0$ because α can be regarded as the risk premium for the price fluctuation in the electricity market. If $\alpha \leq 0$, the power producer has no incentive to hold the power plant to sell to the electricity market.

The power producer determines the holding ratio, r, by maximizing (16). Since (16) is a quadratic function for r, the optimal holding ratio is given by

$$r^*(p) = \begin{cases} 1, & 0 < p \leq p_0, \\ \frac{p_0}{p}, & p > p_0 \end{cases} \tag{18}$$

where $p_0 = \frac{\alpha}{\gamma}$. When $p \leq p_0$, the power producer can obtain the income from electricity selling without selling in the capacity market, whereas when $p > p_0$, the power producer tends to sell a portion in the capacity market. The ratio for selling in the capacity market increases with the market price. By substituting $r = r^*(p)$ into (16), the utility function is given by

$$f_2(p) = \begin{cases} f_1(p), & 0 < p \leq p_0, \\ f_c(p) = Ap + \frac{\alpha^2}{2\gamma} - I, & p > p_0 \end{cases} \tag{19}$$

where $f_1(p)$ is the utility function without the capacity market. Since $f_1(p)$ is an increasing function for $0 < p \leq p_0$ from $p_0 < p^\flat$, $f_2(p)$ is an increasing function for $p > 0$. When a capacity market does not exist, the utility function, $f_1(p)$, decreases due to high price and an increase in the fluctuation risk. However, when the market exists, $f_2(p)$ is an increasing function because the power producer can hedge the fluctuation risk by using the capacity market.

Thus, the optimal investment problem with the capacity market is formulated by

$$V_2(p) = \sup_{\tau \in \mathcal{T}} \mathrm{E}\left(e^{-\rho\tau} f_2(P_\tau) \,\middle|\, P_0 = p\right). \tag{20}$$

3. Comparison of Optimal Investment Policies

In Section 3.1, we derive the optimal investment policy of the problems formulated in Section 2. Based on the results, we show comparative statics and investigate in Section 3.2 how model parameters have effects on the optimal policies.

3.1. Optimal Investment Policies

We first derive the optimal policy of the investment problem without the capacity market. As is well-known in the real option analysis, the quadratic equation

$$\frac{\sigma^2}{2}x^2 + (\mu - \frac{\sigma^2}{2})x - \rho = 0 \tag{21}$$

has two real solutions, $\beta_1 > 1$ and $\beta_2 < 0$, under (13). We define

$$p_1^* = \frac{2\beta_1 I}{(\beta_1 - 1 + \sqrt{D})B_0}, \tag{22}$$

$$p_1^- = \frac{2(-\beta_2)I}{(1 - \beta_2 - \sqrt{D^-})B_0}, \tag{23}$$

$$D = (\beta_1 - 1)^2 - \beta_1(\beta_1 - 2)\theta, \tag{24}$$

$$D^- = (1 - \beta_2)^2 - (-\beta_2)(2 - \beta_2)\theta, \tag{25}$$

$$\theta = \frac{\lambda I}{\eta} \tag{26}$$

where $0 < \theta < 1$ under (13). Thus, $D > 0$ and $D^- > 0$.

Theorem 1. *If the capacity market does not exist, the optimal policy is to invest when $p_1^* \leq p \leq p_1^-$ and continue waiting otherwise, where p denotes the electricity price. The value function of this policy is*

$$V_1(p) = \begin{cases} f_1(p_1^*)(p/p_1^*)^{\beta_1}, & 0 < p < p_1^*, \\ f_1(p), & p_1^* \leq p \leq p_1^-, \\ f_1(p_1^-)(p/p_1^-)^{\beta_2}, & p > p_1^-. \end{cases} \tag{27}$$

Proof. See Appendix A.2. □

As discussed in Section 2.1, utility function $f_1(p)$ in this case is quadratic and concave. Since $V_1(p)$ is smoothly pasted at $p = p_1^*$ and $p = p_1^-$, price $p_1^\flat$ that attains the maximum of $f_1(p)$ satisfies $p_1^* < p_1^\flat < p_1^-$. Value function $V_1(p)$ is increasing for $0 < p < p_1^\flat$ and decreasing for $p > p_1^\flat$. It is optimal not to invest both when the electricity price is low and when the price is high enough since it leads to high risk due to a large price fluctuation.

To state the optimal policy when a capacity market exists, we define

$$p_2^* = \frac{\beta_1 I}{(\beta_1 - 1)A}\left[1 - \frac{(1 - A/B_0)^2}{\theta}\right]. \tag{28}$$

Recall that $p_0 = \alpha/\gamma$, as defined in Section 2.2.

Theorem 2. *If there exists a capacity market, the optimal policy is of a threshold type; that is, there exists a threshold price p^* such that it is optimal to invest when $p \geq p^*$ and continue waiting if $p < p^*$.*
(Case 1) If $p_2^* \leq p_0$, the threshold price $p^* = p_1^*$, and the value function is given by

$$V_2(p) = \begin{cases} f_1(p_1^*)(p/p_1^*)^{\beta_1}, & 0 < p < p_1^*, \\ f_1(p), & p_1^* \leq p < p_0, \\ f_c(p), & p \geq p_0. \end{cases} \tag{29}$$

(Case 2) If $p_2^* > p_0$, the threshold price $p^* = p_2^*$, and the value function is given by

$$V_2(p) = \begin{cases} f_c(p_2^*)(p/p_2^*)^{\beta_1}, & 0 < p < p_2^*, \\ f_c(p), & p \geq p_2^*. \end{cases} \tag{30}$$

Proof. See Appendix A.3. □

In Case 1, if the current electricity price $p < p_1^*$, it is optimal for the power producer to invest when the price first reaches p_1^*. This policy is the same as in Theorem 1, wherein the capacity market does not exist. In this case, the power producer does not use the capacity market. If the current price $p \geq p_1^*$, an immediate investment is optimal. The capacity market is utilized only when the current price $p > p_0$ and $1 - p_0/p$ of the capacity of the power plant is sold in the capacity market to hedge the risk of price fluctuation in the future. Consequently, the value function $V_2(p)$ is increasing for all $p > 0$ compared to $V_1(p)$, which decreases for $p > p^\flat$.

In Case 2, the capacity market is utilized even when the price first reaches to p_2^* and $1 - p_0/p_2^*$ of the capacity is sold in the capacity market. Value function $V_2(p)$ is increasing for $p > 0$ as the payoff function $f_c(p)$ is increasing. From the viewpoint of electricity power producers, a capacity market has a role to hedge risk especially when the price is high.

Figure 2 presents a numerical example where panels (a) and (b) correspond to Cases 1 and 2 in Theorem 2, respectively. We set the parameters as $\mu = 0.02, \rho = 0.04, \sigma^2 = 0.01$, $A = 3, B_0 = 10, I = 5$, and $\lambda = 0.15$ in panel (a). Panel (b) uses the same set of parameters except for $\lambda = 0.25$. In panel (a), the red solid line represents $V_2(p)$ and the blue dashed line represents $f_1(p)$. The two marked points in green and blue correspond to p_0 and p_1^*. In panel (b), the red solid line represents $V_2(p)$, and the blue and green dashed lines represent $f_1(p)$ and $f_c(p)$, respectively. The marked points in green and red correspond to p_0 and p_2^*.

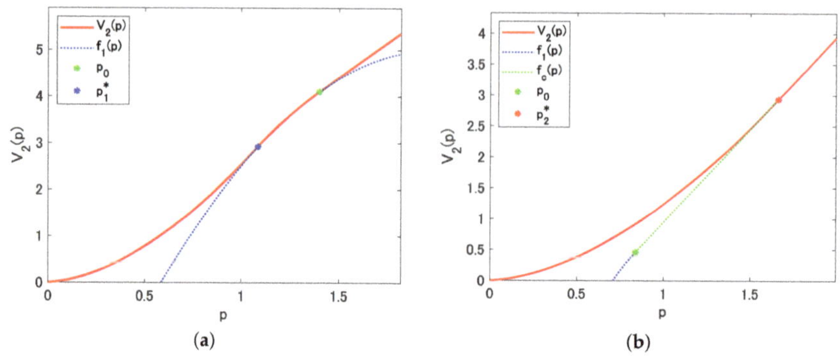

Figure 2. (**a**) Value function $V_2(p)$ in Case 1. (**b**) Value function $V_2(p)$ in Case 2. Note: Red solid line represents $V_2(p)$. For comparison, $f_1(p)$, p_1^*, and p_0 are plotted in (**a**), and $f_1(p)$, $f_c(p)$, p_0, and p_2^* are plotted in (**b**). In (**a**), $\mu = 0.02, \rho = 0.04, \sigma^2 = 0.01, \lambda = 0.15, A = 3, B_0 = 10$, and $I = 5$. In (**b**), $\lambda = 0.25$ and other parameters are the same as in (**a**).

3.2. Comparative Statics

As discussed in Section 3.1, the optimal investment policy is of the threshold type when there is a capacity market. In this section, we show the comparative statics of the threshold price in terms of four important parameters: risk aversion coefficient λ, profit ratio of the capacity market A, electricity price volatility σ^2, and investment cost I. Investigating how these parameters have effects on the threshold price makes it possible to better understand the role of the capacity market in the investment into a new power plant. We also suppose that the remaining parameters μ, ρ, and B (and hence B_0) are fixed.

To identify which cases in Theorem 2 occur, we consider

$$K(\lambda, A, \sigma^2, I) = \frac{p_2^*}{p_0} = \frac{\beta_1\{\theta - (1 - A/B_0)^2\}}{2(\beta_1 - 1)(1 - A/B_0)(A/B_0)}, \tag{31}$$

where θ is given in (26). For presentation simplicity, we will use an abbreviated notation K, rather than $K(\lambda, A, \sigma^2, I)$. Theorem 2 implies that the threshold price $p^* = p_1^*$ if $K \leq 1$ and $p^* = p_2^*$ if $K > 1$. Moreover, if $K > 1$ and the power producer makes an investment at the price $p = p_2^*$, $1 - K^{-1}$ of the capacity of the plant is sold in the capacity market. To satisfy (13), the domain of λ, σ^2, and I are limited from above by

$$\overline{\lambda} = \frac{\eta}{I}, \qquad \overline{\sigma^2} = \frac{\rho - \mu}{\lambda I + \frac{1}{2}} \qquad \text{and} \qquad \overline{I} = \frac{\eta}{\lambda}, \tag{32}$$

respectively. Since the domain of A is $(0, B_0)$, we denote $\overline{A} = B_0$.

Theorem 3. *For any of the parameters $x \in \{\lambda, A, \sigma^2, I\}$, suppose that all other parameters are fixed. Then, there exists a unique x_0 such that $K \leq 1$ for $0 < x \leq x_0$, and $K > 1$ for $x_0 < x < \overline{x}$. Moreover, $\partial K/\partial x > 0$ in the region satisfying $K > 0$.*

Proof. See Appendix A.4. □

Theorem 3 indicates that, when each of the parameters λ, A, σ^2, or I increases, the threshold price p^* switches monotonically from p_1^* to p_2^*. For parameters λ, σ^2, and I, this is a natural consequence since an increase in these parameters also increases the risk of the power producer, inducing an incentive to hedge risk by using the capacity market. When A increases, the capacity market becomes more attractive so that p^* switches from p_1^* to p_2^*.

Theorem 3 also provides an important implication about the utilization of the capacity market. When $K > 1$ and investment is made at price $p = p_2^*$, $1 - K^{-1}$ of the capacity of a plant is sold in the capacity market. From $\partial K/\partial x > 0$, the amount of capacity sold in the capacity market increases as each parameter increases. Thus, an increase in these parameters enhances the utilization of the capacity market. When θ goes to 1 (i.e., each of λ, σ^2 or I goes to the upper limit of the domain), the ratio sold in the capacity market tends to

$$\lim_{\theta \to 1}(1 - K^{-1}) = \frac{(\beta_1 - 2)(A/B_0) + 2}{\beta_1(2 - A/B_0)}. \tag{33}$$

Meanwhile, $1 - K^{-1}$ tends to 1 when A goes to $\overline{A}$, which implies that all of the capacity of the plant is sold in the capacity market if the risk premium contained in the price in the electricity market vanishes.

Subsequently, we further investigate the characteristics of threshold prices. Particularly, we pay attention to whether the threshold price changes monotonically when each parameter increases. The ratio sold in the capacity market is also discussed.

3.2.1. Risk Aversion Coefficient λ

We observe from (22) that p_1^* is a decreasing function of λ if $1 < \beta_1 < 2$, is constant if $\beta_1 = 2$, and is increasing if $\beta_1 > 2$. Case $1 < \beta_1 < 2$ seems somewhat counter-intuitive since, in many applications of real option analysis, threshold price increases as the investor becomes more risk averse. This may be interpreted in the following manner. The payoff function $f_1(p)$ contains a term $-\frac{\gamma}{2}p^2$, which quadratically decreases as the investment price increases. Meanwhile, the value of waiting increases as p^{β_1} from (27). Since γ is a linear function of λ, the former effect becomes dominant for large λ when $1 < \beta_1 < 2$, leading to investment at a lower threshold price.

Unlike p_1^*, (28) shows that p_2^* is an increasing function of λ irrespective of the value of β_1. This is because a risk-sensitive power producer attempts to obtain a higher payoff by increasing the investment threshold. Although larger p_2^* increases the risk of price fluctuation, this risk can partially be hedged when there is a capacity market.

Figure 3 presents a numerical example that shows threshold price p^* as a function of λ. Here and in subsequent numerical examples, we use the same set of parameters in Figure 2b as the base case. In panel (a), red and blue solid lines represent p_1^* and p_2^*, respectively. For comparison, we also plot p_1^* in Case 2 with a red dashed line. Since $\beta_1 = 1.70 < 2$ in this case, p_1^* gradually decreases. The threshold price switches from p_1^* to p_2^* at $\lambda_0 = 0.199$ and starts to increase rather sharply. As a result, an effect of risk aversion on the investment policy is non-monotonic by the existence of the capacity market unlike previous papers such as [17,19]. The previous papers [17,19] show the risk averse firms tend to withhold the investment opportunity due to avoiding the risk. In this paper, however, since the risk could be mitigated by the transaction in the capacity market, the investment opportunity increases even if the risk aversion is relatively large. Intuitively, when the risk aversion is relatively low, the power producer increases the investment opportunity in the energy-only market for lack of need for the capacity market. In contrast, when the risk aversion is relatively high, the selling income in both the electricity and capacity markets might be obtained. Panel (b) shows the ratio sold in the capacity market $1 - K^{-1}$. Evidently, the ratio sharply increases when λ exceeds the threshold λ_0. When $\lambda \to \overline{\lambda} = 0.3$, $1 - K^{-1}$ tends to the limit 0.66 given by (33). When the risk averse for the power producer is relatively small, the income in the energy-only market is obtained, whereas when the risk averse is relatively high, the income is gained in both the electricity and capacity markets for the sake of enough profit. This implies that the risk-averse power producer might use the capacity market for the risk hedge.

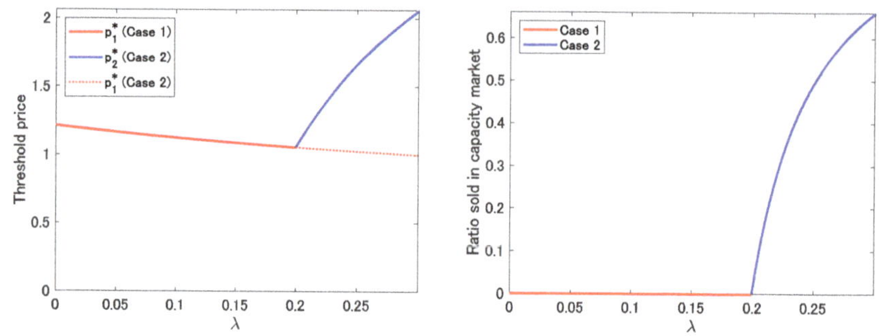

(a) (b)

Figure 3. (**a**) Threshold price and risk aversion coefficient λ. (**b**) Ratio sold in the capacity market and risk aversion coefficient λ. Note: Red and blue solid lines represent Cases 1 and 2, respectively. For comparison, p_1^* in Case 2 is plotted by a red dashed line in (**a**). $\mu = 0.02, \rho = 0.04, \sigma^2 = 0.01$, $A = 3$, $B_0 = 10$, and $I = 5$.

3.2.2. Profit Rate of Capacity Market A

We consider the effect that the profit rate of the capacity market A has on the threshold price p^*. Unlike other parameters, p_1^* does not contain A and hence is constant. Concerning p_2^*, we rewrite p_2^* as

$$p_2^* = \frac{\beta_1 I}{(\beta_1 - 1)\theta}\left(\frac{\theta - 1}{A} + \frac{2}{B_0} - \frac{A}{B_0^2}\right), \tag{34}$$

where θ is defined in (26). p_2^* is the sum of hyperbolic and linear functions of A, each with a negative coefficient since $0 < \theta < 1$ under condition (13). Then, it is not difficult to check that p_2^* is increasing for $0 < A < A_M$ and decreasing for $A_M < A < \overline{A}$, where $A_M = \sqrt{1-\theta}B_0$. This implies that p_2^* is not monotonic in A.

The effects that A has on p_2^* are two-fold. Since the power producer can hedge the price fluctuation risk by selling some amount of capacity in the capacity market, an increase in A will allow the producer to take more risk, and this pushes up p_2^*. As A approaches to $\overline{A}$, immediate investment becomes more attractive since the ratio of the plant sold in the capacity market increases to 1 and the value of waiting comparatively decreases. This pushes down p_2^*. The unimodal shape of p_2^* in terms of A is interpreted as the former effect is dominant for small A, while the latter effect is dominant for larger A. Note also that turning point A_M is a decreasing function of all λ, σ^2, and I. Investment becomes more risky when these parameters increase. Then, p_2^* increases and then decreases for a smaller value of A to avoid risk increase.

When A approaches $\overline{A}$, p_2^* approaches $p_2^*(\overline{A}) = \frac{\beta_1 I}{(\beta_1 - 1)B_0}$. It can be checked that $p_1^* > p_2^*(B_0)$ if and only if $\beta_1 > 2$. If this is the case, p_2^* decreases and goes below p_1^* as A approaches $\overline{A}$.

Figure 4 shows a numerical example of p^* and $1 - K^{-1}$ as a function of A. In panel (a), the red solid line is p_1^*, which is constant, and blue solid line is p_2^*. In Case 2, p_2^* increases rather sharply for $A_0 = 1.45 < A < A_M = 4.08$ and gradually decreases for $A_M < A < \overline{A} = 10$. Since $\beta_1 = 1.70 < 2$ in this case, we observe that $p_2^*(\overline{A}) > p_1^*$. The result indicates that the optimal investment policy depends in the relationship of both prices in electricity and capacity markets. For low capacity price compared to electricity, the investment threshold increases with the profit rate. Intuitively, while the power producers have an incentive to sell the capacity in the market, the investment opportunity decreases because the profit from the electricity market decreases. In contrast, when the profit rate is relatively high, the power producers have a large incentive to invest because enough profit from the capacity market can be obtained. Panel (b) depicts the ratio sold in the capacity market. Similar to Figure 3, the graph starts to increase sharply when A exceeds A_0, and the slope becomes gentle and approaches 1 as A goes to $\overline{A}$.

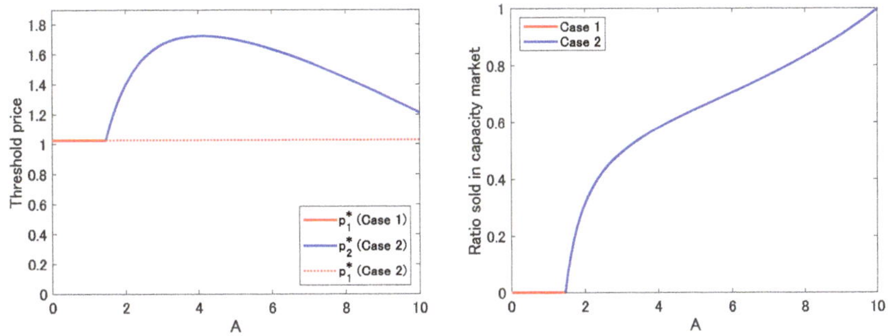

(a) (b)

Figure 4. (**a**) Threshold price and profit rate of capacity market A. (**b**) Ratio sold in the capacity market and profit rate of capacity market A. Note: Red solid line shows Case 1 and blue solid line shows Case 2. For comparison, p_1^* in Case 2 is plotted by a red dashed line in (**a**). $\mu = 0.02$, $\rho = 0.04, \sigma^2 = 0.01, I = 5, B_0 = 10, \lambda = 0.25$.

3.2.3. Electricity Price Volatility σ^2

We consider the relation between price volatility σ^2 and the threshold price, as well as the ratio sold in the capacity market. Since not only θ in (22) but also β_1 is dependent on σ^2, the shape of p_1^* as a function of σ^2 is difficult to analytically derive. Therefore, we investigate it by using a numerical example. Conversely, it is easy to check from (28) that p_2^* increases as σ^2 increases for $\sigma_0^2 = 0.0084 < \sigma^2 < \overline{\sigma^2} = 0.0114$.

Panel (a) of Figure 5 shows p^* as a function of volatility σ^2. In this case, p_1^* first increases gradually and then decreases. After switching from p_1^* to p_2^* at σ_0^2, p_2^* increases rather sharply compared to p_1^*. Panel (b) shows $1 - K^{-1}$, which is quite similar to that in Figure 3. The ratio sold in the capacity market increases sharply when σ^2 exceeds σ_0^2.

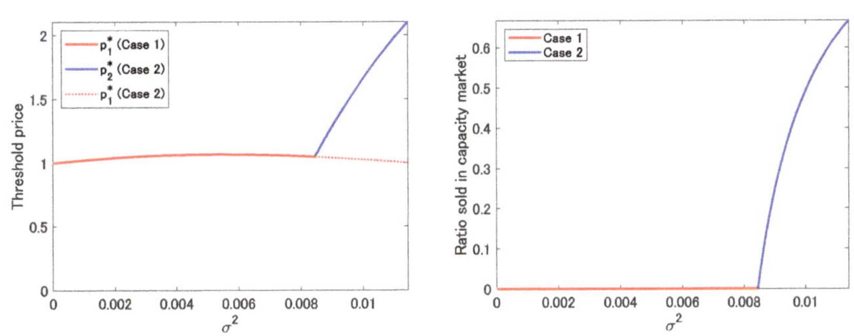

(a) (b)

Figure 5. (**a**) Threshold price and price volatility σ^2. (**b**) Ratio sold in the capacity market and price volatility σ^2. Note: Red and blue solid lines represent Cases 1 and 2, respectively. For comparison, p_1^* in Case 2 is plotted by a red dashed line in (**a**). $\mu = 0.02, \rho = 0.04, I = 5, A = 3, B_0 = 10$, and $\lambda = 0.25$.

3.2.4. Investment Cost I

Finally, we consider the relation between investment cost I and the threshold price, as well as the ratio of the capacity sold in the capacity market. From (22), it is readily checked that p_1^* is an increasing function of I when $\beta_1 \geq 2$. Since p_1^* is rewritten as

$$p_1^* = \frac{\eta(\sqrt{D} - \beta_1 + 1)}{\lambda(2 - \beta_1)}, \tag{35}$$

we observe that p_1^* is an increasing function even when $1 < \beta_1 < 2$. Concerning p_2^*, we observe from (28) that p_2^* is a linear function of I with the slope $\frac{\beta_1}{(\beta_1 - 1)A}$. Thus, threshold price p^* always increases as the investment cost increases. If I is large, the power producer will wait until electricity prices go up to a high threshold to cover the high investment cost.

Figure 6 depicts a numerical example. From panel (a), the threshold price switches from p_1^* to p_2^* at $I_0 = 3.98$. The slope of increase of p_2^* is steeper than that of p_1^*. We observe from panel (b) that the ratio sold in the capacity market starts to increase sharply when I exceeds threshold I_0, as observed in Figures 3–5.

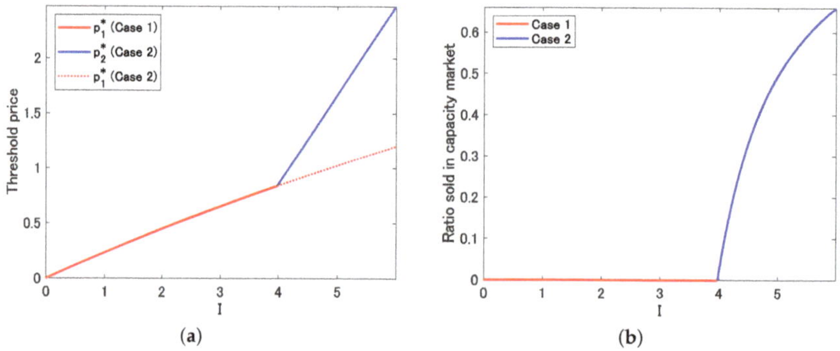

Figure 6. (**a**) Threshold price and investment cost I. (**b**) Ratio sold in the capacity market and investment cost I. Note: Red and blue solid lines represent Cases 1 and 2, respectively. For comparison, p_1^* in Case 2 is plotted by a red dashed line in (**a**). $\mu = 0.02, \rho = 0.04, \sigma^2 = 0.01, A = 3, B_0 = 10$, and $\lambda = 0.25$.

4. Conclusions

This paper focuses on the interaction between the investment decision for the risk-averse power producer and the policy of the capacity market in the framework of real options. In particular, we show how the risk aversion and the profit rate of the capacity market affect the optimal investment policy. The results indicate that the power producer obtains the selling income in both the electricity and capacity markets when the risk aversion is relatively high, the effect of risk aversion on the investment policy is ambiguous unlike the previous work, and the power producers have a large incentive to invest because enough profit from the capacity market can be obtained when the profit rate of the capacity market is relatively high.

Our study contributes to the argument on the 'missing money' problem, and highlights the relations among the investment opportunity, risk aversion, and the profit rate of the capacity market. If policymakers can observe the risk aversion of power producers, the missing money problem would be resolved by increasing the investment opportunity for the optimal policy of the profit rate of the capacity market. Even if the power producers have less investment opportunities due to low risk aversion, the problem might be mitigated by means of a subsidy policy for the investment cost.

We highlight the importance of understanding the role of the capacity market for the risk-averse power producers that have both problems of investment in power plants and missing money. In this paper, it is assumed that the capacity price is correlated with that of electricity. Since in any realistic case there exists no correlation, it is necessary to mitigate the assumption in the model. Hence, as an extension of our work, the capacity price is modeled as a stochastic process different from the electricity price by estimating the real data for the capacity market. The penetration of power generations from renewable energy sources leads to the missing money problem, and then the capacity market is created due to mitigating the problem. This means that not only local problems such as missing money, but also whole market problems are included. Thus, it also would be important to address social welfare in order to examine an effect of the interaction between the penetration of renewable energy and the capacity market. In addition, the relations between the capacity mechanism and risk aversion for power producers should be investigated by empirical analyses because evidence of the results in this study could be obtained.

Author Contributions: Conceptualization, R.T.; methodology, N.M. and R.T.; software, N.M.; validation, R.T.; formal analysis, N.M.; investigation, R.T.; resources, N.M.; data curation, N.M.; writing—original draft preparation, N.M. and R.T.; writing—review and editing, R.T.; visualization, N.M.; supervision, R.T.; project administration, R.T.; funding acquisition, N.M. and R.T. All authors have read and agreed to the published version of the manuscript.

Funding: This work was supported by JSPS KAKENHI Grant Number JP20H00285.

Institutional Review Board Statement: Not applicable.

Informed Consent Statement: Not applicable.

Data Availability Statement: Not applicable.

Conflicts of Interest: The authors declare no conflicts of interest.

Abbreviation

GBM Geometric Brownian motion

Appendix A. Proofs

Appendix A.1. Derivation of (6) and (7)

Let $Q_t = \exp\left[(\mu - \frac{\sigma^2}{2})t + \sigma W_t)\right]$. Then, (6) is obtained from

$$\mathrm{E}(P_s R \mid P_s = p) = p\mathrm{E}(R) = p\int_0^\infty e^{-\rho t}\mathrm{E}(Q_t)dt = \frac{p}{\rho - \mu}. \tag{A1}$$

Since Q_t is a Markov process,

$$\begin{aligned} \mathrm{E}\left(R^2\right) &= 2\int_0^\infty \int_t^\infty e^{-\rho t}e^{-\rho u}\mathrm{E}(Q_t Q_u)dudt \\ &= 2\int_0^\infty \int_t^\infty e^{-\rho t}e^{-\rho u}\mathrm{E}\left(Q_t^2\right)\mathrm{E}\left(\frac{Q_u}{Q_t}\right)dudt \\ &= 2\int_0^\infty \int_t^\infty e^{-\rho t}e^{-\rho u}e^{(2\mu+\sigma^2)t}e^{\mu(u-t)}dudt \\ &= \frac{1}{(\rho-\mu)(\rho-\mu-\frac{\sigma^2}{2})}. \end{aligned} \tag{A2}$$

From $\mathrm{V}(R) = \mathrm{E}(R^2) - \mathrm{E}(R)^2$, (7) is derived from

$$\mathrm{V}(P_s R \mid P_s = p) = p^2\mathrm{V}(R) = \frac{p^2\sigma^2}{2(\rho-\mu)^2(\rho-\mu-\frac{\sigma^2}{2})}. \tag{A3}$$

□

Appendix A.2. Proof of Theorem 1

Based on [25], we first show the solution of the optimal stopping problem with a general payoff function when the underlying process is a GBM. Let $f(p)$ be a function such that $f(p) > 0$ for some $p > 0$ and $\mathrm{E}(e^{-\rho\tau} f(P_\tau) \mid P_0 = p)$ is well-defined for every stopping time $\tau \in \mathcal{T}$, where P_t is a GBM (1). To solve the optimal stopping problem

$$V(p) = \sup_{\tau \in \mathcal{T}} \mathrm{E}(e^{-\rho\tau} f(P_\tau) \mid P_0 = p), \tag{A4}$$

we define

$$g(u) = u^{\frac{\beta_1}{\beta_1 - \beta_2}} f(u^{\frac{1}{\beta_2 - \beta_1}}) \tag{A5}$$

and let $G(u)$ be the smallest non-negative concave majorant of $g(u)$ on $(0, \infty)$. The following is a simplified version of [25] when P_t follows a GBM.

Theorem A1. *Value function of the optimization problem (A4) is given by*

$$V(p) = p^{\beta_1} G(p^{\beta_2 - \beta_1}), \quad p > 0. \tag{A6}$$

Proof. Since the underlying process is a GBM, we plug $\psi(p) = p^{\beta_1}$ and $\phi(p) = p^{\beta_2}$ into Proposition 5.12 of [25]. The value function is given by $V(p) = \psi(p)\widetilde{W}(-p^{\beta_2 - \beta_1})$ for $p > 0$ where, $\widetilde{W}(y)$ is the smallest non-negative concave majorant of $\widetilde{H}(y) = (-y)^{\frac{\beta_1}{\beta_1 - \beta_2}} f((-y)^{\frac{1}{\beta_2 - \beta_1}})$ for $y < 0$. By replacing $-y$ for $y < 0$ with $u > 0$, $g(u)$ and $G(u)$ correspond to $\widetilde{H}(-u)$ and $\widetilde{W}(-u)$, respectively. □

Theorem A1 reduces an optimal stopping problem to find the smallest non-negative concave majorant $G(u)$. The next result is useful to identify $G(u)$ in many applications.

Corollary A1. *If $u^* = \mathrm{argmax}_{u>0}\{g(u)\}$ exists and $\lim_{u\to\infty} \frac{g(u)}{u} = 0$, then the smallest non-negative concave majorant of $g(u)$ is in the form*

$$G(u) = \begin{cases} G_0(u), & 0 < u \le u^*, \\ g(u^*), & u > u^* \end{cases} \tag{A7}$$

where $G_0(u)$ is the smallest non-negative concave majorant of $g(u)$ on $(0, u^*]$.

Proof. It is not difficult to check that $G(u)$ in (A7) is a non-negative concave majorant of $g(u)$ on $(0, \infty)$ and is the smallest on $(0, u^*]$. To prove the minimality on (u^*, ∞), suppose that there exists a non-negative concave majorant $\widehat{G}(u)$ satisfying $\widehat{G}(u^*) = g(u^*)$ and $\widehat{G}(\widehat{u}) < g(u^*)$ for some $\widehat{u} > u^*$. From the concavity of $\widehat{G}(u)$,

$$\widehat{G}(u) \le g(u^*) + \frac{\widehat{G}(\widehat{u}) - g(u^*)}{\widehat{u} - u^*}(u - u^*), \quad \forall u > \widehat{u} \tag{A8}$$

which shows $\limsup_{u\to\infty} \frac{\widehat{G}(u)}{u} \le \frac{\widehat{G}(\widehat{u}) - g(u^*)}{\widehat{u} - u^*} < 0$. Since $\lim_{u\to\infty} \frac{g(u)}{u} = 0$, $\widehat{G}(u) < g(u)$ holds for sufficiently large u, which contradicts to that $\widehat{G}(u)$ is a majorant of $g(u)$. □

We now prove Theorem 1.

Proof of Theorem 1. We define

$$g_1(u) = u^{\frac{\beta_1}{\beta_1 - \beta_2}} f_1(u^{\frac{1}{\beta_2 - \beta_1}}) = -\frac{\gamma}{2} u^{\frac{\beta_1 - 2}{\beta_1 - \beta_2}} + B_0 u^{\frac{\beta_1 - 1}{\beta_1 - \beta_2}} - I u^{\frac{\beta_1}{\beta_1 - \beta_2}}. \tag{A9}$$

(Case A) $\beta_1 > 2$: In this case, it is not difficult to check that $g_1'(u) = 0$ has two positive solutions

$$u_\pm = \left[\frac{(\beta_1 - 1 \pm \sqrt{D})B_0}{2\beta_1 I}\right]^{\beta_1-\beta_2} \tag{A10}$$

where D is defined in (24). Similarly, $g_1''(u) = 0$ has two positive solutions

$$w_\pm = \left\{\frac{[(\beta_1 - 1)(1-\beta_2) \pm \sqrt{D_w}]B_0}{2\beta_1(-\beta_2)I}\right\}^{\beta_1-\beta_2} \tag{A11}$$

where

$$D_w = (\beta_1 - 1)^2(1-\beta_2)^2 - \beta_1(\beta_1 - 2)(-\beta_2)(2-\beta_2)\theta \tag{A12}$$

that is positive under (13). Moreover, $0 < u_- < w_- < u_+ < w_+$ in this case. Then, we obtain

u	0	$\cdots$	u_-	$\cdots$	w_-	$\cdots$	u_+	$\cdots$	w_+	$\cdots$
$g_1(u)$	0	$\searrow$		$\nearrow$		$\nearrow$		$\searrow$		$\searrow$
$g_1'(u)$		$-$	0	$+$	$+$	$+$	0	$-$	$-$	$-$
$g_1''(u)$		$+$	$+$	$+$	0	$-$	$-$	$-$	0	$+$

(A13)

(Case B) $1 < \beta_1 \leq 2$: In this case, both $g_1'(u) = 0$ and $g_1''(u) = 0$ have unique positive solution u_+ in (A10) and w_+ in (A11), respectively. Thus,

u	0	$\cdots$	u_+	$\cdots$	w_+	$\cdots$
$g_1(u)$	$-$	$\nearrow$		$\searrow$		$\searrow$
$g_1'(u)$		$+$	0	$-$	$-$	$-$
$g_1''(u)$		$-$	$-$	$-$	0	$+$

(A14)

In both Cases A and B, $g_1(u)$ attains the maximum at $u = u_+$. Moreover, $\lim_{u\to\infty} \frac{g_1(u)}{u} = 0$. From Corollary A1, the smallest non-negative concave majorant of $g_1(u)$ is in the form

$$G_1(u) = \begin{cases} G_0(u), & 0 < u < u_+, \\ g_1(u_+), & u \geq u_+. \end{cases} \tag{A15}$$

To obtain $G_0(u)$, consider a straight line that goes through the origin and is tangent to $g_1(u)$ at some $u \in (0, u_+)$. The slope of the straight line connecting $(0,0)$ and $(u, g_1(u))$ is $h(u) = g_1(u)/u$. With some algebra, we can observe that $h'(u) = 0$ has two positive solutions:

$$v_\pm = \left[\frac{(1-\beta_2 \pm \sqrt{D^-})B_0}{2(-\beta_2)I}\right]^{\beta_1-\beta_2} \tag{A16}$$

where D^- is given in (25), and that $h'(u) > 0$ for $0 < u < v_-$ and $u > v_+$, while $h'(u) < 0$ for $v_- < u < v_+$. This, along with $\lim_{u\to\infty} h(u) = 0$, imply that slope $h(u)$ is the maximum at $u = v_-$. Moreover, $v_- < u_+$ since $g_1'(v_-) > 0$, and $v_- > w_-$ in Case A since $g_1''(v_-) < 0$. Then, the smallest non-negative majorant for $0 < u < u_+$ is given by

$$G_0(u) = \begin{cases} g_1'(v_-)u, & 0 < u \leq v_-, \\ g_1(u), & v_- < u < u_+. \end{cases} \tag{A17}$$

The value function is now derived from Theorem A1 by noting that $p_1^* = (u_+)^{\frac{1}{\beta_2-\beta_1}}$ and $p_1^- = (v_-)^{\frac{1}{\beta_2-\beta_1}}$. $\square$

Appendix A.3. Proof of Theorem 2

As in the proof of Theorem 1, we define

$$g_2(u) = \begin{cases} g_c(u), & 0 < u < u_0, \\ g_1(u), & u \geq u_0 \end{cases} \tag{A18}$$

where $u_0 = p_0^{\beta_2-\beta_1}$, $g_1(u)$ is defined in (A9) and

$$g_c(u) = u^{\frac{\beta_1}{\beta_1-\beta_2}} f_c\left(u^{\frac{1}{\beta_2-\beta_1}}\right) = \frac{(\beta_1-1)A}{\beta_1}\left(\frac{\beta_1}{\beta_1-1}u^{\frac{\beta_1-1}{\beta_1-\beta_2}} - p_2^* u^{\frac{\beta_1}{\beta_1-\beta_2}}\right). \tag{A19}$$

Here, the second equality follows from (28). It is noted that $g_1(u) < g_c(u)$ for $0 < u < u_0$ since $f_1(p) < f_c(p)$ for $p > p_0$. Further, $g_1(u)$ and $g_c(u)$ are smoothly pasted at $u = u_0$; that is $g_1(u_0) = g_c(u_0)$ and $g_1'(u_0) = g_c'(u_0)$.

From

$$g_c'(u) = \frac{(\beta_1-1)A}{\beta_1-\beta_2}u^{\frac{\beta_2}{\beta_1-\beta_2}}\left(u^{-\frac{1}{\beta_1-\beta_2}} - p_2^*\right), \tag{A20}$$

$$g_c''(u) = -\frac{(\beta_1-1)(1-\beta_2)A}{(\beta_1-\beta_2)^2}u^{\frac{2\beta_2-\beta_1}{\beta_1-\beta_2}}\left[u^{-\frac{1}{\beta_1-\beta_2}} - \frac{(-\beta_2)}{1-\beta_2}p_2^*\right], \tag{A21}$$

$g_c(u)$ is increasing and concave for $u > 0$ if $p_2^* \leq 0$. If $p_2^* > 0$, the shape of $g_c(u)$ is given as

u	0	$\cdots$	x_1	$\cdots$	x_2	$\cdots$
$g_c(u)$	0	$\nearrow$		$\searrow$		$\searrow$
$g_c'(u)$		+	0	−	−	−
$g_c''(u)$		−	−	−	0	+

(A22)

where

$$x_1 = (p_2^*)^{\beta_2-\beta_1}, \tag{A23}$$

$$x_2 = \left[\frac{(-\beta_2)}{1-\beta_2}p_2^*\right]^{\beta_2-\beta_1}. \tag{A24}$$

(Case 1) $p_2^* \leq p_0$: Since $p_2^* \leq p_0$ is equivalent to $u_0 < x_1$, $g_c(u_0) = g_1(u_0) > 0$, which in turn implies $u_0 < u_+$ from (A13) and (A14). Thus, $g_2(u)$ attains the maximum at $u = u_+$. From Corollary A1, the smallest non-negative majorant of $g_2(u)$ is

$$G_2(u) = \begin{cases} g_c(u), & 0 < u \leq u_0, \\ g_1(u), & u_0 < u \leq u_+, \\ g_1(u_+), & u > u_+. \end{cases} \tag{A25}$$

The concavity of $G_2(u)$ can be checked from the concavity of $g_c(u)$, (A13) and (A14), and the smooth pasting conditions. $V_2(p)$ in (29) is obtained from Theorem A1.

(Case 2) $p_2^* > p_0$: Since $u_0 > x_1$ in this case, $g_c'(u_0) = g_1'(u_0) \leq 0$, and thus $u_0 \geq u_+$ from (A13) and (A14). The maximum of $g_2(u)$ is given by $u = x_1$. Since $g_2(u)$ is concave for $0 < u < x_1$, the smallest non-negative majorant is obtained from Corollary A1 as

$$G_2(u) = \begin{cases} g_c(u), & 0 < u \leq x_1, \\ g_c(x_1) & u > x_1. \end{cases} \tag{A26}$$

Theorem A1 now proves $V_2(p)$ in (30). □

Appendix A.4. Proof of Theorem 3

Noting that K can be rewritten as

$$K = \frac{\beta_1}{2(\beta_1 - 1)}\left(\frac{\theta}{1 - A/B_0} - \frac{1-\theta}{A/B_0} + 1\right),$$

clear $\partial K/\partial x > 0$ for $x \in \{\lambda, A, I\}$. For $x \in \{\lambda, I\}$, $\theta \to 0$ when $x \to 0$ and $\theta \to 1$ when $x \to \overline{x}$. From $0 < A < B_0$, we observe that $\lim_{x\to 0} K < 0$, and $\lim_{x\to\overline{x}} K > 1$, which proves the desired result for $x \in \{\lambda, I\}$. Case $x = A$ is also easy to show since $\lim_{A\to 0} K = -\infty$ and $\lim_{A\to\overline{A}} K = \infty$.

The remaining case is $x = \sigma^2$. From (28), $K > 0$ if and only if $\sigma^2 > \sigma_m^2$, where $\sigma_m^2 = (\rho - \mu)\left[\frac{\lambda I}{(1-A/B_0)^2} + \frac{1}{2}\right]^{-1}$. For $\sigma^2 > \sigma_m^2$, $\partial K/\partial\sigma^2 > 0$ since, as σ^2 increases, θ increases, while β_1 decreases. Noting that $\theta \to 1$ as $\sigma^2 \to \overline{\sigma^2}$, $\lim_{\sigma^2\to\overline{\sigma^2}} K > 1$ as before. This completes the proof for σ^2. □

References

1. Joskow, P.L. Capacity payments in imperfect electricity markets: Need and design. *Util. Policy* **2008**, *163*, 159–170. [CrossRef]
2. Green, R.; Staffell, I. Electricity in Europe: Exiting fossil fuels? *Oxf. Rev. Econ. Policy* **2016**, *32*, 282–303. [CrossRef]
3. Woo, C.K.; Ho, T.; Zarnikau, J.; Olson, A.; Jones, R.; Chait, M.; Horowitz, I.; Wang, J. Electricity-market price and nuclear power plant shutdown: Evidence from California. *Energy Policy* **2014**, *73*, 234–244. [CrossRef]
4. Newbery, D. Missing money and missing markets: Reliability, capacity auctions and interconnectors. *Energy Policy* **2016**, *94*, 401–410. [CrossRef]
5. Bublitz, A.; Keles, D.; Zimmermann, F.; Fraunholz, C.; Fichtner, W. A survey on electricity market design: Insights from theory and real-world implementations of capacity remuneration mechanisms. *Energy Econ.* **2019**, *80*, 1059–1078. [CrossRef]
6. Duggan, J.E., Jr. Capacity market mechanism analyses: A literature review. *Curr. Sustain. Energy Rep.* **2020**, *7*, 186–192. [CrossRef]
7. De Vries, L.; Heijnen, P. The impact of electricity market design upon investment under uncertainty: The effectiveness of capacity mechanisms. *Util. Policy* **2008**, *16*, 215–227. [CrossRef]
8. Hach, D.; Spinler, S. Capacity payment impact on gas-fired generation investments under rising renewable feed-in—A real options analysis. *Energy Econ.* **2016**, *53*, 270–280. [CrossRef]
9. Höschle, H.; De Jonghea, C.; Le Cadre, H.; Belmans, R. Electricity markets for energy, flexibility and availability—Impact of capacity mechanisms on the remuneration of generation technologies. *Energy Econ.* **2017**, *66*, 372–383. [CrossRef]
10. Hach, D.; Spinler, S. Robustness of capacity markets: A stochastic dynamic capacity investment model. *OR Spectr.* **2018**, *40*, 517–540. [CrossRef]
11. Rios-Festner, D.; Blanco, G.; Olsina, F. Long-term assessment of power capacity incentives by modeling generation investment dynamics under irreversibility and uncertainty. *Energy Policy* **2020**, *137*, 111185. [CrossRef]
12. de Maere d'Aertryckea, G.; Ehrenmanna, A.; Smeers, Y. Investment with incomplete markets for risk: The need for long-term contracts. *Energy Policy* **2017**, *105*, 571–583. [CrossRef]
13. Petitet, M.; Finon, D.; Janssen, T. Capacity adequacy in power markets facing energy transition: A comparisonof scarcity pricing and capacity mechanism. *Energy Policy* **2017**, *103*, 30–46. [CrossRef]
14. Ousman, A.A.; Hary, N.; Rious, V.; Saguan, M. The impact of investors' risk aversion on the performances of capacity remuneration mechanisms. *Energy Policy* **2018**, *112*, 84–97. [CrossRef]
15. Basak, S.; Chabakauri, G. Dynamic mean-variance asset allocation. *Rev. Financ. Stud.* **2010**, *23*, 2970–3016. [CrossRef]
16. Dixit, A.K.; Pindyck, R.S. *Investment under Uncertainty*; Princeton University Press: Princeton, NJ, USA, 1994.
17. Hugonnier, J.; Morellec, E. Real options and risk aversion. in *Real Options, Ambiguity, Risk and Insurance*; Bensoussan, A., Peng, S., Sung, J., Eds.; IOS Press: Amsterdam, The Netherlands, 2013; pp. 52–65.
18. McDonald, R.L.; Siegel, D.S. The value of waiting to invest. *Q. J. Econ.* **1986**, *101*, 707–728. [CrossRef]
19. Chronopoulos, M.; De Reyck, B.; Siddiqui, A. Optimal investment under operational flexibility, risk aversion, and uncertainty. *Eur. J. Oper. Res.* **2011**, *213*, 221–237. [CrossRef]
20. Chronopoulos, M.; Lumbreras, S. Optimal regime switching under risk aversion and uncertainty. *Eur. J. Oper. Res.* **2017**, *256*, 543–555. [CrossRef]
21. Barbosa, L.; Ferrão, P.; Rodrigues, A.; Sardinha, A. Feed-in tariffs with minimum price guarantees and regulatory uncertainty. *Energy Econ.* **2018**, *72*, 517–541. [CrossRef]
22. Detemple, J.; Kitapbayev, Y. The value of green energy under regulation uncertainty. *Energy Econ.* **2017**, *89*, 104807. [CrossRef]
23. Babich, V.; Lobel, R.; Yücel, S. Promoting solar panel investments: Feed-in-tariff vs. tax- rebate policies. *Manuf. Serv. Oper. Manag.* **2021**, *22*, 1148–1164. [CrossRef]

24. Conejo, A.J.; Carrión, M.; Morales, J.M. *Decision Making under Uncertainty in Electricity Markets*; Springer: New York, NY, USA, 2010.
25. Dayanik, S.; Karatzas, I. On the optimal stopping problem for one-dimensional diffusions. *Stoch. Process. Their Appl.* **2003**, *107*, 173–212. [CrossRef]

energies

Article

Machine Learning-Based Estimation of COP and Multi-Objective Optimization of Operation Strategy for Heat Source Considering Electricity Cost and On-Site Consumption of Renewable Energy

Daishi Sagawa * and Kenji Tanaka

School of Engineering, The University of Tokyo, Tokyo 113-8656, Japan
* Correspondence: sagawa@ioe.t.u-tokyo.ac.jp

Abstract: Air conditioning is a significant consumer of electricity in buildings, accounting for around 40% of the total consumption. While previous studies have focused on planning methods to minimize electricity costs, recent years have seen an increasing need for energy management methods that consider environmental performance, such as CO_2 emissions, alongside economic efficiency. This study proposes a mechanism to support stakeholders' decision-making by calculating Pareto solutions based on the multi-objective optimization of economic and environmental characteristics for entities that own renewable energy generation facilities. Unlike many existing studies that assume a specific equation for COP (Coefficient of Performance) estimation, this study adopts a nonparametric COP estimation method using machine learning, resulting in a more realistic and flexible modeling of the system. The study also presents a model for selecting an operation strategy that balances environmental and economic goals, incorporating a thermal storage facility to improve the renewable energy rate. Specifically, we proposed and compared methods for calculating solutions using only the GA (Genetic Algorithm) and a two-step optimization method combining a GA and gradient-based optimization method, confirming the superiority of the two-step optimization method. The case study unveiled unique operational profiles corresponding to cost-saving, renewable-energy, and balanced orientation points, suggesting the existence of specific strategies tailored to each orientation. The findings of this study can help stakeholders make more informed decisions regarding energy management in air conditioning systems, with benefits for both the environment and the bottom line.

Keywords: machine learning; multi-objective optimization; heat source; renewable energy

1. Introduction

1.1. Background

Air conditioning is a critical component in buildings and consumes a significant amount of electricity, accounting for around 40% of the total energy usage. In central heat source air conditioning, heat generated by the equipment is transferred to air conditioners through pipes, where the air and heat medium exchange heat to cool or heat the air. In large-scale central heat source systems, there are multiple heat source devices, and their operation and shutdown must be scheduled according to the demand for heat. To optimize the efficiency of daily energy equipment operation, energy consumers need to consider various factors, such as fluctuations in renewable energy generation due to weather conditions, fluctuations in equipment efficiency, fluctuations in electricity prices, and the use of energy storage and thermal storage facilities. Furthermore, during the equipment installation phase, it is essential to determine the appropriate scale of the equipment to be installed. With the increasing focus on the introduction of renewable energy and the establishment of decarbonization targets, energy management methods based on multi-objective optimization have gained attention. These methods aim to address not only economic efficiency, but also environmental efficiency, such as reducing CO_2 emissions and achieving other environmental goals. In this study, we propose a multi-objective optimization approach to develop a strategy for the daily operation of existing HVAC (Heating, Ventilation, and Air Conditioning) equipment. The proposed method aims to minimize power costs while maximizing on-site consumption of renewable energy. It takes into account various variable factors such as equipment efficiency and solar power generation to reflect the reality of daily operations. To verify the effectiveness of our proposed approach, we conduct a case study using actual operating data of multiple heat source equipment and thermal storage units at an airport.

Overall, our study presents a method to support stakeholders in decision-making related to energy management in the air conditioning domain. It offers a more realistic and flexible modeling of the system by using a nonparametric COP (Coefficient of Performance) estimation method based on machine learning, which is then incorporated into the optimization equation. By doing so, we can achieve a better balance between the environmental and economic performance in HVAC operations, which is crucial for promoting sustainable energy consumption.

1.2. Related Work

This section focuses on the optimization of heat source equipment, with a particular emphasis on thermal energy storage devices. These devices act as a form of energy storage and can shift the electricity demand, similar to how storage batteries function. As such, this section not only reviews studies on the optimization of heat source equipment, but also considers research on the optimization of electricity management, including storage batteries.

In the operation strategy of heat source equipment, demand shift can be achieved by using thermal energy storage (TES: Thermal Energy Storage) equipment such as ice thermal storage tanks. Existing TES systems are often operated on a fixed schedule and do not take advantage of their load shifting capability; appropriate demand shifting with TES and storage batteries can reduce costs by utilizing the price difference in time-of-use (TOU: Time Of Use) electricity rates and the surplus of renewable energy [1–4] and suppressing annual peaks in electricity demand [5–7].

Studies using multi-objective optimization in the field of energy storage and thermal storage include minimizing the cost of electricity and the probability of power loss through the operation of storage batteries [8–10], respectively; Abdelkader et al. [8], Ould Bilal et al. [9], and Yang et al. [10] have investigated the impact of stand-alone environments on energy prosumers consisting of wind-solar PV systems, storage batteries, and electricity demand. They formulated the cost of electricity and the probability of power loss, and by employing a genetic algorithm, they solved a multi-objective optimization problem aiming to minimize both. This approach facilitated the identification of facility installations that are both economically viable and robust. This optimizes parameters, such as the size of the wind power, solar power, and storage battery installations and the angle at which they are installed, and not for day-to-day operational strategies. Additionally, due to the stand-alone environment, the system's behavior regarding a power purchase from the grid was not considered. Another multi-objective optimization study from the environmental and economic point of view of storage batteries is the multi-objective optimization [11,12], which considers CO_2 emissions and power usage costs. The object being optimized here is not the charging and discharging schedule, but rather the capacity and other parameters of the facility while the heat storage and heat dissipation and the charging and discharging are rule-based and fixed. Therefore, this is not a discussion of day-to-day operational strategies. Thus, in the multi-objective optimization of energy systems, including storage batteries, which are similar to the heat source and thermal storage equipment that are the subject of this study, the scale of the system installation that achieves both environmental and economic performance through multi-objective optimization has been calculated, but the effects were obtained through a multi-objective optimization of operational strategies at a certain installation scale. However, there is still room for research on the effects that can be obtained with a multi-objective optimization of operational strategies at a certain scale of installation.

As for the optimization of the system, including heat source and thermal storage equipment, which is the subject of this study, [13–15], Ascione et al. [13] used a genetic algorithm to minimize the primary energy demand and investment cost in a house with renewable energy, a Pareto solution that minimizes the primary energy demand and investment cost in a house with renewable energy. The system was calculated to support optimal decision making on the balance between the environmental and economic characteristics of the stakeholders. However, this is an optimization for determining the installed capacity of each facility, and there is room for research on the effects that can be obtained through the multi-objective optimization of operational strategies; Navidbakhsh et al. [14] modeled an ice thermal energy storage (ITES) system that incorporated phase change material (PCM) as a partial cold storage material and used ice. A multi-objective optimization of the air conditioning system, including the ice thermal energy storage system, was performed. The objective functions of the multi-objective optimization are exergy efficiency (energy use efficiency) and total cost. In addition, CO_2 emissions are also quantified, but they are not considered as objective functions of the multi-objective optimization. Lee et al. [15] performed an optimization of the installed equipment using the particle swarm algorithm to minimize the life cycle cost of air conditioning equipment, including ice thermal storage systems, and found the ice thermal storage air conditioning system. They studied and analyzed the increase in electricity consumption and CO_2 emissions due to the use of ice thermal storage air-conditioning systems. Here, the parameters related to the operational strategy and equipment installation were optimized simultaneously. However, CO_2 emissions were not the objective function of the optimization, and CO_2 emissions were only quantified for the optimization results in pursuit of economic efficiency. In addition, the efficiency COP of air conditioning equipment is fixed and does not consider changes in the COP due to the external environment; Zhou et al. [16] constructed a multi-objective optimization model based on the improved firefly algorithm (IFA) for the rational allocation of cooling load between chillers and ice thermal storage tanks. Energy consumption loss rate and operating cost are set as objective functions and minimized. The use of renewable energy within the energy system is not considered, and it is assumed that the consideration of the COP of the chiller depends only on the magnitude of the output and can be expressed as a quadratic function of the magnitude of the output.

In reality, the COP of a chiller varies with external factors, such as the season and temperature, and there is room for improvement in terms of realistic modeling. Therefore, a planning method for heat source equipment is required that takes into account the COP characteristics that vary depending on the load ratio and the weather conditions of the day and time. The load ratio indicates the ratio of the actual load to the maximum capacity of the system. Although there is a manufacturer's nominal COP curve for heat source equipment, it shows the relationship between the output and COP under limited conditions, and in order to use it, it is necessary to interpolate the values under the conditions, and since the COP changes as the heat source equipment deteriorates over time, it is necessary to take into account the actual operating data of power consumption and output heat quantity, as well as the weather conditions and settings at that time. Therefore, the data-driven approach, which calculates the COP based on actual operating data of the power consumption and output heat quantity, as well as weather conditions and settings at the time, can calculate the COP in line with actual conditions. As an example of the data-driven approach, a method of operation optimization based on a COP estimation by a multivariate linear regression [17] has been proposed, but the linear model has limitations in expressive power and does not take into account variations in the COP with the load ratio. There is also a study [18] that modeled power consumption using MLP (Multi-Layer Perceptron) and performed operation planning based on particle swarm optimization, but no study of methods for various modeling has been conducted.

Ren et al. [19] proposed a particle swarm optimization of the hourly partial load ratio of the chiller and the cooling ratio of the ice thermal storage tank for the purpose of energy consumption, operating cost, and energy loss. Zheng et al. [20] solved MINLP (mixed integer non-linear programming) by setting up an objective function that linearly combines economic efficiency, environmental impact, and annual EB indices with an entropy weighting method. Lo et al. [21] solved MINLP (mixed integer non-linear programming), which in effect has only one objective function and does not allow stakeholders to select an operational strategy from the Pareto solution, and proposed an optimal dispatch model for thermal storage air conditioning systems to minimize costs under dynamic rates. Again, no consideration was given to the use of renewable energy.

As we have seen, various studies have been conducted to help stakeholders make decisions on the scale of equipment installation based on their own criteria from the Pareto optimal solution at the stage of equipment installation, but there is room for further study on how to support similar decision making in operational strategies. Therefore, this study establishes a mechanism to support stakeholders' decision-making by calculating Pareto solutions based on a multi-objective optimization of economic and environmental characteristics for entities that own renewable energy generation facilities. In addition, while many studies assume that a COP estimation is a quadratic equation, this study aims to achieve a more realistic and flexible modeling of the system by adopting a nonparametric COP estimation method using machine learning and incorporating the estimation function into the optimization equation.

1.3. Contribution of This Paper

The novelty of this study is to develop a mechanism to support stakeholders' decision-making through the calculation of Pareto solutions by multi-objective optimization of economic and environmental characteristics for entities that own renewable energy generation facilities. In particular, we adopt the ratio of renewable energy to energy consumption (RE ratio) as an indicator of environmental performance, taking into account the recent trend toward decarbonization, and the cost of electricity as an indicator of economic performance. This enables the selection of an operational strategy that considers the balance between how much the on-site consumption of renewable energy can be increased and how much the cost of electricity can be reduced. In other words, by selecting daily operational strategies from the Pareto front, the economic and environmental use efficiency of existing facilities can be improved in line with stakeholder intentions. This approach proves beneficial as the Pareto front represents a set of points where RE ratio and cost are in equilibrium. Hence, by choosing a point from this set that aligns with their preferred balance, users can devise equipment operation plans that cater as closely as possible to their intent.

1.4. Structure of This Paper

Section 2 describes COP estimation; Section 3 describes the method for optimizing the operating strategy in the case of only heat source equipment considering only economic efficiency as a preparation for multi-objective optimization; Section 4 describes the multi-objective optimization method for economic and environmental efficiency; Section 5 describes the case study.

2. Methodology

2.1. Machine Learning Based COP Estimation

COP is an indicator of the energy efficiency of a device and is generally expressed as "the amount of energy output from the device"/"the amount of energy input to the device". In this context, COP can be defined as COP = "heat output of the heat source equipment"/"power consumption of the heat source equipment". Since the value of COP varies depending on weather factors, such as temperature and the magnitude of the output of the heat source equipment, it was necessary to determine the COP function while taking into account the weather conditions of the day and time. Therefore, a machine learning model was constructed to estimate the COP using the load ratio "r" to the maximum output and the temperature "T" as inputs. The load ratio "r" was a key parameter in our model, representing the proportion of the current output of the heat source equipment relative to its maximum possible output. This ratio allowed us to normalize the output across different heat source equipment, accounting for variations in capacity. By considering this ratio in conjunction with the current temperature "T", our model could estimate the COP under various weather conditions and output levels. In the optimization process, we adjusted the value of "r" to find the optimal balance between output and energy efficiency while the temperature "T" was set using forecast data, as shown in Equation (1).

$$COP(T,r) = f_T(r) \tag{1}$$

We used RBF kernel of SVR (Support Vector Regressor), polynomial kernel of SVR, MLP (Multi-Layer Perceptron), K-Neighbors Regressor (K-nearest neighbor method), and Random Forest as representative machine learning methods. The reproduction accuracy of these methods was compared in a case study. In addition to the accuracy, the shape of the function f_T that composed them was considered to be an important factor in the subsequent optimization calculations. This was because if this COP function was a multimodal function, it might affect the convergence of the optimization. In more detailed modeling, it was desirable to input coolant temperature and other parameters, but it was easy to increase the number of input parameters in machine learning, and a similar approach should be sufficient.

2.2. *Optimization of Operating Strategies for Economic Efficiency in the Case of Heat Source Equipment Only*

This section describes a method for optimizing the operation of a heat source equipment for the purpose of economic efficiency only in the absence of a heat storage device as a prerequisite knowledge for the next section: multi-objective optimization method for operational strategy considering economic and environmental efficiency.

Using the COP model estimated in the previous step, the optimization equation could be written as follows. Equation (2) was the objective function, which aimed to minimize the amount of electric energy consumed by the operation of the heat source equipment (refrigerator) for cooling. Equations (3) and (4) were the constraints. Equation (3) required that the heat demand could be met while Equation (4) defined a lower limit for the load ratio because there was a minimum load ratio for the chiller.

$$\text{minimize} \sum_{i=1}^{N} u_{i,t} \frac{P_i r_{i,t}}{COP_i(T_t,\ r_{i,t})} \tag{2}$$

$$Subject\ to$$

$$\sum_{i=1}^{N} u_{i,t} P_i r_{i,t} \geq D_t \tag{3}$$

$$r_{lower} \leq r_{i,t} \leq 1 \tag{4}$$

Each symbol was defined as follows:
P_i: Maximum output of chiller i [GJ]
T_t: Temperature at time t [°C]
$u_{i,t} \in [0,\ 1]$: Whether or not refrigerator i was operating at time t
$r_{i,t}$: Output ratio of chiller i at time t
$COP_i(T_t,\ r_{i,t})$: COP at temperature T_t, output ratio $r_{i,t}$ for refrigeration unit i
D_t: Heat demand to be supplied [GJ]
r_{lower}: Lower limit of settable output ratio
N: Number of heat source devices

The solution was calculated using the flow shown in Figure 1. The thermal storage system was not taken into account here, so the plan could be formulated independently for each time. The 0 and 1 variables for availability were prepared in advance, and $r_{i,t}$ was calculated for each combination, turning the mixed integer problem into a nonlinear programming problem. The total number of combinations with and without operation was 2^N, which was O (2^N). However, since the number of N refrigeration units in an individual air-conditioning system was assumed to be up to about 10 at most, the calculation was assumed to be realistic as far as the air-conditioning system was concerned. In the calculation of $r_{i,t}$, Sequential Least Squares Programming (SLSQP) was used. In general, $\frac{P_i r_{i,t}}{COP_i(T_t, r_{i,t})}$. in the objective function may have been a non-convex function because it depended on the COP function to be estimated, but if we assumed that increasing the output $r_{i,t}$ will increase the power consumption and $\frac{P_i r_{i,t}}{COP_i(T_t, r_{i,t})}$. was a monotonically increasing function with respect to $r_{i,t}$, it could be optimized with a gradient-based method.

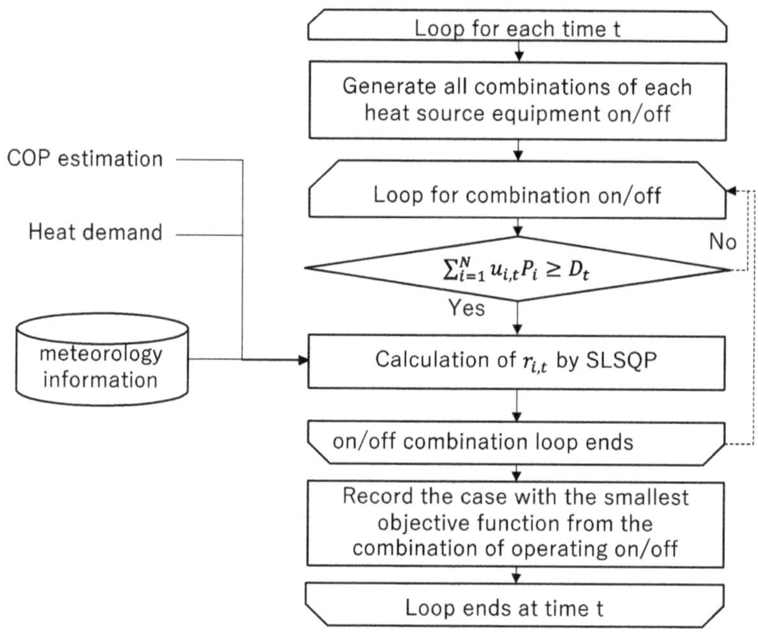

Figure 1. Calculation flow of operating strategies.

2.3. Multi-Objective Optimization Method Reflecting Economic and Environmental

This paper describes a method for developing an operation strategy that reflects economic and environmental concerns when, in addition to heat source equipment, the operation of a thermal storage device that enables a shift in the timing of electricity consumption is included. Thermal storage devices use electricity to produce hot water or ice, which is then stored and used to shift the timing of electricity consumption by extracting heat at an arbitrary point in time. In the case where the surplus of solar power generation could be sold to an electric power retailer, we formulated Equations (5) to (14) as a multi-objective optimization that took into account both the economic efficiency and the self-consumption of solar power generation to the maximum extent possible. Equations (5) and (6) were the objective functions of the multi-objective optimization. Equation (6) maximized the amount of solar power generation consumed on site. Equation (7) showed the constraints of meeting the heat demand, and Equation (8) showed the relationship between the amount of heat storage and the hourly heat storage and heat release. Equation (10) represented the upper and lower limits of the heat storage capacity.

$$\text{minimize} \sum_{t}^{M} p_t^{retail}\left(e_t - s_t^{used}\right) + s_t^{used} p_t^{pv\_cost} - s_t^{surplus} p_t^{pv\_sell} \tag{5}$$

$$\text{Maximize} \sum_{t}^{M} s_t^{used} \tag{6}$$

$$\textit{Subject to}$$

$$\sum_{i=1}^{N} P_i r_{i,t} + |b_t|^- \geq D_t \tag{7}$$

$$I_{t+1} = b_t + I_t \tag{8}$$

$$0 \leq I_t \leq I_{upper} \tag{9}$$

$$-P_{melt} \leq b_t \leq P_{freeze} \tag{10}$$

Here,

$$e_t = K\left\{\sum_{i=1}^{N} \frac{P_i r_{i,t}}{COP_i(T_t, r_{i,t})} + \frac{|b_t|^+}{COP_{freeze}} + \frac{|b_t|^-}{COP_{melt}}\right\} \tag{11}$$

$$s_t = s_t^{used} + s_t^{surplus} \tag{12}$$

$$|b_t|^+ = \begin{cases} b_t \; if \; b_t > 0 \\ 0 \; else \end{cases} \tag{13}$$

$$|b_t|^- = \begin{cases} -b_t \; if \; b_t < 0 \\ 0 \; else \end{cases} \tag{14}$$

Each symbol was defined as follows:
s_t: Self-owned solar power generation at time t [kWh]
s_t^{used} : Amount of s_t consumed at home [kWh]
$s_t^{surplus}$: Amount of electricity sold as surplus out of s_t [kWh]
COP_{freeze}: COP at thermal storage (constant)
COP_{melt}: COP at heat dissipation (constant)
b_t: Amount of heat stored and dissipated (+ indicates heat stored—indicates heat dissipated) [GJ].
I_t: Heat storage at the start of time t [GJ]
I_{upper}: Upper limit of heat storage [GJ].
K: Unit conversion factor for converting [GJ] to [kWh]
p_t^{retail}: Unit price of electricity purchased from retailers [yen/kWh]
$p_t^{pv\_sell}$: Unit price for selling PV power surplus [yen/kWh].
$p_t^{pv\_cost}$: Cost of PV power generation [yen/kWh].
P_{freeze}: Upper limit output for heat storage [GJ].
P_{melt}: Upper output limit for heat dissipation [GJ].

In the optimization calculations, the solution without the heat storage device for each time was first calculated as the initial solution using the approach in Section 3, and then, the multi-objective optimization genetic algorithm NSGAII (Elitist Non-dominated Sorting Genetic Algorithm) [22] was used to compute the solution. Among the constraints, Equations (7) and (9) were added to the objective function as penalty terms. Equations (8) and (10) were incorporated as relationships between design variables.

In addition to the method of determining the heat storage and heat radiation plans and the output plan of each heat source component using only the GA in Figure 2, a two-step optimization method was proposed and verified by generating only the heat storage and heat radiation plans using the GA and determining the output plan of each heat source component using SLSQP for the generated heat storage and heat radiation plans, as shown in Figure 3. The first method using only a GA was a two-step optimization method that used a GA to generate only the heat storage and heat dissipation plans, as in Figure 3. The former method, using only GA, simultaneously searched both b_t: heat release/storage and r_(i,t): output ratio of chiller i at time t using GA and obtained a Pareto solution for the electricity cost and the amount of renewable energy consumed on-site. In this method, the constraints in Equations (8) and (10) could be expressed in terms of relationships among design variables while the constraints in Equations (7) and (9) were added to the objective function as penalties. Therefore, a search was conducted for solutions that did not satisfy the two constraints, which made the search inefficient. On the other hand, in the latter two-stage optimization method, after the heat storage and heat dissipation plans were formulated by the GA, the output plan of each heat source component was optimized by SLSQP to satisfy the constraint conditions in Equation (7) so that a solution that satisfied the constraint conditions in Equation (7) was always output. The constraint condition in Equation (9) was added to the objective function as a penalty in the same way. The b_t: heat release and heat storage was calculated by GA, the required heat demand was recalculated, and r_(i,t): output of chiller i at time t was calculated by the flow on the right side of Figure 3. The flow on the right side of Figure 3 was almost the same as the output calculation in the base model, and the bifurcation condition took the heat dissipation into account. Equation (3) was used as the constraint for SLSQP. The latter approach had the advantage that the search space of the GA could be narrowed because the GA only searched for the heat storage and heat dissipation plans, and the output plan of each heat source component was calculated using SLSQP. In both approaches, the initial solution without heat storage devices was first calculated for each time using the base model method, and then, the multi-objective optimization GA, NSGAII (Elitist Non-dominated Sorting Genetic Algorithm) [22], was used for the calculation.

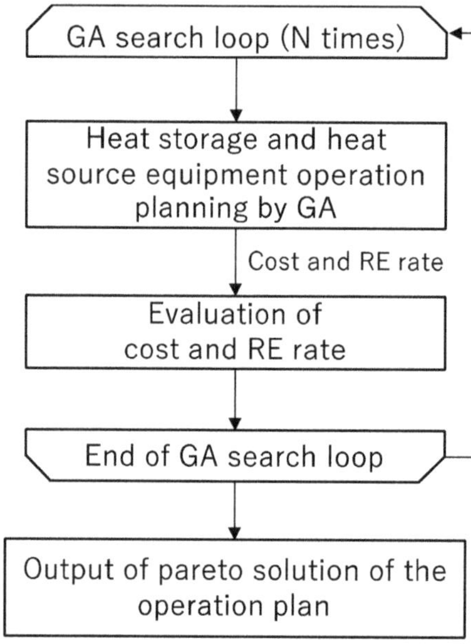

Figure 2. Multi-objective optimization using only GA.

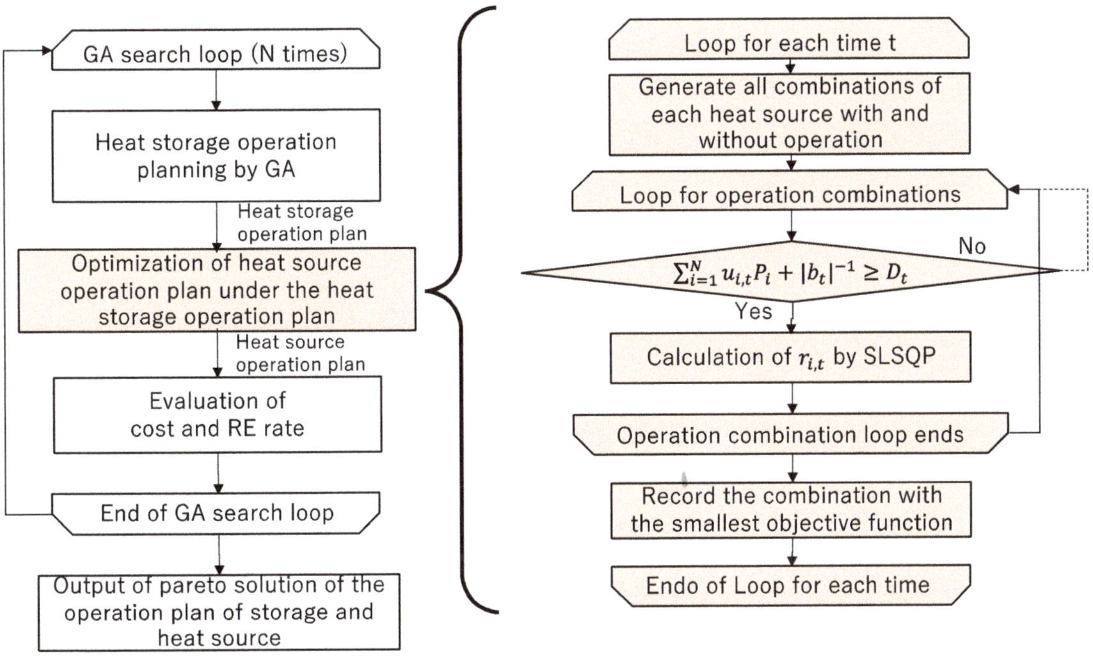

Figure 3. Two-step multi-objective optimization with GA and SLSQP.

3. Case Study

3.1. Data Condition Setting

A case study of the method described above is conducted using actual operating data of air-conditioning equipment in a large-scale facility. An overview of the data used is shown in Tables 1 and 2.

Table 1. Data period and number of devices.

Data Period	1 January 2021 to 31 December 2021
Number of chillers	4
Number of thermal storage unit	1

Table 2. Specifications of heat source equipment and thermal storage devices.

Chiller No.	Maximum Cooling Capacity [GJ/h]
Chiller 1	25.3
Chiller 2	25.3
Chiller 3	25.3
Chiller 4	15.2
Thermal storage unit	12.7

3.2. Case Studies of Methods for Estimating COP Based on Actual Operational Data

The COP is estimated for four chillers using one year of operating data. Chillers 1 and 2 are turbo chillers, and chillers 3 and 4 are inverter turbo chillers. The latter is characterized by a maximum COP at a load ratio of 30 to 40% when the temperature is low.

These four chillers were modeled and evaluated with k-partition cross-validation (k = 10), and the results are shown in Tables 3–6, respectively. The error indices are MAE: Mean Absolute Error, RMSE: Root Mean Squared Error, and MAPE: Mean Percentage Error. Overall, the accuracy of SVR's RBF kernel is relatively good. Next is the K-nearest neighbor method. Figures 4 and 5 show the results of estimating the COP of chiller 4 using these two methods: Figure 4 with the SVR:RBF kernel shows a smooth line while Figure 5 with the K-nearest neighbor shows a line where the COP is not stable with respect to the output ratio. Although this function is used as a part of the objective function, the SVR: RBF kernel is used here because it is preferable to use smoother lines because unstable lines are more likely to lead to local solutions since the gradient information of the objective function is used to solve the optimization problem.

Table 3. List of COP estimation errors for each method: Chiller 1 (in ascending order of MAE).

Model	MAE	RMSE	MAPE
SVR(C = 1, kernel = 'rbf')	0.189	0.246	0.030
K-Neighbors Regressor (n_neighbors = 20)	0.191	0.249	0.030
Random Forest Regressor (max_depth = 10)	0.201	0.259	0.032
SVR (C = 1, kernel = 'poly')	0.201	0.258	0.032
MLP	0.352	0.411	0.055

Table 4. List of COP estimation errors for each method: Chiller 2 (in ascending order of MAE).

Model	MAE	RMSE	MAPE
SVR (C = 1, kernel = 'poly')	0.201	0.255	0.030
K-Neighbors Regressor (n_neighbors = 20)	0.209	0.264	0.031
SVR (C = 1, kernel = 'rbf')	0.210	0.263	0.031
Random Forest Regressor (max_depth = 10)	0.221	0.286	0.033
MLP	0.340	0.407	0.050

Table 5. List of COP estimation errors for each method: Chiller 3 (in ascending order of MAE).

Model	MAE	RMSE	MAPE
MLP	0.615	0.757	0.065
SVR (C = 1, kernel = 'rbf')	0.618	0.757	0.064
K-Neighbors Regressor (n_neighbors = 20)	0.635	0.778	0.066
Random Forest Regressor (max_depth = 10)	0.636	0.789	0.067
SVR (C = 1, kernel = 'poly')	0.914	1.088	0.093

Table 6. List of COP estimation errors for each method: Chiller 4 (in ascending order of MAE).

Model	MAE	RMSE	MAPE
SVR (C = 1, kernel = 'rbf')	0.586	0.738	0.056
K-Neighbors Regressor (n_neighbors = 20)	0.603	0.760	0.058
Random Forest Regressor (max_depth = 10)	0.605	0.763	0.058
MLP	0.622	0.783	0.059
SVR (C = 1, kernel = 'poly')	1.016	1.244	0.112

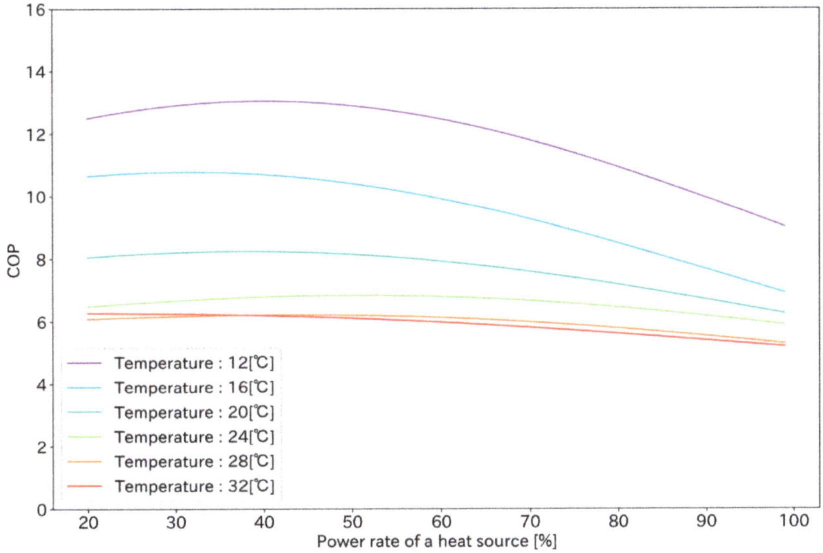

Figure 4. COP curve of chiller 4 with SVR: RBF kernel.

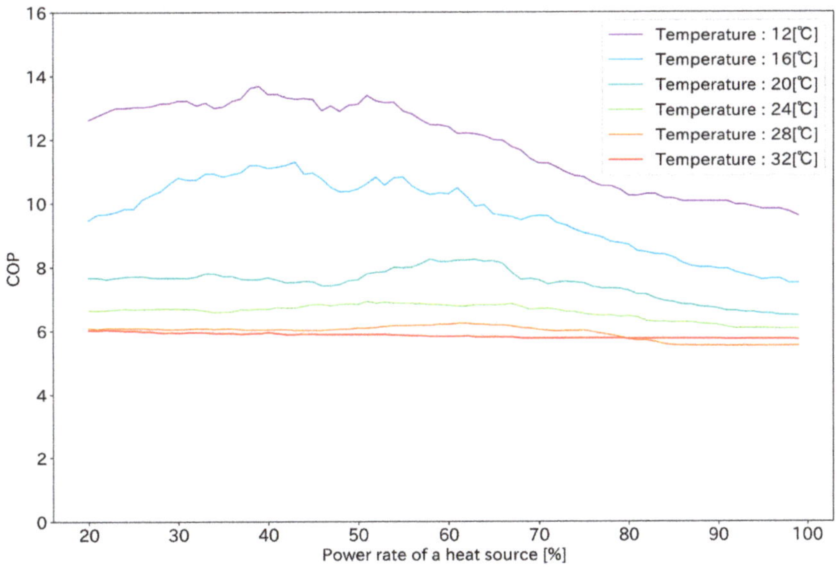

Figure 5. COP curve of chiller 4 with K-nearest neighbor.

Assuming that increasing the output ratio $r_{i,t}$ increases the energy consumption, we mentioned that $\frac{r_{i,t}}{COP_i(T_t, r_{i,t})}$ in the objective function is a monotonically increasing function with respect to $r_{i,t}$, but in reality, based on the estimated COP, when $\frac{r_{i,t}}{COP_i(T_t, r_{i,t})}$ was plotted, a monotonically increasing line was obtained as shown in Figure 6. Optimization can be performed for this COP function using a gradient-based optimization method.

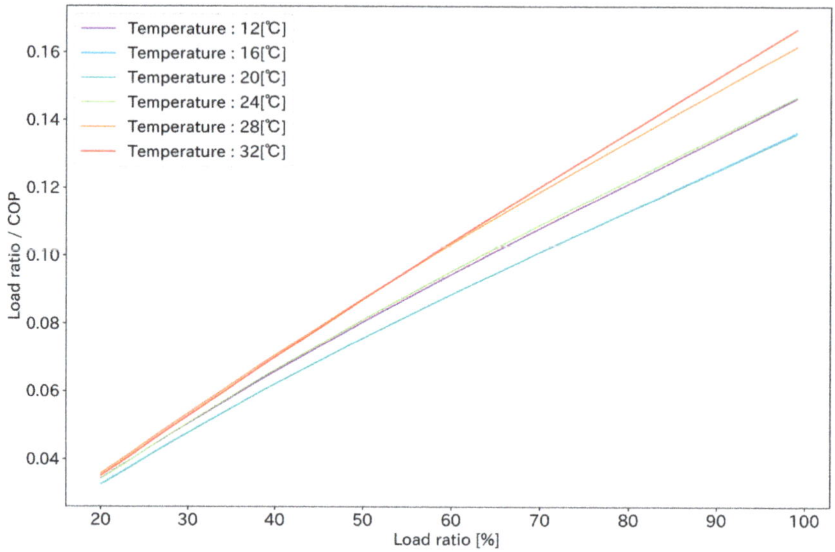

Figure 6. Output ratio/COP of chiller 1.

3.3. *Case Study of a Heat Source Equipment Operation Planning Method Considering COP Variation*

The actual operation history is compared with the cost of operation using the planning method, which takes into account the variation of the COP using the proposed method, to verify how much cost reduction can be achieved by the proposed method. The learned COP model estimated previously is used, and Table 7 shows the results of power consumption reduction by the proposed method in the case of the COP model SVR. The actual power consumption of the actual chiller operation ("Actual" in the table) and the power consumption based on the estimated COP model ("Actual: COP Adjusted" in the table) are set as the comparison targets. "Actual after COP Adjustment" is prepared. "Actual after COP Adjusted" is intended to ensure the equality of the comparison since the energy consumption from the proposed method is calculated based on the estimated COP. For example, if the estimated COP function is inaccurate, and the COP is estimated to be higher than it should be, the power consumption in the "Proposed Method" will be improperly evaluated as small. To solve this problem, the "Actual after adjustment for COP" is calculated by dividing the chiller output by the estimated COP and comparing it with the "Proposed Method". Although PV power generation forecasts, heat demand forecasts, and energy price forecasts are required to develop an actual operational strategy, here, for the sake of proof of concept, forecast values are assumed to be actual values for simplicity. The results for the actual and COP-adjusted results also show that the proposed method can reduce the electricity consumption by approximately 4% per year. Calculated at 14 yen/kWh, this amounts to about 4.1 million yen per year. Table 8 shows the results of power consumption reduction by the proposed method in the case of the COP model KNN. As shown in Table 9, KNN takes about three times longer to calculate the plan than the COP model SVR. This may be due to the fact that the COP estimation function with KNN is multimodal, which causes the gradient-based SLSQP calculation to fall into a local solution. It is also thought that the computation took longer for the same reason.

Table 7. Comparison of proposed and actual electricity consumption and electricity prices (for COP model SVR).

Item	Actual	Actual after Adjustment for COP	Proposed Method
Electricity consumption [MWh]	7326	7320	7030
Electricity cost [thousand yen]	102,600	102,500	98,400
Reduction rate [%] (vs. actual)	-	-	4.04
Reduction rate [%] (vs. actual after COP adjustment)	-	-	3.96

Table 8. Comparison of proposed and actual electricity consumption and electricity prices (for COP model K-nearest neighbor).

Item	Actual	Actual after Adjustment for COP	Proposed Method
Electricity consumption [MWh]	7326	7302	7092
Electricity cost [thousand yen]	102,600	102,200	99,300
Reduction rate [%] (vs. actual)	-	-	3.20
Reduction rate [%] (vs. actual after COP adjustment)	-	-	2.88

Table 9. Difference in computation time of planning methods by COP estimation model.

COP Estimation Model	Computing Time [s]
SVR: RBF	445
KNN(k = 20)	1154

The specific operational schedules generated are examined here. Figure 7 presents the operational schedule for a given day while Figures 8 and 9 depict the schedules proposed by our methodology with the SVR and KNN COP models, respectively, for the same day. The color scheme in these figures indicates the ratio of operation intensity to maximum output. Figures 10 and 11 show the heat demand and temperature variations, respectively, for the same day. As can be inferred from Figure 7, the actual operation bases itself on chiller No.3, adjusting the operation of other chillers in line with the heat demand changes. However, the proposed operation schedule based on the SVR COP model, as shown in Figure 8, operates primarily on chillers No.2 and No.3. The main difference appears in the comparatively smaller demand period of late night to early morning. While the actual operations mainly involve chillers No.3 and No.4, with chiller No.1 serving as a supplement, the proposed methodology instead primarily operates chillers No.2 and No.3, with No.4 serving in a supplementary capacity. Figure 12 presents the COP curves of each chiller at a temperature of 26 degrees. Here, chiller No.3 consistently exhibits the highest COP, followed closely by No.2. Hence, based on the COP estimated from the data, it is rational under these weather conditions to prioritize chiller No.2 over No.4. This reasoning is reflected in the schedule generated by our proposed methodology. Moreover, chiller No.4 is more efficient than

No.1 when the output ratio is small, but this is reversed around a 75% output ratio. Thus, under these weather conditions, there is a need to switch between chillers No.1 and No.4 according to the heat demand size. Our proposed method prioritizes chiller No.1 when the daytime heat demand is high and then switches to chiller No.4 in the evening when the heat demand decreases, demonstrating rational decision-making. It is difficult to consider both weather conditions and output ratio simultaneously in operations based on intuition and experience. Therefore, a scheduling methodology, such as our proposed one, which is based on an equipment efficiency model derived from actual data, is effective.

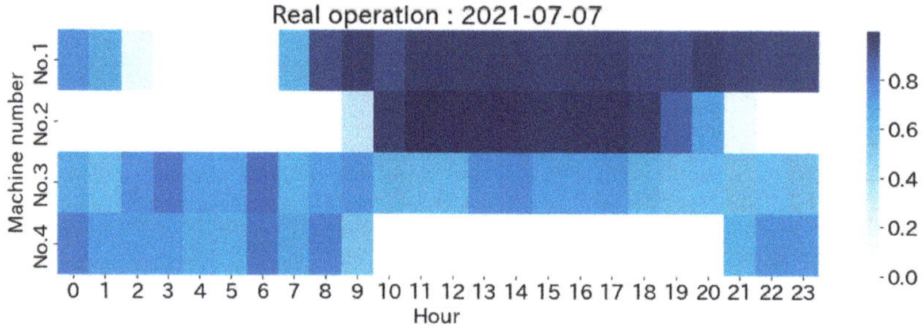

Figure 7. Actual operating schedule for a certain day.

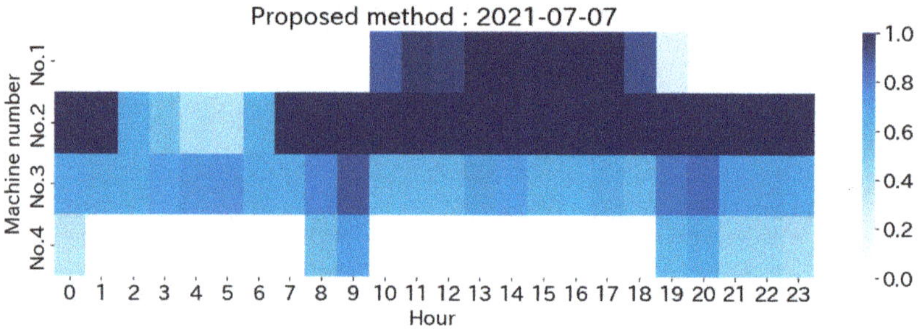

Figure 8. Operating schedule based on the proposed method for a certain day (COP model: SVR).

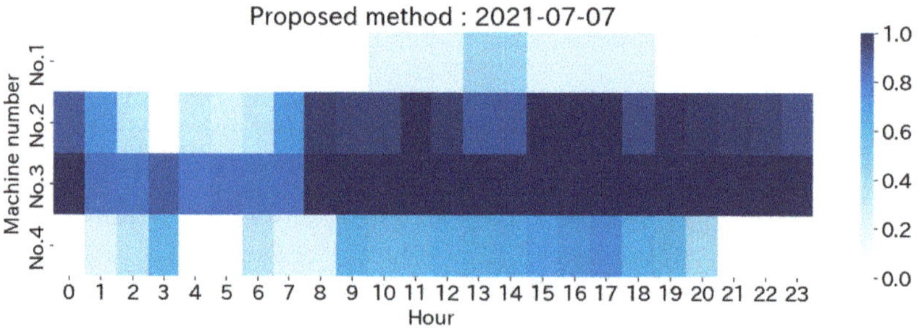

Figure 9. Operating schedule based on the proposed method for a certain day (COP model: KNN).

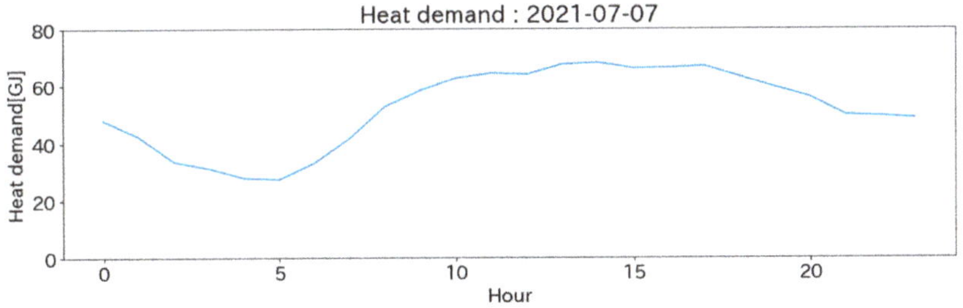

Figure 10. Heat demand on a certain day.

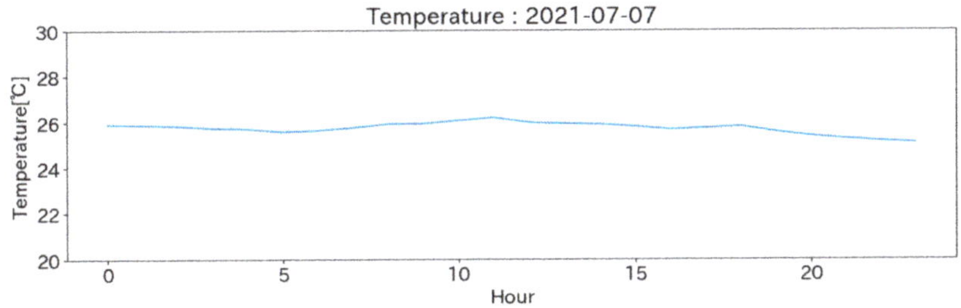

Figure 11. Outside temperature on a certain day.

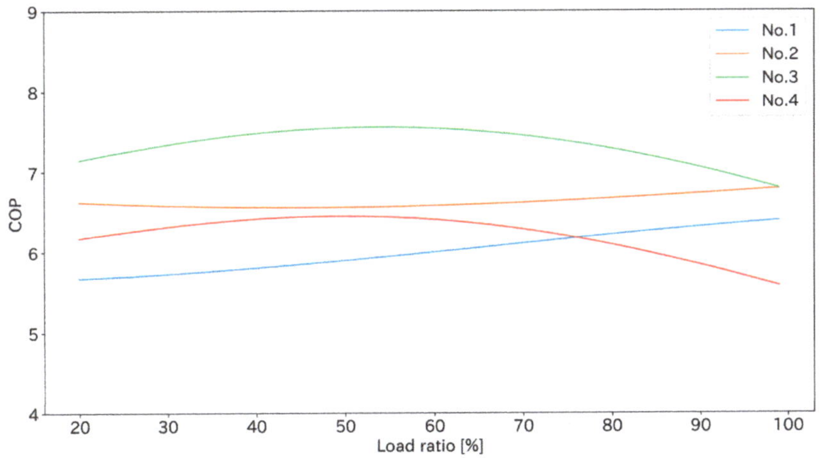

Figure 12. Estimated COP curves for each chiller at an outside temperature of 26 °C (for SVR).

3.4. Case Study of an Operational Planning Method Capable of Exploring Trade-Offs between Environmental and Economic Performance

In this section, we conduct a multi-objective optimization comparison using two methods. The first method, depicted in Figure 2, uses the GA exclusively to decide on the heat storage and release plan, along with the output plan for each heat source equipment. The second method, represented in Figure 3 employs a two-stage optimization approach. Table 10 lists the set values of the constants used in this case study. We assume a scenario where the price of power purchased from the retail business remains consistent at 14 JPY/kWh, regardless of the time of day. As for the rate when selling surplus power generated by the PV system, we have considered the purchase price by a retail business for non-FIT (Feed-In Tariff) power. We assumed that 12 MW of PV was installed.

Table 10. Setting Constants in Case Studies.

COP_{freeze}: COP (constant) for heat storage (Calculated from actual operation data)	4.48
COP_{melt}: COP at heat dissipation (Calculated from actual operation data)	23.8
I_{upper}: Capacity of heat storage [GJ].	200
I_t: Heat storage capacity at the start of time t [GJ].	0
p_t^{retail}: Unit price of electricity purchased from retailers [yen/kWh]	14
$p_t^{pv\_sell}$: Unit price for selling PV power surplus [yen/kWh].	12
$p_t^{pv\_cost}$: Cost of PV power generation [yen/kWh].	8
P_{freeze}: Upper limit output for heat storage [GJ].	38.0
P_{melt}: Upper output limit for heat dissipation [GJ].	38.0

We examine how the Pareto front changes when altering the number of exploratory steps for two optimization techniques. Figure 13 depicts the evolution of the Pareto front using a method that solely utilizes the Genetic Algorithm (GA). Multiple heat storage and release plans, along with operation plans for a given day, are plotted with the Renewable Energy (RE) rate on the y-axis and the cost ratio (minimized power consumption cost divided by total cost) on the x-axis. As the number of explorations increases, the Pareto front shifts towards the upper left, indicating an improvement with higher RE rates and lower costs. Until 3000 explorations, the Pareto front significantly improves, and after this point, the improvements gradually continue as the number of explorations increases. Beyond 7000 explorations, however, there seems to be little improvement. Figure 14 illustrates the evolution of the Pareto front using a two-stage optimization method. Compared to Figure 13, the Pareto front is positioned more towards the upper left overall, indicating superior exploratory performance. Additionally, until 3000 explorations, the Pareto front significantly improves, then it converges, demonstrating the two-stage optimization method's efficiency per exploration.

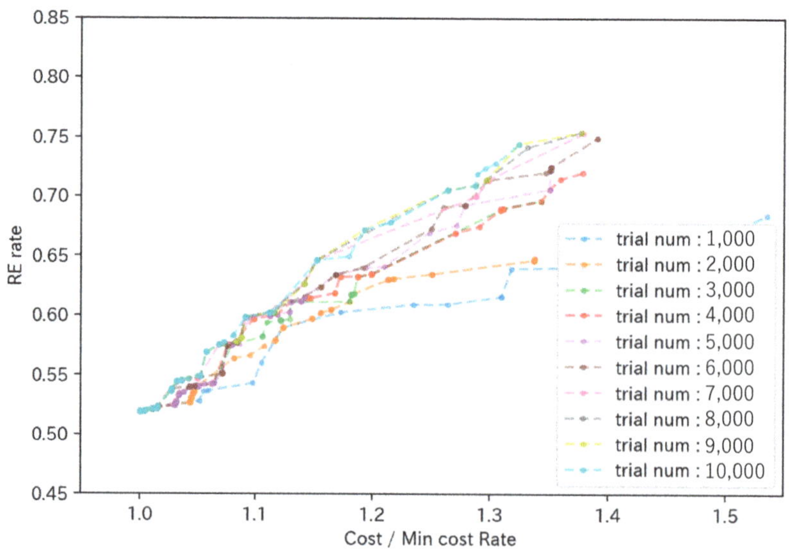

Figure 13. Pareto front transition for methods using only GA.

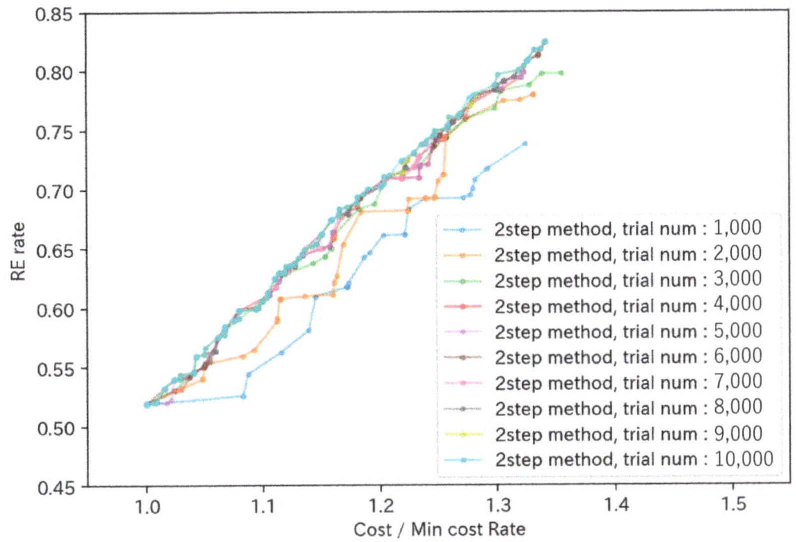

Figure 14. Pareto Front Transition for Two-Step Optimization Method.

To compare and evaluate the differences between the Pareto solutions calculated by the two methods at the same search count, Figure 15 illustrates the Pareto solutions for both methods for search counts ranging from 1000 to 10,000. As depicted in the figure, the two-stage optimization method (labeled "2 step method" in the figure) generally positions the Pareto curve to the upper left more so than the method using only the GA (labeled "Only GA" in the figure). This demonstrates that the two-stage optimization method can achieve a higher RE ratio at a lower cost.

Figure 16 illustrates the Pareto frontiers for the method using only the Genetic Algorithm (GA) after 10,000 searches (represented by the black dotted line in the figure), and the two-stage optimization method after 1000 (blue), 2000 (orange), 3000 (green), and 10,000 (light blue) searches. As can be observed, the Pareto frontier obtained after 3000 searches using the two-stage optimization method (represented by the solid green line in the figure) is positioned further to the upper left compared to the frontier achieved after 10,000 searches using the GA-only method. This indicates that the two-stage optimization method can generate superior Pareto frontiers with fewer search iterations.

In the method using only the GA, 10,000 searches take 5766 s, averaging 0.577 s per search. Conversely, the two-stage optimization method takes 12,293 s for 10,000 searches, averaging 1.23 s per search. This is approximately 2.13 times the time per search. This increased time is due to the two-stage processing of the algorithm, involving GA search and optimization through Sequential Least Squares Programming (SLSQP). However, as shown in Figure 16, a better Pareto frontier is reached with 3000 searches using the two-stage optimization method than with 10,000 searches using only GA. In Table 11, 10,000 searches using the GA-only method take 5766 s while 3000 searches using the two-stage optimization method take 3761 s. This suggests that the two-stage optimization method can reach a superior Pareto frontier in just 65.2% of the time needed by the GA-only method. Despite the increased processing time per search, the two-stage optimization method demonstrates superior efficiency in a Pareto frontier search per unit of time because it can arrive at better solutions with fewer searches compared to the GA-only method.

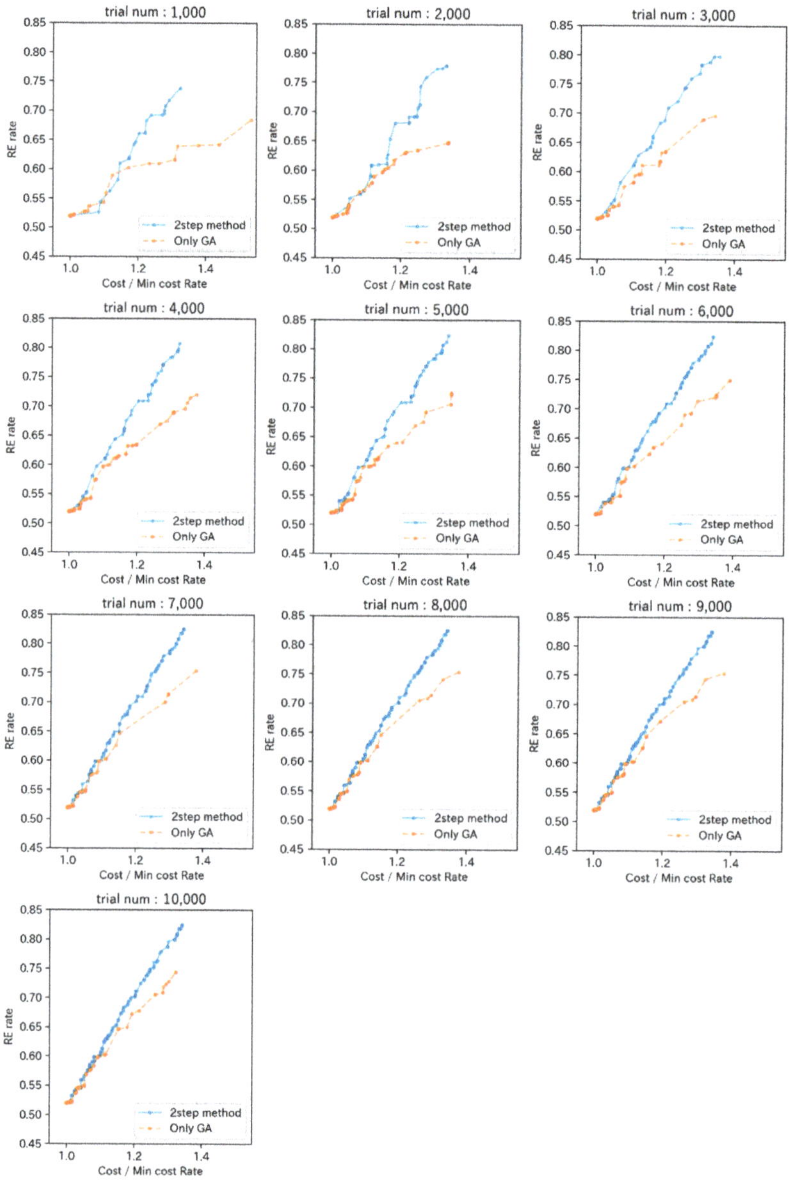

Figure 15. Comparison of Pareto fronts of the two methods by number of searches.

Next, we will examine the details of the solutions produced by the two-stage optimization method. Figure 17 includes a plot of the results of 10,000 searches using the two-stage optimization method, including both Pareto-optimal solutions and sub-optimal solutions (those that are not Pareto-optimal). The color of the points represents red points for Pareto-optimal solutions that satisfy the constraints, light red points for Pareto-optimal solutions that do not satisfy the constraints, blue points for sub-optimal solutions that satisfy the constraints, and light blue points for sub-optimal solutions that do not satisfy the constraints. Here, the points representing the non-use of heat storage devices indicate the lowest cost solutions. This suggests that under these price settings, running the heat storage device is generally unprofitable due to the low Coefficient of Performance (COP) of the heat storage device. Furthermore, under this setting, where the price of electricity purchased from the retailer remains constant regardless of the time of day, there is no cost reduction effect through demand shifting by using heat storage devices. Similarly, under these pricing conditions (electricity sale price and purchase price), selling surplus PV power generated by the user to the retailer is more economical than storing it. Hence, there is no cost reduction effect through the utilization of surplus PV. Therefore, the pattern of not operating the heat storage device is the most economical. In the Pareto-optimal solutions that satisfy the constraints (represented by red points), the cost and RE rate rise at almost a constant ratio. When the RE rate reaches around 82%, there are no longer any Pareto-optimal solutions that satisfy the constraints. From the Pareto-optimal solutions satisfying the constraints (red points), one would select the heat storage and dissipation schedule and the operation schedule for each heat source device. The Pareto point with the lowest cost is (cost ratio, RE rate) = (1.00, 0.52) and the Pareto point with the highest RE rate is (cost ratio, RE rate) = (1.34, 0.82). In other words, it is found that if a cost increase of 34% is permitted, the RE rate can be improved by about 30%.

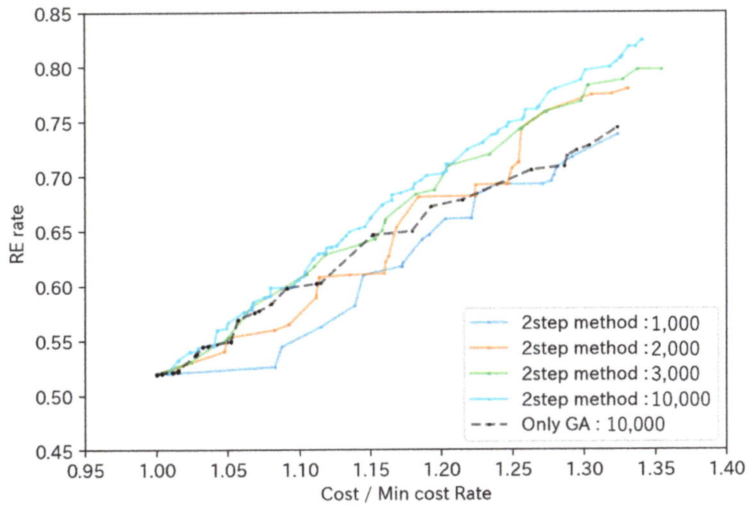

Figure 16. Comparison of the Pareto front of the two methods at different times.

Table 11. Execution time of the two methods per number of searches.

Number of Search [Times].	Execution Time [s]	
	Only GA	2 Step Method
1000	470	1239
2000	960	2493
3000	1473	3761
4000	2013	4964
5000	2582	6156
6000	3178	7320
7000	3795	8502
8000	4424	9702
9000	5079	10,950
10,000	5766	12,293

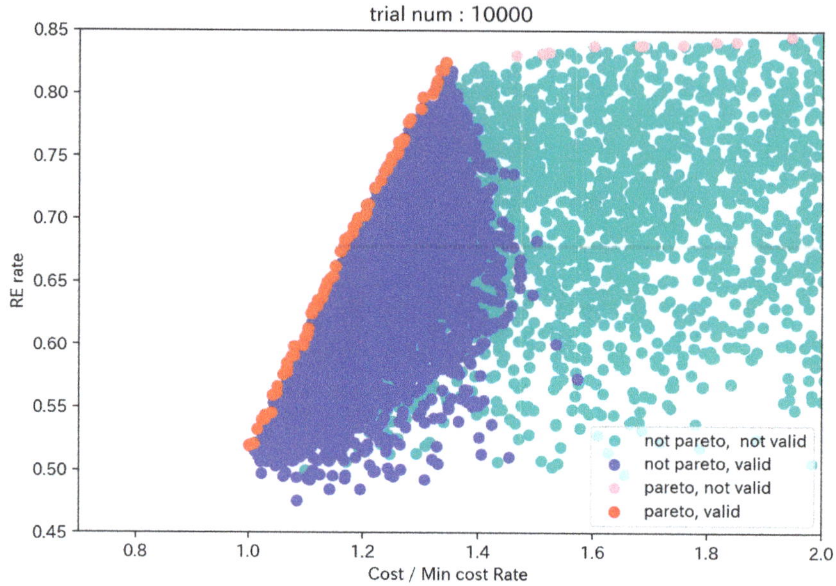

Figure 17. Search result by 2-step optimization method: 10,000 times search.

Figure 18 shows the power consumption, heat supply, and heat storage in the operation plan for the cost-saving oriented point (cost ratio, RE rate) = (1.00, 0.52). Figure 19 shows the operating schedule of the heat source equipment No.1 to 4 and the heat storage device under the same cost-saving oriented plan. There is no heat storage or dissipation, and heat demand is met solely through the operation of the heat source equipment.

Figure 20 shows the power consumption, heat supply, and heat storage in the operation plan for the renewable-energy oriented point (cost ratio, RE rate) = (1.34, 0.82). It can be seen that surplus PV during the day is stored up to the maximum amount of ice that can be stored, and heat dissipation begins from the evening when the surplus power generation ceases. Although a slight amount of heat storage can be observed around 1–2 am, it does not contribute to improving the renewable energy ratio, nor does it contribute to cost reduction. Therefore, this is not an optimal strategy and suggests that there is room for further exploration of solutions. Figure 21 shows the operating schedule of the heat source equipment No.1 to 4 and the heat storage device for the same Pareto point. Considering the amount of heat supplied by heat dissipation from evening to night, the operating level of the heat source equipment is confirmed to be lower than that in Figure 19. Moreover, Figure 22 presents the power consumption, heat supply, and heat storage in the operation plan for a point with a balance between the cost and renewable energy orientation (cost ratio, RE rate) = (1.15, 0.65). While it absorbs surplus renewable energy during the day, it does not store as much as the renewable-energy oriented plan. Figure 23 shows the operating schedule of the heat source equipment No.1 to 4 and the heat storage device for the same cost-renewable energy balance oriented plan. The amount of heat dissipation at night is not as significant as in the renewable-energy oriented plan.

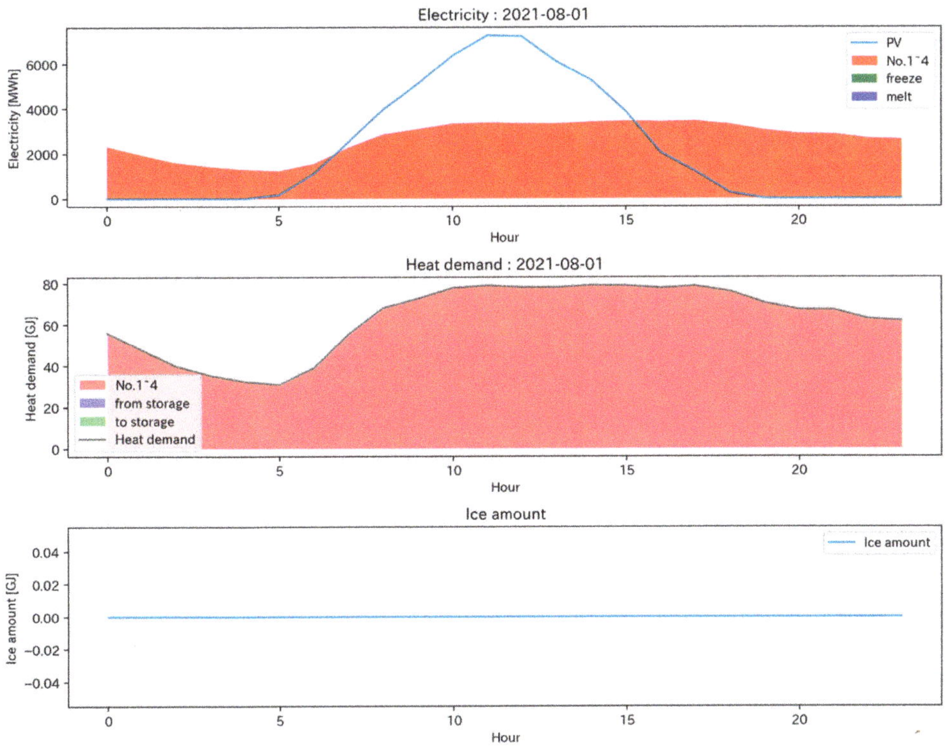

Figure 18. Electricity consumption, heat supply, and heat storage through cost-saving oriented operation planning.

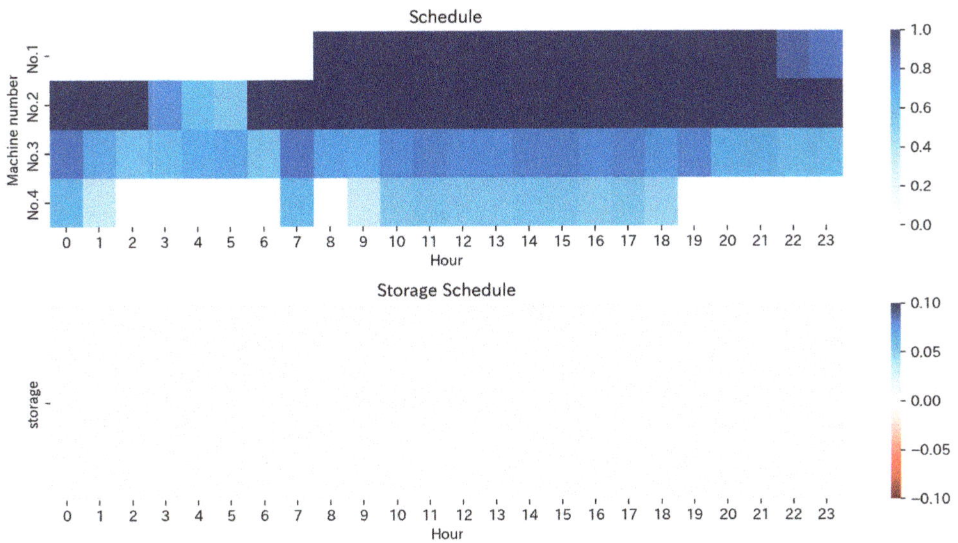

Figure 19. Cost-saving oriented operation plan.

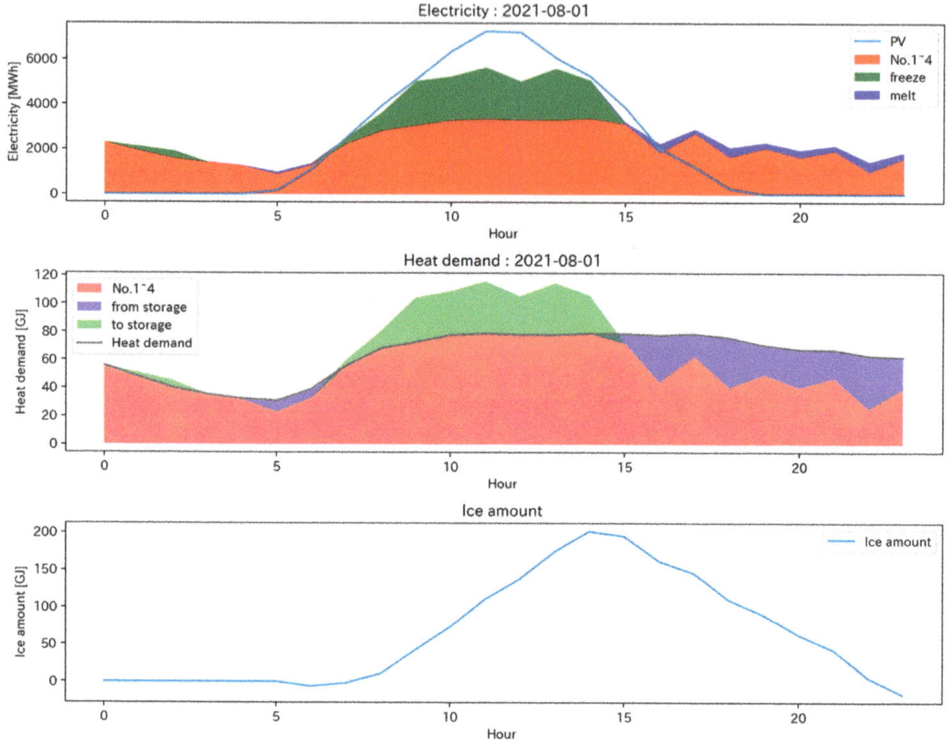

Figure 20. Electricity consumption, heat supply, and heat storage based on an RE oriented operating plan.

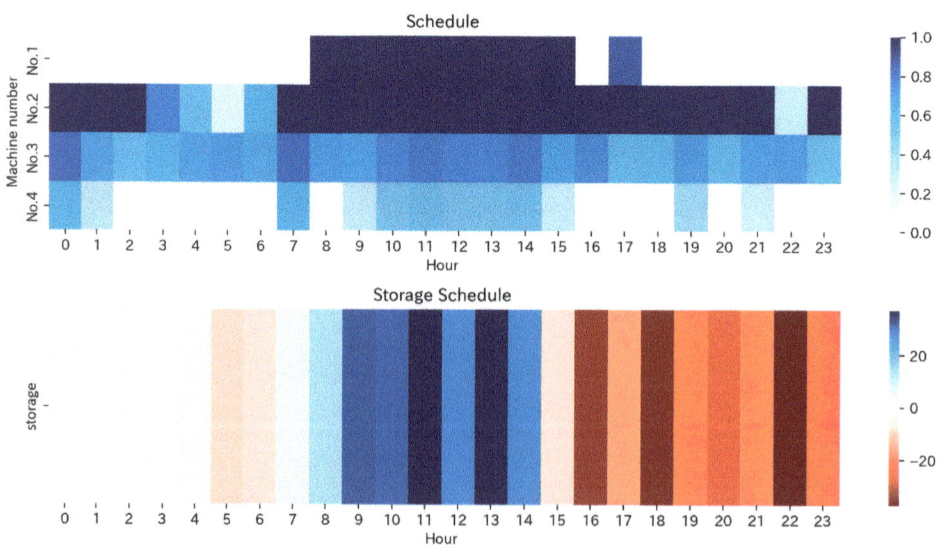

Figure 21. RE oriented operation plan.

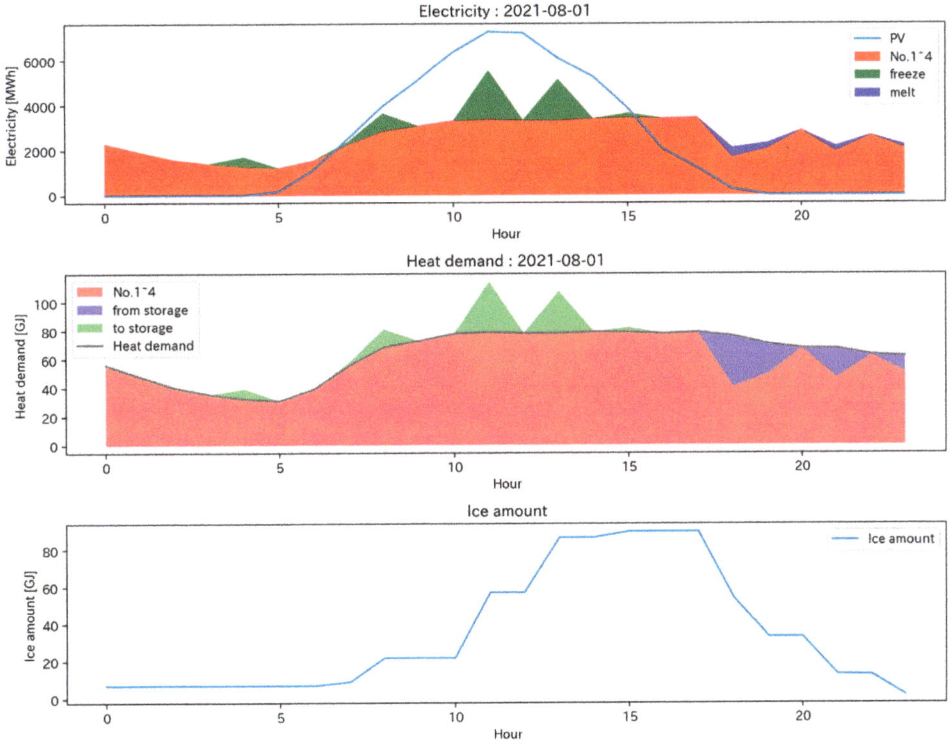

Figure 22. Electricity consumption, heat supply, and heat storage based on cost and RE balanced operation plan.

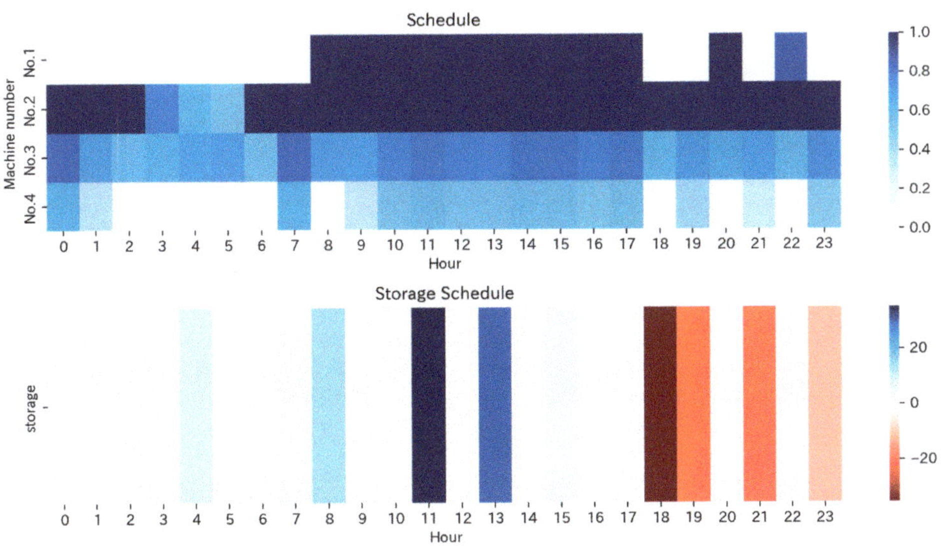

Figure 23. Cost and RE balanced operation plan.

By deriving Pareto solutions from multi-objective optimization, it is possible to determine the heat storage and dissipation strategy and the operation plan of the heat source equipment that an entity should adopt according to the entity's orientation, be it cost-saving, renewable-energy, or cost-renewable energy balance oriented.

4. Conclusions

In this study, we constructed a system to support the decision-making process of entities owning renewable energy generation facilities by calculating Pareto solutions through a multi-objective optimization that balances economic and environmental objectives. We adopted a non-parametric COP estimation method using machine learning, and by integrating the estimation function into the optimization formula, we achieved a more realistic and flexible system modeling, demonstrating that an approximate annual cost reduction of 4% is possible from the current operation. Moreover, when adding a thermal storage facility, we set the proportion of renewable energy in power consumption as the RE rate, an environmental index and the electricity bill as an economic index.

By solving the multi-objective optimization of operation strategies with these as objective functions, we were led to Pareto solutions and constructed a model for selecting operation strategies considering the balance of objective functions. Specifically, we proposed and compared methods for calculating solutions using only the GA and a two-step optimization method combining GA and SLSQP, confirming the superiority of the two-step optimization method. The case study unveiled unique operational profiles corresponding to cost-saving, renewable-energy, and balanced orientation points, suggesting the existence of specific strategies tailored to each orientation. As a result, we built a model that allows stakeholders to select daily operation strategies according to their preferences.

It should be noted that the optimization method used here is based on the genetic algorithm, and its optimality is not guaranteed; hence, there is a possibility that better Pareto solutions may exist.

5. Future Work

One limitation of this method is that it may require approximately a year's worth of data to train a model that can estimate the Coefficient of Performance (COP). However, the real value of this approach becomes evident when the manufacturer's COP curve starts to diverge from reality over time due to deterioration or other factors. It is assumed that by this point, sufficient data will have been accumulated. If sufficient data cannot be prepared to train the COP model, it is still possible to carry out a subsequent optimization using the manufacturer's COP curve as an alternative. In such a case, a future task would be to determine the extent to which this approach deviates from reality.

In addition, we aim to delve deeper into the optimization of parameters for our non-MLP models, including the SVR model. The parameters for these models were kept fixed in our current study. However, we recognize the potential impact that parameter optimization could have on the performance of these models. Specifically, for the SVR model, we will look into optimizing parameters like the regularization parameter 'C' and the kernel functions, which could significantly affect the model's accuracy. We also plan to explore different kernels beyond the 'rbf' kernel used in this study. Moreover, we plan to extend our investigations to other machine learning models and optimization techniques. This would allow us to compare and contrast the effectiveness of various models and strategies, potentially leading to more robust and efficient solutions.

Lastly, we anticipate refining our data collection and handling methods. This would include obtaining more precise information about data quality and sensor accuracy, which can further enhance our model's predictive ability and overall reliability.

Author Contributions: Conceptualization, D.S.; methodology, D.S.; software, D.S.; writing—original draft preparation, D.S.; project administration; K.T. All authors have read and agreed to the published version of the manuscript.

Funding: Grant-in-Aid for Scientific Research (A) 20H00285 from the Japan Society for the Promotion of Science (JSPS).

Data Availability Statement: Restrictions apply to the availability of these data. The data may be available from the authors, but only if permission is granted by Kansai Airports and Kansai International Airport's Heating and Cooling Supply. However, obtaining this permission is not guaranteed.

Acknowledgments: This research project was carried out in cooperation with Kansai Airports. We would like to express our deepest gratitude to Kansai Airports for their helpful advice and to Kansai International Airport Heating & Cooling Supply for provision of their data. The second author is supported by Grant-in-Aid for Scientific Research (A) 20H00285 from the Japan Society for the Promotion of Science (JSPS).

Conflicts of Interest: The authors declare no conflict of interest.

References

1. Hajiah, A.; Krarti, M. Optimal control of building storage systems using both ice storage and thermal mass-Part I: Simulation environment. *Energy Convers. Manag.* **2012**, *64*, 499–508. [CrossRef]
2. Hajiah, A.; Krarti, M. Optimal controls of building storage systems using both ice storage and thermal mass-Part II: Parametric analysis. *Energy Convers. Manag.* **2012**, *64*, 509–515. [CrossRef]
3. Lee, D.; Ooka, R.; Ikeda, S.; Choi, W.; Kwak, Y. Model predictive control of building energy systems with thermal energy storage in response to occupancy variations and time-variant electricity prices. *Energy Build.* **2020**, *225*, 110291. [CrossRef]
4. Kim, D.; Wang, Z.; Brugger, J.; Blum, D.; Wetter, M.; Hong, T.; Piette, M.A. Site demonstration and performance evaluation of MPC for a large chiller plant with TES for renewable energy integration and grid decarbonization. *Appl. Energy* **2022**, *321*, 119343. [CrossRef]
5. Kircher, K.J.; Zhang, K.M. Model predictive control of thermal storage for demand response. In Proceedings of the 2015 American Control Conference (ACC), Chicago, IL, USA, 1–3 July 2015; IEEE: New York, NY, USA, 2015; pp. 956–961.
6. Oldewurtel, F.; Ulbig, A.; Morari, M.; Andersson, G. Building control and storage management with dynamic tariffs for shaping demand response. In Proceedings of the 2011 2nd IEEE PES International Conference and Exhibition on Innovative Smart Grid Technologies, Manchester, UK, 5–7 December 2011; IEEE: New York, NY, USA, 2011; pp. 1–8.
7. Reynders, G.; Nuytten, T.; Saelens, D. Potential of structural thermal mass for demand-side management in dwellings. *Build. Environ.* **2013**, *64*, 187–199. [CrossRef]
8. Abdelkader, A.; Rabeh, A.; Ali, D.M.; Mohamed, J. Multi-objective genetic algorithm based sizing optimization of a stand-alone wind/PV power supply system with enhanced battery/supercapacitor hybrid energy storage. *Energy* **2018**, *163*, 351–363. [CrossRef]
9. Ould Bilal, B.; Sambou, V.; Ndiaye, P.A.; Kébé, C.M.F.; Ndongo, M. Optimal design of a hybrid solar–wind–battery system using the minimization of the annualized cost system and the minimization of the loss of power supply probability (LPSP). *Renew. Energy* **2010**, *35*, 2388–2390. [CrossRef]
10. Yang, H.; Zhou, W.; Lu, L.; Fang, Z. Optimal sizing method for stand alone hybrid solar-wind system with LPSP technology by using genetic algorithm. *Sol. Energy* **2008**, *82*, 354–367. [CrossRef]
11. Dufo-López, R.; Bernal-Agustín, J.L.; Yusta-Loyo, J.M.; Domínguez-Navarro, J.A.; Ramírez-Rosado, I.J.; Juan Lujano, J.; Ismael Aso, I. Multi-objective optimization minimizing cost and life cycle emissions of stand-alone PV wind–diesel systems with batteries storage. *Appl. Energy* **2011**, *88*, 4033–4041. [CrossRef]
12. Saif, A.; Gad Elrab, K.; Zeineldin, H.H.; Kennedy, S.; Kirtley, J.L. Multi-objective capacity planning of a PV-wind-diesel-battery hybrid power system. In Proceedings of the IEEE International Conferences, Manama, Bahrain, 18–22 December 2010.
13. Ascione, F.; Bianco, N.; De Masi, R.F.; De Stasio, C.; Mauro, G.M.; Vanoli, G.P. Multi-objective optimization of the renewable energy mix for a building. *Appl. Therm. Eng.* **2016**, *101*, 612–621. [CrossRef]
14. Navidbakhsh, M.; Shirazi, A.; Sanaye, S. Four E analysis and multi-objective optimization of an ice storage system incorporating PCM as the partial cold storage for air-conditioning applications. *Appl. Therm. Eng.* **2013**, *58*, 30–41. [CrossRef]
15. Lee, W.S.; Chen, Y.T.; Wu, T.H. Optimization for ice-storage air-conditioning system using particle swarm algorithm. *Appl. Energy* **2009**, *86*, 1589–1595. [CrossRef]
16. Zhou, X.; Yu, J.; Zhang, W.; Zhao, A.; Zhou, M. A multi-objective optimization operation strategy for ice-storage air-conditioning system based on improved firefly algorithm. *Build. Serv. Eng. Res. Technol.* **2022**, *43*, 161–178. [CrossRef]
17. Zhao, T.; Wang, J.; Xu, M.; Li, K. An online predictive control method with the temperature based multivariable linear regression model for a typical chiller plant system. *Build. Simul.* **2020**, *13*, 335–348. [CrossRef]
18. Wei, X.; Xu, G.; Kusiak, A. Modeling and optimization of a chiller plant. *Energy* **2014**, *73*, 898–907. [CrossRef]
19. Ren, Y.; Yu, J.; Zhao, A.; Jing, W.; Ran, T.; Yang, X. A multi-objective operation strategy optimization for ice storage systems based on decentralized control structure. *Build. Serv. Eng. Res. Technol.* **2021**, *42*, 62–81. [CrossRef]
20. Zheng, X.; Wu, G.; Qiu, Y.; Zhan, X.; Shah, N.; Li, N.; Zhao, Y. A MINLP multi-objective optimization model for operational planning of a case study CCHP system in urban China. *Appl. Energy* **2018**, *210*, 1126–1140. [CrossRef]

21. Lo, C.C.; Tsai, S.H.; Lin, B.S. Ice storage air-conditioning system simulation with dynamic electricity pricing: A demand response study. *Energies* **2016**, *9*, 113. [CrossRef]
22. Deb, K.; Pratap, A.; Agarwal, S.; Meyarivan, T.A.M.T. A fast and elitist multiobjective genetic algorithm: NSGA-II. *IEEE Trans. Evol. Comput.* **2002**, *6*, 182–197. [CrossRef]

energies

MDPI

Article

Simulation Analysis of Electricity Demand and Supply in Japanese Communities Focusing on Solar PV, Battery Storage, and Electricity Trading

Mika Goto [1,*], Hiroshi Kitamura [2], Daishi Sagawa [3], Taichi Obara [3] and Kenji Tanaka [3]

[1] School of Environment and Society, Tokyo Institute of Technology, Tokyo 108-0023, Japan
[2] NEC Corporation, Tokyo 108-8001, Japan; kitamura@nec.com
[3] School of Engineering, The University of Tokyo, Tokyo 113-8656, Japan; sagawa@ioe.t.u-tokyo.ac.jp (D.S.); obara-taichi@g.ecc.u-tokyo.ac.jp (T.O.); tanaka@tmi.t.u-tokyo.ac.jp (K.T.)
* Correspondence: goto.m.af@m.titech.ac.jp

Abstract: This study analyzes how the electricity demand and supply constitutions affect electricity independence and power trading within a community and between a community and a grid through simulation analysis. To that aim, we create a simulation model equipped with a community-building function and trading capability. We first construct a community consisting of various types of residential and industrial consumers, and renewable power plants deployed in the community. Residential and industrial consumers are characterized by a state of family/business and ownership and the use of energy equipment such as rooftop solar PV and stationary battery storage in their homes/offices. Consumers' electricity demand is estimated from regression analyses using training data. Using the hypothetical community constructed for the analysis, the simulation model performs rule-based electricity trading and provides outputs comprising the total electricity demand in the community, the state of use of battery storage and solar PV, the trading volume, and the electricity independence rate of the community. From the simulation results, we discuss policy implications on the effective use of renewable energy and increasing electricity independence by fully utilizing battery and trading functions in a community.

Keywords: simulation; community; renewable energy; energy trading; battery

1. Introduction

Awareness of risks related to climate change and global warming issues is increasing every year worldwide. The Paris Agreement was adopted in December 2015 at the United Nations (UN) Climate Change Conference (COP21) held in Paris and entered into force in November 2016. The long-term goal of the agreement is to hold "the increase in the global average temperature to well below 2 °C above pre-industrial levels" and pursue efforts "to limit the temperature increase to 1.5 °C above pre-industrial levels" (https://unfccc.int/event/cop-21, accessed on 25 March 2023). Since that, policymakers, corporate leaders, researchers, and society have recognized common concerns of climate change problems and the necessity of future carbon neutrality more than ever. In response, many countries stipulated carbon neutrality by 2050 in law or announced in public that should be achieved.

In Japan, former Prime Minister Suga declared Japan's goal of becoming carbon neutral by 2050 in October 2020 in his policy speech, and in April 2021, he further announced at Japan's Global Warming Prevention Headquarters and the Leaders' Summit on Climate hosted by the United States, "Consistent with the 2050 goal, we aim to achieve an ambitious 46% reduction in greenhouse gas emissions in Fiscal Year (FY) 2030 from FY2013 levels. Furthermore, we will continue to take on the challenge of reaching the 50% mark." Along with the carbon neutrality goal, the Japanese government is moving forward with green society initiatives through the advancement of innovation, energy policy, green growth strategy, the promotion of changes in lifestyle, regional decarbonization, sustainable finance, a reduction in food loss, and various related events (https://www.kantei.go.jp/jp/headline/tokushu/green.html, accessed on 25 March 2023).

To reduce greenhouse gas (GHG) emissions effectively in society, an increase in renewable energy sources such as solar photovoltaic (PV) and wind power generation is a promising measure [1]. In line with the global trend of decarbonization, Japan's local governments are also interested in reducing GHG emissions. An example is the Tokyo Metropolitan Government (TMG), which established a program that requires certain small- and medium-sized new buildings, including houses, to ensure environmental performance and reinforces a green building program for new buildings, a local energy planning program, and an energy environment planning program to strengthen effective efforts to realize a decarbonized society [2]. These programs will start coming into effect from 1 April 2025, etc., and related rules and regulations will be developed at the same time. In particular, since the TMG is a leading example for all the other local governments in Japan, it is of interest as a policy guidance that newly built individual residential homes are required to install solar PV and battery facilities under certain conditions.

Meanwhile, we understand it is not straightforward to drastically increase solar PV and wind power generation because they suffer from intermittency due to the various weather conditions and would bring new challenges in transmission and distribution network operation [3,4]. To remedy the inconvenience, it is suggested to deploy and use batteries combined with this intermittent renewable generation as one of the promising measures. However, it has not been sufficiently discussed what capacities and combinations of these facilities should be deployed in communities that are characterized by different consumer compositions with different characteristics, preferences, and electricity demands for the sake of fully utilizing renewable energy resources. Under the consideration of the combined use of renewable energy sources and battery storage, it is critical to know what battery capacities should be installed and used in what ways [5]. Concretely, it is an essential task to examine how to use a battery in charging and discharging to respond to various changes in demand and supply conditions brought by different community settings. In particular, we emphasize a viewpoint of making the most effective use of renewable energy in a community. The goal of this study is to investigate and accumulate knowledge on an effective strategy of installation and a combination of energy facilities such as solar PV and batteries and the use of electricity trading from simulation results to achieve higher energy independence in a community.

The simulation-based analysis is a solid and useful measure that provides fundamental information to facilitate policy discussions on the electricity demand and supply of a community with the installation of distributed energy resources (DERs). The community is characterized by consumers and DERs, which would basically consist of residential and industrial consumers who hold various types of preferences on energy use, renewable power plants deployed in the community, solar PV facilities installed in residential homes and industrial offices, and the installation and use of battery storage. As a measure to increase the efficient use of DERs while keeping the balance of the power system in a community, peer-to-peer (P2P) electricity trading is of interest to researchers and practitioners. Indeed, the literature on P2P electricity trading has been gradually gaining attention in recent years. Previous studies on P2P electricity trading include [6–31] that investigated, e.g., the efficiency of power systems with DERs, the preferences of energy buyers and sellers and their matching mechanisms, the bidding strategies of prosumers and their influence on electricity trading with/without the use of storage capabilities, and a comprehensive survey of electricity trading and markets including P2P.

The literature, using the agent algorithm [6], examined a bidding agent in a P2P continuous electricity trading market. The agent optimizes and automatically performs electricity trading through bids based on electricity demand and generation forecasts, user preferences for renewable energy (renewable energy-oriented or economically oriented), and assets such as electric vehicles (EVs). As shown in the study, a simulation model approach is a promising tool to examine desirable conditions to effectively use PV generation and battery facilities. Therefore, we employ a similar approach and create a simulation model that is equipped with a flexible community-building capability comprising various consumers and generation facilities and an electricity trading function inside the community as well as between the community and the grid.

Using a hypothetical community with various characteristics created in the simulator, we investigate what combinations of solar PV generation and battery storage usage facilitate power trading for the efficient use of electricity in the community. The efficient use of energy improves the community's energy independence by increasing the self-sufficiency of electricity consumption from renewable power produced in the community. Under the different prefixed conditions of a community, we can observe different patterns of trading that would occur, and thereby, we examine the impacts of different combinations of renewable energy generation, the status of battery holding, and the usage of the electricity trading between the community and the grid on the efficient use of energy.

On the other hand, this study significantly differs from [6] in that we employ a rule-based power allocation model incorporating the predetermined preferences of consumers. That is, the model in this study does not perform optimization calculations for each trading entity or agent (consumer); instead, the community manager performs the rule-based power allocation within the community or between the community and the grid, thus reducing the computation burden and allowing for power exchanges in a large-scale community. Thus, one of the key distinctions of our study lies in the scale of transactions we handle. Unlike existing studies [6], which focused on transactions among 5 users, this study significantly expands the scope by conducting transactions among more than 500 entities. Further, [7] tried to provide a comprehensive understanding of relevant consumer-centric electricity markets from a literature survey by classifying the P2P markets into a fully P2P market, a community-based market, and a hybrid P2P market. According to the classification of [7], the simulation model used in this study can be viewed as a community-based market if we define the community as the entity that establishes the rules for a rules-based transaction. The new contribution of this study in the community-based market approach is that we first incorporate consumers' preferences into the electricity trading and allocation system among agents that implicitly influence agents' trading behavior through the background principle applied to rule-based electricity trading. In addition, we use Japanese real data for parameter estimates. This study also stands out in its ability to generate a community's power demand with simple parameter settings. This feature allows us to recreate a variety of communities in a simplified manner. The flexibility of our simulation model makes it a powerful tool for understanding and planning energy use in diverse community settings. These are the novel contributions of this study.

The remainder of this paper is organized as follows. Section 2 describes the framework and functions of the simulation model used in this study. Section 3 explains simulation cases with different settings and presents the computation results. Section 4 concludes this study and addresses the remaining issues and future extensions.

2. Framework of Simulation

2.1. Brief Description of Simulation

As mentioned in Section 1, this study belongs to the community-based markets model. A unique characteristic that differentiates this study from other community-based market studies is that our model is equipped with internal (within a community) and external (between community and grid) trading functions particularly based on consumers' preferences. Using the trading function and the battery operation, we examine the conditions of the community's efficient electricity consumption that would lead to the community's electricity independence or self-sufficiency.

Figure 1 depicts a brief overview of our simulation. This simulation aims to examine the conditions that lead to efficient power consumption in communities, including how to facilitate the use of EVs, batteries, and renewable power generation facilities owned by industries and households, and those installed in a community. Each household and industry holds three preferences regarding power consumption.

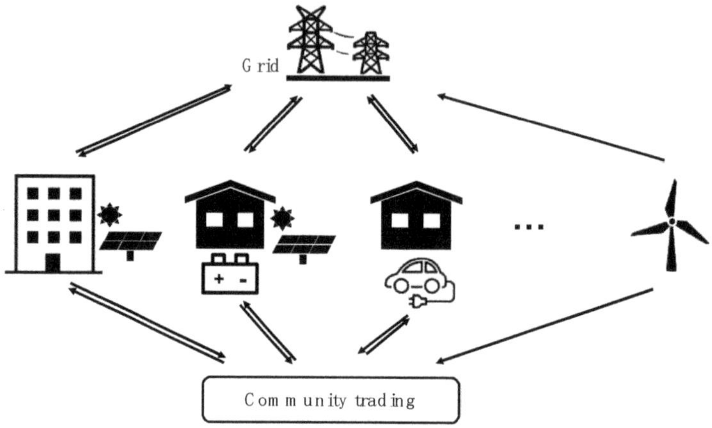

Figure 1. Overview of the simulation.

This simulation was created using Ubuntu 20.04LTS as the operating system, PostgreSQL 12 as the database, and Python 3.8.5 as the programming language. The data used in the simulation are listed in Table 1.

Table 1. List of data used for the simulation.

No.	Data Name	Description
1	Power Demand Data within Tokyo Electric Power Company Jurisdiction	Hourly power demand data of Tokyo Electric Power Company in the past (FY2013). http://www.tepco.co.jp/forecast/html/download-j.html [32], accessed on 25 March 2023
2	Actual Meteorological Information in Tokyo Area	Past meteorological information from the Japan Meteorological Agency. Meteorological elements include temperature, precipitation, humidity, wind speed, wind direction, and solar radiation. http://www.data.jma.go.jp/gmd/risk/obsdl/index.php [33], accessed on 25 March 2023
3	Monthly Electricity Consumption by Household Type	Tokyo Metropolitan Government, Report on the Actual Conditions of Energy Consumption Trends in Households (pp. 30–32). https://www.kankyo.metro.tokyo.lg.jp/climate/home/energy.files/syouhidoukouzittaityousa26honpen_3.pdf [34], accessed on 25 March 2023
4	SII Power Demand Data of Businesses/Building Area and Attribute Data of Businesses	Obtained from EMS OPENDATA of The Environmental Co-creation Initiative. https://www.ems-opendata.jp/ [35], accessed on 25 March 2023

2.2. *Community Generation*

2.2.1. Business Demand Generation

The power demand of each business type (or industry type) is calculated as a product of the power demand per square meter (Figure 2) and the business area, as shown in Figure 3. The flow of the calculation is described as follows.

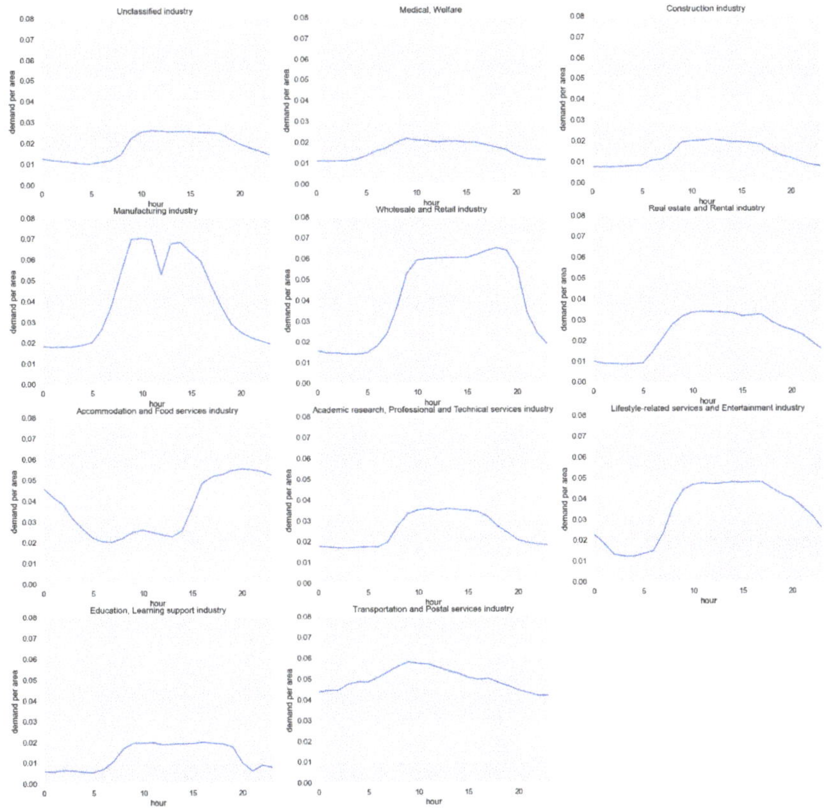

Figure 2. Average demand per area (kWh/m^2) for each business type.

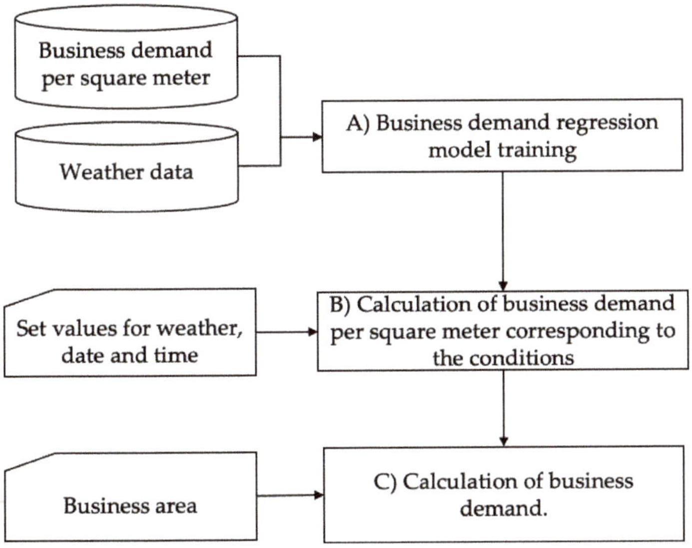

Figure 3. Flow of Business Demand Calculation.

- A model that reproduces the power consumption per square meter for each business type is trained by applying a random forest. The explanatory variables used in this study are as follows, from i to vii. The demand data are based on SII (sustainable open innovation initiative) data, and the model predicts the value of each business type per day.
- Business type;
- Day of the week;
- Time;
- Month;
- Precipitation;
- Temperature;
- Humidity.

The data from SII used in this study include 11 different business types. Table 2 shows these 11 types.

Table 2. List of business types.

Unclassified industry
Medical, welfare
Construction industry
Manufacturing industry
Wholesale and retail industry
Real estate and rental industry
Accommodation and food services industry
Academic research, professional and technical services industry
Lifestyle-related services and entertainment industry
Education, learning support industry
Transportation and postal services industry

- The trained model is provided with weather data corresponding to the simulation target date to calculate the power consumption per square meter for each business type, considering the weather conditions. We provide the trained model with the weather data for the simulation target period as an input, and then we obtain the basic load curve for each business type as an output.
- The power demand for each business is calculated by multiplying the basic load curve by the number of square meters of the business.

2.2.2. Household Demand Generation

The power demand is calculated by the product of the temperature influence coefficient, the proportion of the demand at the specified time within the basic load curve of one day, and the daily power demand for the number of people in the household (apartment/detached house). The flow of the calculation is shown in Figure 4.

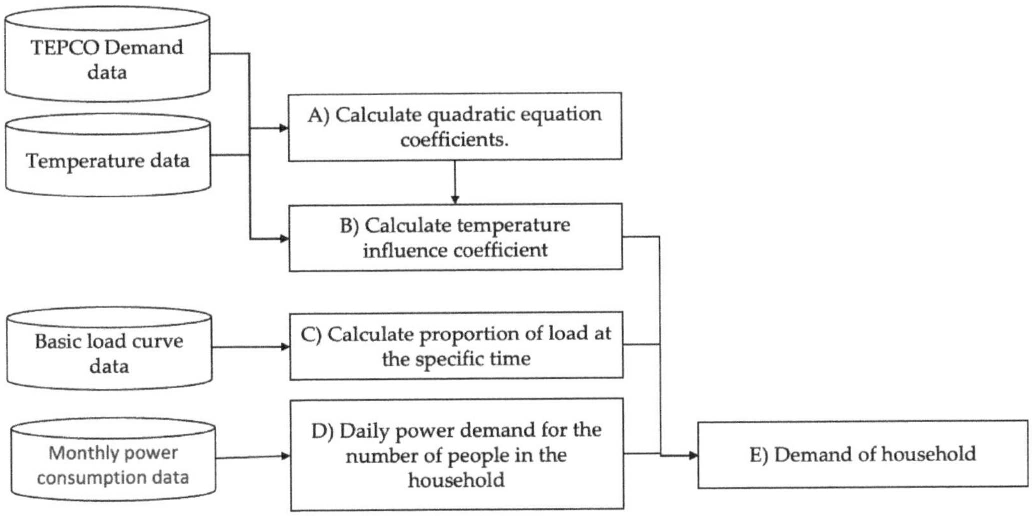

Figure 4. Flow of household demand calculation.

Quadratic Equation Coefficient

We create a quadratic regression equation between power demand and temperature to calculate the coefficient for the difference in power demand due to temperature. Separate regression equations are created for weekdays/weekends and times, and the estimated coefficients are stored. Coefficients of the quadratic regression (a, b, and c in Equation (1)) are calculated using power demand and temperature (T) data in the Tokyo electric power area by the least squares method. The d_{pred} is the estimated demand.

$$d_{pred} = aT^2 + bT + c \tag{1}$$

Temperature Influence Coefficient

The temperature influence coefficient ($C_{T,m,t}$) is calculated in Equation (2) from the estimated demand of the quadratic equation divided by the average demand ($d_{ave,m,t}$) for that month (m) and time (t).

$$C_{T,m,t} = \frac{aT^2 + bT + c}{d_{ave,m,t}} \tag{2}$$

Proportion of demand at the specified time within the basic load curve of one day

The basic load curve specifies the percentage of demand for a given time period (demand rate). In this case, five basic demand curves were established in Figure 5. Of these basic load curves, pattern 1 has low daytime demand, patterns 2 and 3 have high morning and evening demand, pattern 4 has high daytime demand, and pattern 5 is relatively flat.

The proportion of demand at the specified time of one day is calculated for each basic load curve. The shape of the load curve is based on the previous literature [36].

Daily power demand for the number of people in the household (apartment/detached house).

The average monthly power demand in Tokyo for apartments/detached houses and household sizes is converted into daily power demand values.

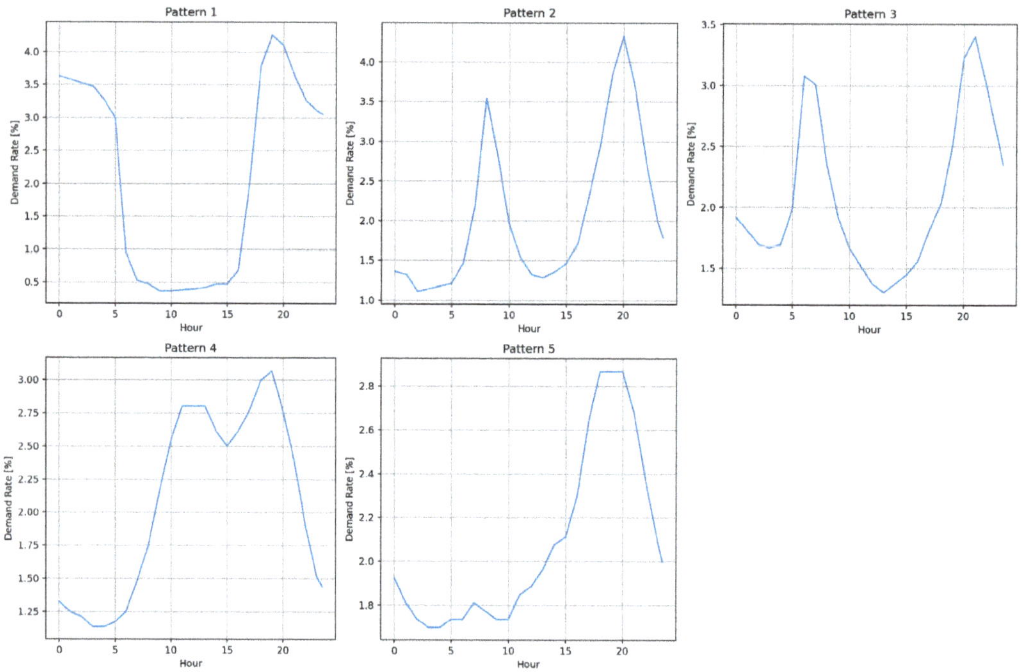

Figure 5. Basic load curves of households.

2.2.3. EV Travel Data Generation

This simulator employs a Python framework "emobpy" [37] to incorporate EV into the simulation model. However, in the present case study, EVs are not considered as industry/household assets. As such, a detailed explanation pertaining to consumers' EV holding and use is omitted from this discussion but could be used for future extension of this study.

2.2.4. Solar Power Generation Data Generation

Solar power generation is calculated by multiplying the solar irradiance [kWh/m^2] by a constant coefficient (*PV_GENERATION_RATIO*) and the system capacity. *PV_GENERATI ON_RATIO* is set to 0.75, which is a power generation of 1 kW solar panel per irradiance [kWh/m^2]. We can set the number and the capacity of solar PV power generation plants depending on the assumptions of a community.

2.2.5. Wind Power Generation Data Generation

The wind power generation is calculated as a product of the wind speed, a constant coefficient, and the system capacity, ensuring that the maximum output is not exceeded: Min (maximum output, wind speed $\times$ coefficient $\times$ maximum output). We can set the number and the capacity of wind power generation plants depending on the assumptions of a community.

2.3. *Power Trading*

In this study, the electricity trading process is divided into two phases: Phase 1 and Phase 2. An overview of the power trading process is described in Figure 6. Initially, in Phase 1, the demand and generation within each entity are calculated, the amount of charging and discharging within each entity is temporarily determined, and the surplus and deficit at each entity are computed. Phase 2 is split into two parts. In the first part, the surplus and deficit of each entity are aggregated to compute the energy surplus or deficit for the community as a whole. In the next part, allocation from entities with surplus to those with deficits occurs, taking into account the preferences of the members and the charging and discharging of batteries.

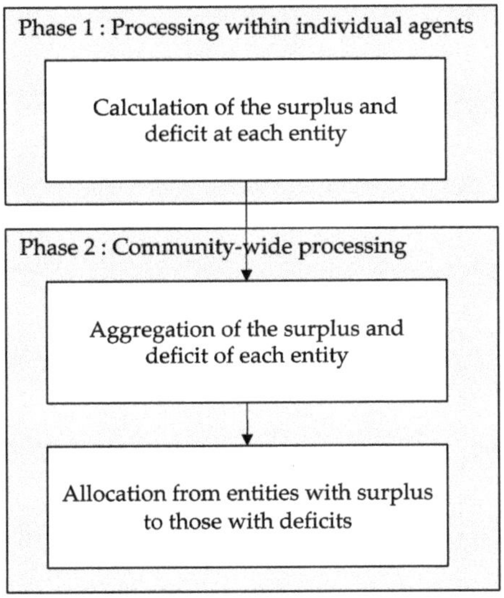

Figure 6. Overview of power trading process.

2.3.1. Phase 1: Processing within Individual Agents

The simulation model calculates the electricity surplus and deficit based on individual agent preferences. In particular, battery operation is influenced by agent preferences. For each entity, considering their preferences, the model calculates the surplus and shortage of electricity after performing battery charging and discharging operations. Table 3 describes the definitions of the variables used in the model, the summarized flow is described in Figure 7, and the detailed process of the consumers (individual agents) is shown in Table 4. In Figure 7, each "Process" step represents multiple actions including calculating, setting variables, checking conditions, etc. Due to the complexity and nested conditions of the pseudocode, it is challenging to represent everything accurately in a flowchart. To fully understand the logic, one needs to look into each "Process" step, where further decisions and actions are taken.

Table 3. Variable definitions.

Defined Variables (in Italics)	Description	Unit
demand	The total energy demand.	kWh
pv_generation	The energy generated from PV (photovoltaic) panels.	kWh
pure_demand	The demand remaining after accounting for PV and wind generation.	kWh
wind_generation	The energy generated from wind turbines.	kWh
pv_surplus	The surplus energy generated from PV panels.	kWh
wind_surplus	The surplus energy generated from wind turbines.	kWh

Table 3. *Cont.*

Defined Variables (in Italics)	Description	Unit
trading_preference	The preference for energy trading strategy: self-consumption, renewable-oriented, or cost-priority.	kWh
ev_dischargable_amount	The amount of energy that can be discharged from the EV.	kWh
ev_soc	The state of charge of the EV battery.	kWh
shortage	The energy deficit that needs to be met after considering generation, battery discharge, and EV charging.	kWh
ev_consumption	The energy consumed by the EV.	kWh
bat_dischargable_amount	The amount of energy that can be discharged from the battery.	kWh
ev_chargable_amount	The amount of energy that can be charged to the EV.	kWh
ev_soc_kwh	The state of charge of the EV battery in kilowatt-hours.	kWh
bat_soc_kwh	The state of charge of the battery in kilowatt-hours.	kWh
bat_chargable_amount	The amount of energy that can be charged to the battery.	kWh
surplus_pv_all	The total surplus energy generated from PV panels across all agents.	kWh
surplus_wind_all	The total surplus energy generated from wind turbines across all agents.	kWh
shortage_all	The total energy deficit across all agents.	kWh
shortage_all_re_preference	The total energy deficit for agents with a "Renewable-oriented" trading preference.	kWh
shortage_all_self_consuming	The total energy deficit for agents with a "Self-consumption" trading preference.	kWh
shortage_all_cost_priority	The total energy deficit for agents with a "Cost-priority" trading preference.	kWh
all_agent_list	The list of all agents participating in the energy trading system.	-

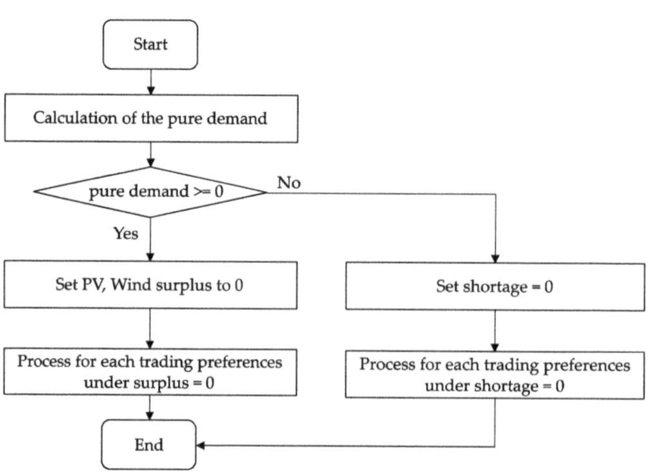

Figure 7. Abstracted flow of phase 1: calculation of surplus and shortage of electricity.

Table 4. Pseudocode of Phase 1: Calculation of surplus and shortage of electricity.

1. Calculate *pure_demand*: *pure_demand = demand-pv_generation-wind_generation*
2. If *pure_demand* >= 0 (generation does not cover consumption):
    2.1. Set *pv_surplus* = 0 and *wind_surplus* = 0
    2.2. If *trading_preference* in ["Self-consumption", "Renewable-oriented"]:
        2.2.1. Discharge as much as possible from the battery
        2.2.2. If EV is at home base (discharging only possible during this time):
            2.2.2.1. If not owning an EV: set *ev_dischargable_amount* = 0
            2.2.2.2. Else, if owning an EV:
                2.2.2.2.1. Calculate EV state of charge
                2.2.2.2.2. If *ev_soc* <= 0.5: charge the EV
                2.2.2.2.3. Else, discharge the EV
            2.2.2.3. Calculate *shortage*
        2.2.3. Else, if EV is driving:
            2.2.3.1. Calculate EV consumption
            2.2.3.2. Calculate *shortage* (without considering *ev_discharge_amount*)
        2.2.4. Update battery state of charge
        2.2.5. Update EV state of charge
    2.3. ElseIf *trading_preference* is "Cost-priority":
        2.3.1. Set *bat_dischargable_amount* = None (decide battery discharge in Phase 2 considering cost)
        2.3.2. If EV is at home base:
            2.3.2.1. If not owning an EV: set *ev_soc* = None and *ev_chargable_amount* = 0
            2.3.2.2. Else, if owning an EV:
                2.3.2.2.1. Calculate EV state of charge
                2.3.2.2.2. If *ev_soc* <= 0.5: charge the EV as much as possible
                2.3.2.2.3. Else: set *ev_chargable_amount* = None (decide EV discharge in Phase 2 considering cost)
            2.3.2.3. If *ev_chargable_amount* is not None: update shortage and *ev_soc_kwh*
            2.3.2.4. Else: set *shortage* = *pure_demand*
        2.3.3. Else, if EV is in use (driving):
            2.3.3.1. Calculate EV consumption
            2.3.3.2. Set shortage = *pure_demand*
3. Else (if the demand is fully met by generation):
    3.1. Set *shortage* = 0
    3.2. If *trading_preference* is "Self-consumption" or "Renewable-oriented":
        3.2.1. Calculate battery charge amount
        3.2.2. If EV is at home base:
            3.2.2.1. Calculate EV charge amount
            3.2.2.2. Calculate PV surplus
        3.2.3. Else (if EV is in use):
            3.2.3.1. Calculate EV consumption
            3.2.3.2. Calculate PV surplus
        3.2.4. Update battery and EV state of charge:
            3.2.4.1. Update *bat_soc_kwh*
            3.2.4.2. Update *ev_soc_kwh* (assuming PHEV and running on gasoline if below 0)
        3.2.5. Set wind surplus: *wind_surplus* = *wind_generation* (all wind generation is surplus as it does not exist at household/business level)
    3.3. ElseIf *trading_preference* is "Cost-priority":
        3.3.1. Calculate PV surplus
        3.3.2. Set wind surplus: *wind_surplus* = *wind_generation* (all wind generation is surplus as it does not exist at household/business level)
        3.3.3. Set *bat_chargable_amount* = None (decide to charge in Phase 2)
        3.3.4. If EV is at home base:
            3.3.4.1. If not owning an EV: set *ev_soc* = None and *ev_chargable_amount* = 0
            3.3.4.2. Else, if owning an EV:
                3.3.4.2.1. Calculate EV state of charge
                3.3.4.2.2. If *ev_soc* <= 0.5: charge the EV as much as possible
                3.3.4.2.3. Else: set *ev_chargable_amount* = None
        3.3.5. Else (if EV is in use):
            3.3.5.1. Calculate EV consumption
        3.3.6. If *ev_chargable_amount* is not None:
            3.3.6.1. Set shortage = max(*ev_chargable_amount*, 0)
            3.3.6.2. Update *ev_soc_kwh*
        3.3.7. Else:
            3.3.7.1. Set *shortage* = 0

Note: This study does not consider wind power generation and EV ownership and use in the community, although the simulator itself is capable of incorporating them for the analysis.

2.3.2. Phase 2: Community-Wide Processing

Phase 2: Community-wide processing aggregates individual agent surpluses and allocates electricity to agents who hold deficits based on their preferences. The distribution of surplus electricity within the community is determined by considering each agent's preferences. Concretely, in order to determine the priority for allocating surplus renewable energy within the community, a price difference based on preference is set as described in Table 5. This ensures that individuals with a preference for renewable energy (renewable-oriented) can secure the community's renewable energy by purchasing it at a higher price. Note that Table 5 is applicable only when the total demand of the community exceeds the total generation of the community. Otherwise, it is possible to purchase electricity at 21 JPY regardless of any preference type. These entities can also trade power with the grid. Table 6 presents the transaction prices with the power grid. The price of 22 JPY for grid power purchasing in Table 5 assumes the power purchased from outside the community where we assume most of the electricity is produced from conventional power plants such as fossil fuel and nuclear power plants.

Table 5. Price table for each preference.

Preferences	Type of Generations	Buying Prices
Cost-oriented	PV	21
Renewable-oriented	PV	25
Self-consumption	PV	23

Note: Although the simulator can incorporate wind power generation plants as a community's renewable power sources, this study focuses on solar PV generation and does not assume community-deployed wind power generation.

Table 6. Transaction price from the power grid.

Type	Prices
Grid power purchasing	22
PV sales to the grid	15

Phase 2 of the process is bifurcated into two main parts. The initial part, as outlined in Table 7, involves the aggregation of surplus and shortage. This step calculates the total surplus and shortage for each community entity at every time interval, using the results derived from Phase 1. The subsequent part, detailed in Table 8, is the allocation of surplus. Here, the previously calculated surplus is distributed to entities that have shortages, in line with their individual preferences. Given the complexity of the entire process, a comprehensive description could be overwhelming. Therefore, we have simplified the process into pseudocode, presented in Tables 7 and 8, to provide a more digestible overview.

Table 7. Process of the surplus and shortage aggregation.

1. Initialize *surplus_pv_all*, *surplus_wind_all*, *shortage_all*, *shortage_all_re_preference*, *shortage_all_self_consuming*, and *shortage_all_cost_priority* to 0
2. For each agent in *all_agent_list*:
   - 2.1. Execute *trade_phase1_time* (dt) for the agent
   - 2.2. Add agent's *pv_surplus* to *surplus_pv_all*
   - 2.3. Add agent's *wind_surplus* to *surplus_wind_all*
   - 2.4. Add agent's *shortage* to *shortage_all*
   - 2.5. If agent's *trading_preference* is " *Renewable-oriented*":
     - 2.5.1. Add agent's shortage to *shortage_all_re_preference*
   - 2.6. ElseIf agent's *trading_preference* is "Self-consumption":
     - 2.6.1. Add agent's shortage to *shortage_all_self_consuming*
   - 2.7. ElseIf agent's *trading_preference* is "Cost-priority":
     - 2.7.1. Add agent's shortage to *shortage_all_cost_priority*
3. EndFor

Table 8. Process of the surplus allocation.

1. If there is no surplus PV and wind generation in the community and there is an overall shortage: 1.1. For each agent in the community: 1.1.1. Determine the agent's *shortage* 1.1.2. If the agent's *trading_preference* is "Cost-oriented": 1.1.2.1. If there is a *shortage*: 1.1.2.1.1. Discharge the battery up to its discharge speed or until the shortage is met 1.1.2.1.2. If the agent's EV is at home, discharge the EV up to its discharge speed or until the remaining shortage is met 1.1.2.1.3. Update the agent's battery and EV states of charge, and chargeable amounts 1.1.2.1.4. Update the agent's grid purchase amount, shortage, and community PV and wind purchase amount to zero 1.1.2.2. If there is no *shortage*: 1.1.2.2.1. Do not discharge the battery or EV 1.1.2.2.2. Update the agent's grid purchase amount, shortage, and community PV and wind purchase amount to zero 1.1.3. If the agent's *trading_preference* is "Self-consumption" or "Renewable-oriented": 1.1.3.1. Since these agents have already been discharged in phase 1, do not discharge them further 1.1.3.2. Update the agent's grid purchase amount to meet the shortage, and set the shortage, community PV, and community wind purchase amounts to zero 2. If there is surplus PV and wind generation in the community, but it is not sufficient to cover the needs of all agents: 2.1. Calculate the *pure_demand* for all agents 2.2. Allocate surplus PV and wind generation to agents based on their trading preferences: 2.2.1. Calculate the amount of surplus PV and wind generation that can be allocated to each preference group (Renewable-oriented, Self-consumption, and Cost-oriented) 2.2.2. Calculate the allocation rates for each preference group based on their shortages and surplus PV and wind generation 2.2.3. For each agent in the community: 2.2.3.1. Assign the allocation rate based on the agent's trading preference 2.2.3.2. Set the PV and wind prices based on the agent's trading preference 2.2.3.3. Update the agent's shortage based on the allocated energy 2.2.3.4. Determine the amount of energy supplied from PV and wind based on the surplus PV and wind generation 2.2.3.5. If the agent's *trading_preference* is "Cost-oriented": 2.2.3.5.1. Discharge the battery if it has not been discharged yet, and update the battery state of charge, shortage, and charge amount 2.2.3.5.2. If the agent's EV is at home and has not been discharged yet, discharge the EV, and update the EV state of charge, shortage, and charge amount 2.2.3.5.3. Update the agent's grid purchase amount to cover the remaining shortage, and set the shortage to zero. 3. If there is enough surplus PV and wind generation to cover the needs of all agents in the community: 3.1. Satisfy the energy demands of all agents 3.2. Calculate the amount of energy that can be absorbed by the batteries and EVs of all agents 3.3. Set the PV and wind prices to the standard price 3.4. Set the *shortage* to zero 3.5. For each agent in the community: 3.5.1. Calculate the supplied energy based on the agent's shortage and surplus PV and wind generation 3.5.2. Assign the energy supplied from PV and wind based on the surplus PV and wind generation 3.5.3. Calculate the total chargeable amount for each agent's battery and EV 3.5.4. Update the total chargeable amount for all agents' batteries and EVs 3.6. Calculate the allocation rate for charging based on the total chargeable amount for all agents' batteries and EVs and the remaining surplus energy 3.7. For each agent in the community: 3.7.1. Update the battery charged amount and calculate the chargeable amount for phase 2 3.7.2. Calculate the amount of energy charged in phase 2 based on the allocation rate 3.7.3. Update the battery state of charge (SOC) and total charge amount 3.7.4. If the agent's EV is at home, perform the same calculations for the EV 3.7.4.1. Update the EV charged amount and calculate the chargeable amount for phase 2 3.7.4.2. Calculate the amount of energy charged in phase 2 based on the allocation rate 3.7.4.3. Update the EV state of charge (SOC) and total charge amount 3.7.5. If the agent's EV is not at home, set the charged amount in phase 2 to zero 3.7.6. Allocate the energy charged in phase 2 to the community PV and wind generation based on the surplus PV and wind generation

For further clarity, we have also included Figures 8 and 9. These figures offer abstracted flow diagrams of the processes described in Tables 7 and 8, respectively. Note that these diagrams are simplified representations intended to aid understanding. For a more detailed understanding of the processes, we recommend referring directly to Tables 7 and 8.

Figure 8. Abstracted flow of phase 2: process of the surplus and shortage aggregation.

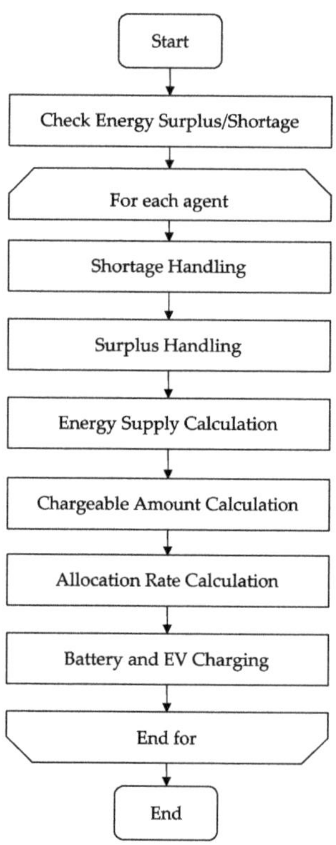

Figure 9. Abstracted flow of phase 2: process of the surplus allocation.

3. Simulation Cases and Results

3.1. Simulation Cases

First, we describe pre-specified conditions and parameters that characterize a community as follows.

The number of industry offices is 50, that of residential homes is 500, and that of community-deployed solar PVs is 4. Although the simulator can incorporate wind power generation plants, this study does not assume the community-deployed wind power generation in order to focus on solar PV generation.

Community solar PV plants: We assume two types of solar PV plants with different capacities, 100 kW (type 1) and 150 kW (type 2). The simulation model is able to set any ratios of these plants for analysis. This study's case sets 0.3 for type 1 and 0.7 for type 2. Note that we assume there are no conventional power plants, such as fossil-fuel and nuclear power plants in a hypothetical community; however, we assume there are conventional power plants outside the community. Therefore, the electricity generated by conventional power plants comes into the community through the power grid.

Residential composition: [Residence 1] 0.5, [Residence 2] 0.1, [Residence 3] 0.2, [Residence 4] 0.2. Residence 1 assumes a one-person household living in an apartment complex, Residence 2 is a two-person household living in a detached house, Residence 3 is a three-person household living in an apartment complex, and Residence 4 is a four-person household living in a detached house. The demand levels depend on the types of residence as described in Section 2.2.2.

Industry composition: 0.1 for each industry from [Industry 1] to [Industry 10]. These industries are assumed from 1 to 10 as follows (see also Table 2): 1: real estate and goods-leasing business, 2: manufacturing industry, 3: lifestyle-related services and entertainment industry, 4: accommodation industry and food services industry, 5: construction industry, 6: education industry and learning support industry, 7: academic research, professional and technical services industry, 8: wholesale and retail industry, 9: transportation industry and postal industry, 10: medical care industry and welfare industry. Each industry is given electricity consumption per square meter of office (electricity intensity), which is used to derive electricity consumption by multiplying the office's square meters as described in Section 2.2.1. This study's case sets 200 m^2 for industry 1 and 1000 m^2 for the others.

Consumer preference: We assume three types of consumer preferences. Preference 1 characterizes a cost-oriented consumer, preference 2 is a renewable-oriented consumer, and preference 3 is a self-consumption-oriented consumer, as shown in Table 5. We assume these three types of consumers exist at the rate of 0.3, 0.2, and 0.5, for each.

Second, we classify the simulation cases into three categories, A to C, by degrees of solar PV capacity installed in homes and offices, with which three battery capacity levels are matched to form individual cases. We set the base case that assumes no PV and battery capacities. Based on that, category A is for large-size PV capacity installed in homes/offices, B is for medium-size, and C is for small-size. Each category is further extended to three cases depending on the battery capacity; large, medium, and small. It should be noted here that these sizes are invariantly applied both to home and office PVs and battery capacities in cases from A to C. They are summarized in Table 9.

Table 9. Description of simulation cases.

Cases	Home/Office PV Capacity	Home/Office Battery Capacity
Base	None	None
A-1	Large	Small
A-2	Large	Medium

Table 9. *Cont.*

Cases	Home/Office PV Capacity		Home/Office Battery Capacity	
A-3	Large		Large	
B-1	Medium		Small	
B-2	Medium		Medium	
B-3	Medium		Large	
C-1	Small		Small	
C-2	Small		Medium	
C-3	Small		Large	
	Home PV	Office PV	Home Battery	Office Battery
D-1	Extra-small	Extra-large	Small	Extra-large
D-2	Extra-small	Extra-large	Medium	Extra-large
D-3	Extra-small	Extra-large	Large	Extra-large

Note: (a) small, medium, and large for PV capacity for home and office indicate 2 kW, 5 kW, and 7 kW, while those for battery capacity for home and office indicate 1 kW, 1.5 kW, and 2 kW. (b) Extra-small for home PV capacity indicates 1 kW, extra-large office PV capacity is 70 kW, and extra-large office battery is 120 kW. The settings used in this simulation for the power plants can be freely changed depending on the assumptions of a community.

Furthermore, to examine the impacts of a combination of larger battery capacity and smaller home PV size, we assume a case in which offices hold extra-large batteries and PVs with a capacity of 120 kW and 70 kW, respectively, and homes hold extra-small PVs with a capacity of 1 kW. We call this case D, which expands to D-1, D-2, and D-3 depending on the combined home battery sizes from small to large. Moreover, we conducted simulation analyses for different consumer preferences as in Table 5; cost-oriented (preference 1), renewable-oriented (preference 2), and self-consumption-oriented (preference 3) ones. These three types of preferences are also examined by simulations for each case in Table 9, referred to as A-1-1 (preference 1 for all home and office consumers), A-1-2 (preference 2 for all home and office consumers), and A-1-3 (preference 3 for all home and office consumers) for the A-1 case as an example. Three types of preferences are also applied to the D-series simulations.

3.2. Results

We present a summary of the simulation results in Table 10. The output measures from the simulation results listed in the table are described as follows.

- [Demand] is the total sum of consumed electricity by all homes and offices in the community (kWh).
- [PV generation] is the total generated electricity by all solar PV generation facilities (community-deployed solar PV plants and consumer-owned solar PV facilities) in the community (kWh).
- [Grid purchase amount] (kWh) is the total electricity that homes and offices purchase from the grid and use in the community.
- [Grid purchase cost] (JPY) is an expenditure for the purchased electricity from the grid, which is calculated from the system average price multiplied by the [Grid purchase amount].
- [Independence rate] (IndR) (%) is a ratio of community-deployed PV generation to community demand. It is calculated as {[Demand]–[Grid purchase amount] + battery net electricity storage}/{[Demand] + battery net electricity storage}. This is a focused core measure in this study because we consider that the higher the rate is, the higher the self-dependency on electricity in the community is achieved, which means being efficient in the use of renewable electricity in the community and also being resilient to critical events that would be caused by supply disruption of grid electricity.

- [PV sell amount to grid] (kWh) is the total amount of electricity sold to the grid that is produced by community-deployed PV facilities.
- [PV sell revenue to grid] (JPY) is revenue earned from electricity produced by community-deployed PVs and sold to the grid, which is calculated from the PV sell price multiplied by the [PV sell amount to grid].
- [Battery capacity] (kWh) is the total capacity installed in the community.
- [Battery loss] (kWh) is the total lost electricity due to the charge/discharge operation of the battery.
- [Community PV trading amount] (kWh) is the total amount of electricity traded (sell/buy) in the community.
- [Community PV trading revenue] (JPY) is revenue/cost earned/paid by electricity traded in the community.
- [Trade rate] (%) is the ratio of community PV power trading to demand, which is calculated by [Community PV trading amount]/[Demand].
- [Battery rate] (%) is the ratio of total battery charge amount to demand, which is calculated by total net electricity charge amount/[Demand].
- [Battery operation rate] is the ratio of total battery charge amount to battery capacity.

From the simulation results in Table 10, this study points out six findings and discusses policy implications. First, the total demand for electricity in the community ranges around 338,000 kWh for all cases. This is found by the pre-defined conditions pertaining to the creation of the hypothetical community for the simulations. Although this study examines a limited number of combinations in Table 10 among electricity demand and supply, generation technologies, home- and industry-owned energy facilities, and their capacities, we can flexibly change those levels and combinations depending on what we want to analyze in various cases.

Second, looking into the core measure [Independence rate] (IndR), all cases are classified into four groups based on the degrees of IndR; (1) base case with 12% IndR, (2) a group with around 30% (C series), (3) a group with around 50% (A and B series), and (4) a group with around 70% (D series). The higher IndR indicates that the community is more electricity-independent from the power grid because the community can procure their consumed electricity in a manner that their electricity demand is fulfilled by solar PV generation installed in the community through plants and home/office facilities. The higher IndR of a community is consistent with the carbon-neutral status of the society because it indicates that the community fulfills a higher rate of its electricity demand from renewable electricity produced in the community. In addition, the lower percentage of the base case is reasonable because no PVs and battery storage are deployed in homes and offices in that case, so that they cannot generate and store electricity for their own convenience and efficient use. On the other hand, the D series assume offices hold a very large battery capacity (120 kW) and PV capacity (70 kW) and a very small PV capacity (1 kW) for homes compared with the other (A to C) simulation series, thereby they can efficiently consume generated electricity using battery storage options. This indicates offices significantly influence the efficient use of electricity because of the large size of capacities. This provides a practical implication to achieve higher IndR, which proposes prioritizing policy institutions that facilitate the large-scale installation of PV and battery assets for industries over homes with relatively small sizes and impacts.

Figure 10 depicts example patterns of demand (Demand all) and generation (PV generation) to represent four groups from (1) to (4) mentioned in the previous paragraph. It is noted that "Demand all" is a net amount that includes the charge/discharge amount. Figure 11 also shows example patterns in which "Original demand" and composite demand (Demand with battery) after adjusting battery charge/discharge amount are separately depicted. Note that "Demand all" in Figure 10 mostly corresponds to "Demand with battery" in Figure 11, although they are slightly different due to the existence of battery loss and conversion loss amounts.

Table 10. Summary of simulation results.

Cases	Demand	PV Generation	Grid Purchase Amount	Grid Purchase Cost	Independence Rate (IndR)	PV Sell Amount to Grid	PV Sell Revenue to Grid	Battery Capacity	Battery Loss	Community PV Trading Amount	Community PV Trading Revenue	Trade Rate	Battery Rate	Battery Operation Rate
Base	338,862	41,025	299,889	6,597,552	12%	0	0	0	0	38,974	961,132	12%	0%	-
A-1	338,258	326,796	151,640	3,336,086	55%	123,354	1,850,308	1100	873	141,377	3,018,140	42%	2%	7.49
A-1-1	338,236	325,007	149,900	3,297,796	56%	119,761	1,796,415	1100	818	148,098	3,110,065	44%	2%	7.00
A-1-2	338,511	325,007	152,549	3,356,076	55%	122,379	1,835,680	1100	902	138,568	2,992,221	41%	3%	7.75
A-1-3	338,258	325,007	152,996	3,365,914	55%	123,077	1,846,153	1100	903	137,818	2,932,895	41%	3%	7.76
A-2	338,035	326,796	148,914	3,276,108	56%	120,633	1,809,502	1650	1292	141,677	3,026,465	42%	4%	7.38
A-2-1	338,234	325,007	146145	3,215,200	57%	115,679	1,735,184	1650	1227	152,180	3,195,786	45%	3%	7.00
A-2-2	338,376	325,007	150,488	3,310,744	55%	120,262	1,803,925	1650	1342	137,188	2,961,467	41%	4%	7.68
A-2-3	338,201	325,007	150,868	3,319,096	55%	120,815	1,812,226	1650	1343	136,583	2,908,509	40%	4%	7.69
A-3	338,173	326,796	145,519	3,201,413	57%	116,866	1,752,986	2200	1723	143,023	3,054,790	42%	5%	7.39
A-3-1	338,248	325,007	142,449	3,133,870	58%	111,638	1,674,570	2200	1636	156,223	3,280,678	46%	5%	7.00
A-3-2	338,369	325,007	148,276	3,262,072	56%	117867	1,768,009	2200	1780	136,097	2,939,624	40%	5%	7.64
A-3-3	338,274	325,007	148,660	3,270,525	56%	118,340	1,775,097	2200	1785	135,568	2,887,727	40%	5%	7.66
B-1	338,183	245,077	161,515	3,553,333	52%	55,702	835,537	1100	852	138,547	2,980,972	41%	2%	7.31
B-1-1	338,214	243,485	158,811	3,493,851	53%	51,248	768,722	1100	818	146,395	3,074,301	43%	2%	7.00
B-1-2	338,375	243,485	162,831	3,582,292	52%	55,390	830,854	1100	877	135,259	2,952,826	40%	2%	7.52
B-1-3	338,266	243,485	163,213	3,590,694	52%	55,879	838,190	1100	878	134,754	2,887,359	40%	2%	7.53
B-2	337,923	245,077	158,736	3,492,197	53%	52,966	794,493	1650	1273	138,856	2,989,854	41%	4%	7.28
B-2-1	338,222	243,485	155,040	3,410,886	54%	47,139	707,081	1650	1227	150,506	3,160,618	44%	3%	7.00
B-2-2	338,366	243,485	161,411	3,551,050	52%	53,806	807,085	1650	1305	133,312	2,929,506	39%	4%	7.46
B-2-3	338,273	243,485	161,786	3,559,290	52%	54,273	814,089	1650	1306	132,835	2,854,487	39%	4%	7.47
B-3	338,097	245,077	156,886	3,451,499	53%	50,733	760,996	2200	1687	138,594	2,989,651	41%	5%	7.23
B-3-1	338,206	243,485	151,366	3,330,044	55%	43,151	647,264	2200	1636	154,493	3,244,343	46%	5%	7.00
B-3-2	338,407	243,485	160,253	3,525,573	52%	52,445	786,673	2200	1727	131,136	2,879,279	39%	5%	7.41
B-3-3	338,338	243,485	160,695	3,535,296	52%	52,960	794,393	2200	1724	130,622	2,807,462	39%	5%	7.39
C-1	338,012	122,498	221,788	4,879,327	34%	112	1672	1100	572	89,718	2,096,405	27%	2%	4.82
C-1-1	338,206	121,201	222,612	4,897,462	34%	0	0	1100	57	93,615	1,965,907	28%	0%	0.26
C-1-2	338,411	121,201	223,806	4,923,731	34%	328	4918	1100	765	86,163	2,134,756	25%	2%	6.53
C-1-3	338,214	121,201	223,556	4,918,237	34%	275	4130	1100	765	86,223	1,969,624	25%	2%	6.53
C-2	338,233	122,498	221,900	4,881,803	34%	0	0	1650	845	87,391	2,045,261	26%	2%	4.75
C-2-1	338,217	121,201	222,374	4,892,238	34%	0	0	1650	71	93,615	1,965,907	28%	0%	0.18
C-2-2	338,320	121,201	223,785	4,923,260	34%	300	4504	1650	1141	82,691	2,047,948	24%	3%	6.49
C-2-3	338,219	121,201	223,631	4,919,885	34%	249	3732	1650	1140	82,750	1,889,751	24%	3%	6.49

Table 10. *Cont.*

Cases	Demand	PV Generation	Grid Purchase Amount	Grid Purchase Cost	Independence Rate (IndR)	PV Sell Amount to Grid	PV Sell Revenue to Grid	Battery Capacity	Battery Loss	Community PV Trading Amount	Community PV Trading Revenue	Trade Rate	Battery Rate	Battery Operation Rate
C-3	338,370	122,498	222,015	4,884,340	34%	0	0	2200	1101	84,957	1,991,712	25%	3%	4.63
C-3-1	338,199	121,201	222,108	4,886,385	34%	0	0	2200	85	93,615	1,965,907	28%	0%	0.13
C-3-2	338309	121,201	223,760	4,922,720	34%	193	2902	2200	1514	79,308	1,963,318	23%	4%	6.46
C-3-3	338,206	121,201	223,614	4,919,512	34%	153	2301	2200	1512	79,350	1,813,432	23%	4%	6.45
D-1	338,548	338,047	90,420	1,989,243	73%	66,119	991,790	13,000	8542	73,657	1,554,804	22%	24%	6.24
D-1-1	338278	336,849	84,061	1,849,348	75%	58,148	872,225	13,000	8010	108,252	2,273,287	32%	22%	5.85
D-1-2	338,371	336,849	96,584	2,124,849	71%	71,798	1,076,967	13,000	8914	49,016	1,039,403	14%	25%	6.52
D-1-3	338,306	336,849	96,484	2,122,640	71%	71,731	1,075,971	13,000	8945	49,032	1,034,849	14%	25%	6.54
D-2	338,455	338,047	87,122	1,916,675	74%	62,739	941,082	13,500	8888	75,376	1,590,744	22%	25%	6.25
D-2-1	338,262	336,849	80,676	1,774,872	76%	54,480	817,201	13,500	8382	111,919	2,350,291	33%	24%	5.89
D-2-2	338,371	336,849	93,364	2,054,006	72%	68,455	1,026,828	13,500	9269	49,472	1,048,988	15%	26%	6.52
D-2-3	338,276	336,849	93,195	2,050,297	72%	68,385	1,025,775	13,500	9276	49,503	1,044,743	15%	26%	6.53
D-3	338,494	338,047	83,944	1,846,769	75%	59,332	889,985	14,000	9234	77,481	1,635,096	23%	26%	6.26
D-3-1	338,328	336,849	77,374	1,702,232	77%	50,811	762,164	14,000	8754	115,575	2,427,081	34%	25%	5.93
D-3-2	338,447	336,849	89,871	1,977,155	73%	64,730	970,946	14,000	9622	50,888	1,078,727	15%	27%	6.53
D-3-3	338,288	336,849	90,074	1,981,637	73%	65,122	976,830	14,000	9611	50,537	1,066,469	15%	27%	6.52

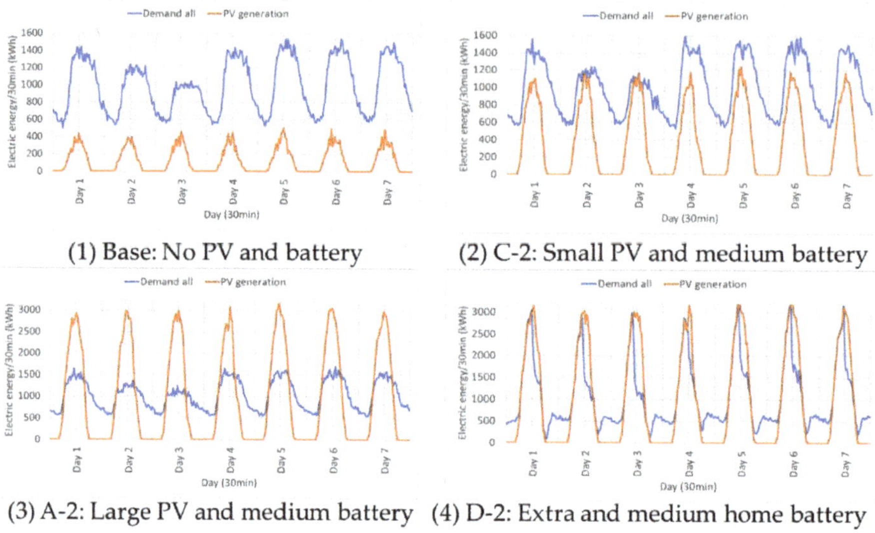

Figure 10. Example patterns of demand and generation. Note: (a) See Table 10 for the corresponding example cases from (**1**)–(**4**). (b) The "Demand all" includes charge/discharge amount, battery loss, and conversion loss. (c) No PV in (**1**) indicates there is no home/office-installed solar PV, only community solar PV plants generate electricity. (d) Extra in (**4**) indicates extra-small PV for home and extra-large PV and battery for office.

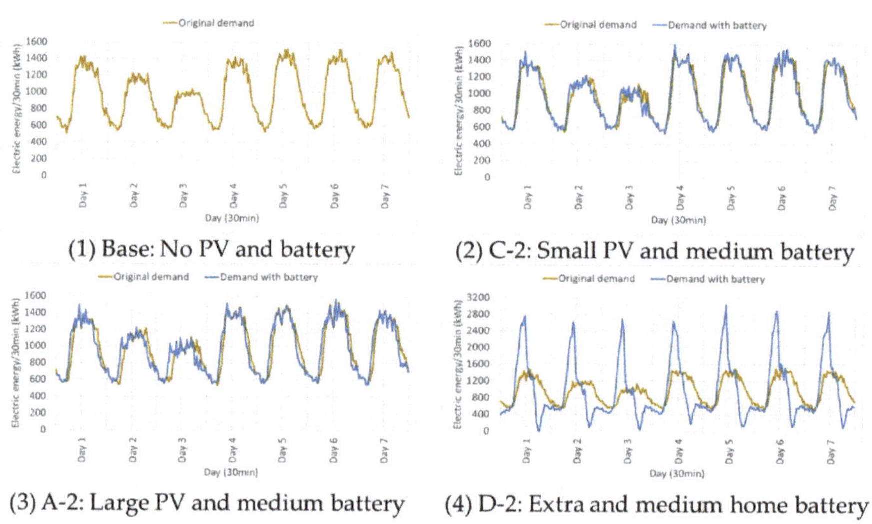

Figure 11. Example patterns of demand and charge amounts. Note: (a) See Table 10 for the corresponding example cases from (**1**)–(**4**). (b) The "Original demand" is consumers' electricity use volume, and "Demand with battery" includes charge/discharge amount, but battery loss and conversion loss are excluded. (c) Extra in (**4**) indicates extra-small PV for home and extra-large PV and battery for office.

Regarding Figure 10, panels (1) and (2) present insufficient solar PV generation compared to the demand, while panel (3) is the opposite in that the solar PV generation shows a large surplus compared to the demand. All of these are not desirable from the perspective of the efficient use of community-produced renewable electricity. On the other hand, the PV generation and demand are well matched in panel (4), receiving balancing benefits from the charge and discharge operations of batteries. Similarly, the active battery operations are also confirmed in Figure 11, panel (4), in which the shape of the original demand is very different from the composite demand (Demand with battery) after charge/discharge adjustments. In other words, the battery operation has been able to follow the demand changes to fill the gap between demand and supply, which consequently achieves the efficient use of electricity in the community.

Further to note, Figures 10 and 11 indicate that a large amount of PV generation compared to the demand in the community is effective to increase IndR. This is presented with IndR levels that increase in the order of (1) < (2) < (3) < (4). In particular, sufficient and balanced PV generation to meet the whole community demand is beneficial for higher IndR as shown in (4) with the D-2 case. Concretely, when homes hold extra-small PV capacity and medium battery capacity, and offices hold extra-large PV capacity and extra-large battery capacity under the pre-defined conditions of this study, IndR is high enough to provide benefits for the whole community in a way that means that solar PV is efficiently used in the community. This is, again, to show that these combinations of facilities deployment are consistent with carbon neutrality because electricity generated by fossil fuel power plants is transmitted into the community only through trading imported from the grid in this study.

Third, related to the second finding, it is clearly presented that higher IndR for the D series is brought by the effective use of battery charge/discharge capability. Concretely, the [Battery rate] for D series ranges from 22% (D-1-1) to 27% (D-3-2, D-3-3), while those for the other series are lower, ranging from 0% (Base, C-1-1, C-2-1, C-3-1) to 5% (A-3, A-3-1, A-3-2, A-3-3, B-3, B-3-1, B-3-2, B-3-3). On the other hand, the [Trade rate] for the D series is relatively low compared to the other cases. Let us remind here that the [Battery rate] is the ratio of total battery charge amount to demand, and the [Trade rate] is the ratio of the community's PV power trading amount to the demand. Thus, the combination of the higher [Battery rate] and the lower [Trade rate] as observed with the D series presents higher reliance on battery use for adjusting the electricity demand and supply and which leads to less use of power trading as an alternative measure. Both measures are consistent with reducing carbon dioxide emissions because they increase the efficient use of renewable electricity produced in the community.

Fourth, it is interesting to note that the [Battery operation rate] in Table 10 ranges over the high level of around six for almost all cases. The exceptions are C-1-1 (with 0.26), C-2-1 (with 0.18), and C-3-1 (with 0.13), which all assume consumers (homes and offices) hold small solar PV capacities and have preference 1 (cost orientation). It is reasonable to see these very small numbers in [Battery operation rate] in exceptional cases in the C series because this study assumes that cost-oriented consumers charge their batteries only when there is a surplus of generation in the community as a whole, and such a surplus situation is rare in the community settings in the C series under the provision of small solar PV capacities. In other words, since the PV generation capacities are very limited in these cases and consumers have a cost-oriented preference, they actually do not have any chances and proactive incentives to use the function of battery storage. Although the batteries do not operate very well in these exceptional cases, electricity is not sold to the grid because of internal community electricity sharing, so electricity is delivered where it is needed. In contrast, self-consumption-oriented and renewable-oriented types charge and discharge when there is a surplus or shortage within each agent (consumer). Thus, we understand battery operations are influenced by consumers' preferences. These characteristics of active and effective battery usage are also presented when we compare C-2-1 and C-2-2 in Figure 12, for example, in which charge and discharge battery functions are evident in C-2-2 (see changes in "Demand with battery") while that for C-2-1 is very limited in the simulation.

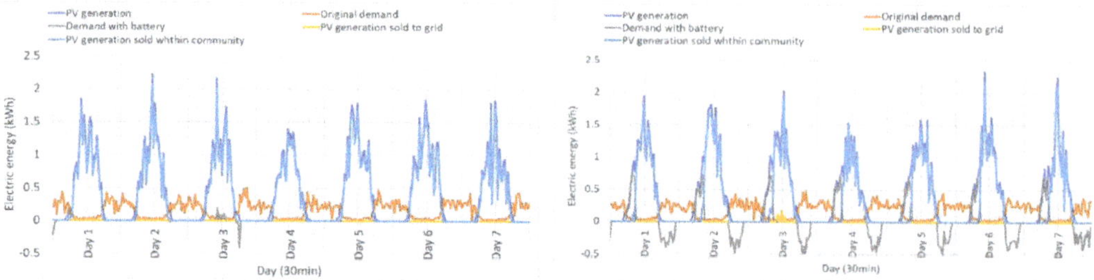

C-2-1: Small PV and medium battery with cost-orientation

C-2-2: Small PV and medium battery with renewable-orientation

Figure 12. Comparison between C-2-1 and C-2-2. Note: (a) See Tables 9 and 10 for the corresponding example cases.

Fifth, differences in the three types of consumer preference do not significantly influence the simulation results. The exceptions are the differences in battery operations in the same cases addressed in the fourth finding, as depicted in Figure 12. This would suggest most combinations of solar PV and battery capacities under different consumer preferences work properly in effective electricity use in a community.

Sixth, the IndR for the A to C series is relatively low compared to the D series, as mentioned in the second finding. This lower IndR for A to C is attributed to the lower [Battery rate]. Meanwhile, the IndR for A to C is supported by a higher [Trade rate], which to some extent compensates for the lower [Battery rate]. For example, the IndR of A-3-1 is 58%, the [Trade rate] is 46%, and the [Battery rate] is 5%, while those of D-3-1 are 77%, 34%, and 25%, respectively. This comparison means the community's internal and external power trading and the use of battery storage are alternative measures to increase the IndR for the community, although the battery seems to be a more powerful tool for this purpose. Therefore, promoting both battery use and power trading is the most desirable measure to achieve higher IndR, but in the case of the battery capacity not being enough for homes/offices to meet their demand under specific preferences, the electricity trading volume would naturally increase to compensate for the limited capacity of the battery storage. This result is interesting from a policy perspective that would suggest a combination of appropriate size of battery installation in homes/offices and conducting power trading as a supplemental tool to provide us with better performance for efficient energy use in a community.

4. Conclusions

This study aimed to examine how the electricity demand and supply constitutions affect a community's electricity independence through the power trading within a community, as well as between a community and a power grid, using the simulation analysis. The demand and supply are influenced by the installation of renewable power plants, consumers' energy facilities (solar PV and battery storage), and their preferences for energy use. For this purpose, we first constructed a hypothetical community consisting of various types of residential and industrial consumers, and renewable power plants focusing on solar PV generation. Residential and industry consumers were characterized by family/business size, the state of ownership, and the use of energy equipment, e.g., rooftop solar PV and battery storage in their homes/offices with various combinations and sizes. Once the community for the analysis was set for the analysis, this study performed the simulations equipped with the rule-based electricity trading capability and obtained outputs comprising total electricity demand in the community, the use statuses of battery storage and solar PVs, and electricity self-sufficiency represented by the IndR of the community. From the simulation results, this study summarized six findings. We suggested it can be effective to achieve higher electricity independence in a community by installing sufficient and balanced solar PV generation and battery storage capacities in combination with functions of power trading.

We note three limitations of this study that should be solved in future studies. First, this study examined a limited number of cases with respect to the constitution of a community, e.g., only incorporating solar PV as a source of generation and employing a fixed composition of different types of consumers. The other DERs, renewable power plants such as wind power generation, and different compositions of consumers can be investigated in the simulation. Second, although EVs are expected to function as battery storage and would provide flexible services to the power grid to arrange the demand and supply of electricity, this study did not include EVs in the simulation. These additional DERs and consumers' compositions are already implemented in the simulation model but were not applied in this study. Third, this study used real data from a Japanese electric power company and weather information in a specific summer period. We can extend the application to other areas and periods, although we need to prepare data for the analysis. These extensions are future tasks of this study.

Author Contributions: Conceptualization, M.G., H.K. and D.S.; methodology, D.S. and H.K.; software, D.S., T.O. and H.K.; validation, M.G., H.K. and D.S.; formal analysis, M.G. and D.S.; investigation, M.G. and D.S.; resources, M.G.; data curation, D.S. and T.O.; writing—original draft preparation, M.G. and D.S.; writing—review and editing, M.G., H.K., D.S. and K.T.; visualization, M.G., H.K. and D.S.; supervision, M.G. and K.T.; project administration, M.G. and K.T.; funding acquisition, M.G. All authors have read and agreed to the published version of the manuscript.

Funding: This work was supported by a Grant-in-Aid of Tokyo Tech Academy of Energy and Informatics.

Data Availability Statement: The data used in this study are openly available in sources presented in Table 1.

Conflicts of Interest: The authors declare no conflict of interest.

References

1. IEA. Energy Technology Perspectives 2023. Available online: https://iea.blob.core.windows.net/assets/a86b480e-2b03-4e25-bae1-da1395e0b620/EnergyTechnologyPerspectives2023.pdf (accessed on 25 March 2023).
2. Tokyo Metropolitan Government. Ordinance Revision to Halve Carbon Emissions (Carbon Half) 2022. Available online: https://www.kankyo.metro.tokyo.lg.jp/en/climate/solar_portal/document.files/ordinance_revision.pdf (accessed on 25 March 2023).
3. Helm, C.; Mier, M. Steering the energy transition in a world of intermittent electricity supply: Optimal subsidies and taxes for renewables and storage. *J. Environ. Econ. Manag.* **2021**, *109*, 102497. [CrossRef]
4. Asiaban, S.; Kayedpour, N.; Samani, A.E.; Bozalakov, D.; De Kooning, J.D.M.; Crevecoeur, G.; Vandevelde, L. Wind and solar intermittency and the associated integration challenges: A comprehensive review including the status in the Belgian power system. *Energies* **2021**, *14*, 2630. [CrossRef]
5. Le, T.S.; Nguyen, T.N.; Bui, D.K.; Ngo, T.D. Optimal sizing of renewable energy storage: A techno-economic analysis of hydrogen, battery and hybrid systems considering degradation and seasonal storage. *Appl. Energy* **2023**, *336*, 120817. [CrossRef]
6. Sagawa, D.; Tanaka, K.; Ishida, F.; Saito, H.; Takenaga, N.; Nakamura, S.; Aoki, N.; Nameki, M.; Saegusa, K. Bidding agents for PV and electric vehicle-owning uses in the electricity P2P trading market. *Energies* **2021**, *14*, 8309. [CrossRef]
7. Sousa, T.; Soares, T.; Pinson, P.; Moret, F.; Baroche, T.; Sorin, E. Peer-to-peer and community-based markets: A comprehensive review. *Renew. Sustain. Energy Rev.* **2019**, *104*, 367–378. [CrossRef]
8. Abdella, J.; Shuaib, K. Peer to Peer Distributed Energy Trading in Smart Grids: A Survey. *Energies* **2018**, *11*, 1560. [CrossRef]
9. Morstyn, T.; Teytelboym, A.; McCulloch, M.D. Bilateral Contract Networks for Peer-to-Peer Energy Trading. *IEEE Trans. Smart Grid* **2019**, *10*, 2026–2035. [CrossRef]
10. Sorin, E.; Bobo, L.; Pinson, P. Consensus-Based Approach to Peer-to-Peer Electricity Markets with Product Differentiation. *IEEE Trans. Power Syst.* **2019**, *34*, 994–1004. [CrossRef]
11. Mengelkamp, E.; Gärttner, J.; Rock, K.; Kessler, S.; Orsini, L.; Weinhardt, C. Designing microgrid energy markets: A case study: The Brooklyn Microgrid. *Appl. Energy* **2018**, *210*, 870–880. [CrossRef]
12. Shrestha, A.; Bishwokarma, R.; Chapagain, A.; Banjara, S.; Aryal, S.; Mali, B.; Thapa, R.; Bista, D.; Hayes, B.P.; Papadakis, A.; et al. Peer-to-Peer Energy Trading in Micro/Mini-Grids for Local Energy Communities: A Review and Case Study of Nepal. *IEEE Access* **2019**, *7*, 131911–131928. [CrossRef]

13. Liu, T.; Tan, X.; Sun, B.; Wu, Y.; Guan, X.; Tsang, D.H.K. Energy management of cooperative microgrids with P2P energy sharing in distribution networks. In Proceedings of the 2015 IEEE International Conference on Smart Grid Communications (SmartGridComm), Miami, FL, USA, 2–5 November 2015; pp. 410–415.
14. Longo, A.; Markandya, A.; Petrucci, M. The internalization of externalities in the production of electricity: Willingness to pay for the attributes of a policy for renewable energy. *Ecol. Econ.* **2008**, *67*, 140–152. [CrossRef]
15. Zori´c, J.; Hrovatin, N. Household willingness to pay for green electricity in Slovenia. *Energy Policy* **2012**, *47*, 180–187. [CrossRef]
16. Tabi, A.; Hille, S.L.; Wüstenhagen, R. What makes people seal the green power deal?—Customer segmentation based on choice experiment in Germany. *Ecol. Econ.* **2014**, *107*, 206–215. [CrossRef]
17. Dagher, L.; Harajli, H. Willingness to pay for green power in an unreliable electricity sector: Part 1. The case of the Lebanese residential sector. *Renew. Sustain. Energy Rev.* **2015**, *50*, 1634–1642. [CrossRef]
18. Morstyn, T.; McCulloch, M.D. Multiclass Energy Management for Peer-to-Peer Energy Trading Driven by Prosumer Preferences. *IEEE Trans. Power Syst.* **2019**, *34*, 4005–4014. [CrossRef]
19. Park, D.-H.; Park, J.-B.; Lee, K.Y.; Son, S.-Y.; Roh, J.H. A Bidding-Based Peer-to-Peer Energy Transaction Model Considering the Green Energy Preference in Virtual Energy Community. *IEEE Access* **2021**, *9*, 87410–87419. [CrossRef]
20. Park, D.-H.; Park, Y.-G.; Roh, J.-H.; Lee, K.Y.; Park, J.-B. A Hierarchical Peer-to-Peer Energy Transaction Model Considering Prosumer's Green Energy Preference. *Int. J. Control. Autom. Syst.* **2021**, *19*, 311–317. [CrossRef]
21. Montakhabi, M.; Zobiri, F.; van der Graaf, S.; Deconinck, G.; Orlando, D.; Ballon, P.; Mustafa, M. An ecosystem view of peer-to-peer electricity trading: Scenario building by business model matrix to identify new roles. *Energies* **2021**, *14*, 4438. [CrossRef]
22. Zepter, J.M.; Luth, A.; del Granado, P.C.; Egging, R. Prosumer integration in wholesale electricity markets: Synergies of peer-to-peer trade and residential storage. *Energy Build.* **2019**, *184*, 163–176. [CrossRef]
23. Iria, J.; Soares, F.; Matos, M. Optimal bidding strategy for an aggregator of prosumers in energy and secondary reserve markets. *Appl. Energy* **2019**, *238*, 1361–1372. [CrossRef]
24. Lee, J.; Cho, Y. Estimation of the usage fee for peer-to-peer electricity trading platform: The case of South Korea. *Energy Policy* **2020**, *136*, 111050. [CrossRef]
25. Vagropoulos, S.I.; Bakirtzis, A. Optimal Bidding Strategy for Electric Vehicle Aggregators in Electricity Markets. *IEEE Trans. Power Syst.* **2013**, *28*, 4031–4041. [CrossRef]
26. Capper, T.; Gorbatcheva, A.; Mustafa, M.; Bahloul, M.; Schwidtal, J.M.; Chitchyan, R.; Andoni, M.; Robu, V.; Montakhabi, M.; Scott, I.J.; et al. Peer-to-peer, community self-consumption, and transactive energy: A systematic literature review of local energy market models. *Renew. Sustain. Energy Rev.* **2022**, *162*, 112403. [CrossRef]
27. Zhou, K.L.; Chong, J.; Lu, X.H.; Yang, S.L. Credit-based peer-to-peer electricity trading in energy blockchain environment. *IEEE Trans. Smart Grid* **2022**, *13*, 678–687. [CrossRef]
28. Li, Z.G.; Chen, S.; Zhou, B.X. Electric vehicle peer-to-peer energy trading model based on SMES and blockchain. *IEEE Trans. Appl. Supercond.* **2021**, *31*, 5700804. [CrossRef]
29. Reis, I.F.G.; Goncalves, I.; Lopes, M.A.R.; Antunes, C.H. Towards inclusive community-based energy markets: A multiagent framework. *Appl. Energy* **2022**, *307*, 118115. [CrossRef]
30. Kim, S.J.; Chu, Y.O.; Kim, H.J.; Kim, H.T.; Moon, H.S.; Sung, J.H.; Yoon, Y.T.; Jin, Y.G. Analyzing various aspects of network losses in peer-to-peer electricity trading. *Energies* **2022**, *15*, 686. [CrossRef]
31. Chung, M.H. Comparison of economic feasibility for efficient peer-to-peer electricity trading of PV-equipped residential house in Korea. *Energies* **2020**, *13*, 3568. [CrossRef]
32. TEPCO Electricity Forecast. Available online: https://www.tepco.co.jp/forecast/html/download-j.html (accessed on 13 February 2023).
33. Japan Meteorological Agency. Available online: https://www.data.jma.go.jp/gmd/risk/obsdl/index.php (accessed on 13 February 2023).
34. Arpak Report. Available online: https://www.kankyo.metro.tokyo.lg.jp/climate/home/energy.files/syouhidoukouzittaityousa26honpen_3.pdf (accessed on 13 February 2023).
35. Energy Management System Open Data. Available online: https://www.ems-opendata.jp/ (accessed on 13 February 2023).
36. Kwac, J.; Flora, J.; Rajagopal, R. Household Energy Consumption Segmentation Using Hourly Data. *IEEE Trans. Smart Grid* **2014**, *5*, 420–430. [CrossRef]
37. Gaete-Morales, C.; Kramer, H.; Schill, W.P. An open tool for creating battery-electric vehicle time series from empirical data, emobpy. *Sci. Data* **2021**, *8*, 152. [CrossRef]

energies

MDPI

Article

Credibility Theory-Based Information Gap Decision Theory to Improve Robustness of Electricity Trading under Uncertainties

Xin Zhao [1], Peng Wang [1], Qiushuang Li [1], Yan Li [1], Zhifan Liu [1], Liang Feng [2,*] and Jiajia Chen [2]

1 Economic & Technology Research Institute, State Grid Shandong Electric Power Company, Jinan 250021, China; suazhaoxin@163.com (X.Z.); wpsgcc@163.com (P.W.)
2 School of Electrical and Electronic Engineering, Shandong University of Technology, Zibo 255000, China; jjchen@sdut.edu.cn
* Correspondence: liangfeng@sdut.edu.cn

Abstract: In the backdrop of the ongoing reforms within the electricity market and the escalating integration of renewable energy sources, power service providers encounter substantial trading risks stemming from the inherent uncertainties surrounding market prices and load demands. This paper endeavors to address these challenges by proposing a credibility theory-based information gap decision theory (CTbIGDT) to improve robustness of electricity trading under uncertainties. To begin, we establish credibility theory as a foundational risk assessment methodology for uncertain price and load, incorporating both necessity and randomness measures. Subsequently, we advance the concept by developing the CTbIGDT optimization model, grounded in the consideration of expected costs, with the primary aim of fortifying the robustness of electricity trading practices. The ensuing model is then transformed into an equivalent form and solved using established standard optimization techniques. To validate the efficacy and robustness of our proposed methodology, a case study is conducted utilizing a modified IEEE 33-node distribution network system. The results of this study serve to underscore the viability and potency of the CTbIGDT model in enhancing the effectiveness of electricity trading strategies in an uncertain environment.

Keywords: electricity service providers; credibility; information gap decision theory; robustness; uncertainties management

1. Introduction

In light of the global imperative to achieve "carbon neutrality," particularly in China, and acknowledging the pivotal role of electricity as a primary secondary energy source, the low-carbon operation of electricity has become a critical concern [1]. While renewable energy sources offer the promise of clean and abundant power, challenges persist, notably encompassing the issues of substantial market volatility and significant deviations in predictive accuracy [2]. The modern power industry, propelled by deregulation and the evolution of electricity market dynamics, has engendered a competitive landscape for global power entities. This transformation not only augments the integration of renewable energy into consumption patterns but also propels enhancements in system efficiency and the quality of power supply [3]. Within the backdrop of an increasingly permeable renewable energy market, the task of ensuring the secure and economically viable operation of the electricity market, replete with myriad energy transactions, becomes notably intricate [4]. This complexity is starkly manifested in heightened uncertainty in electricity price fluctuations, directly impacting market transactions [5]. Serving as the intermediary between end users and power generation enterprises, electricity service providers (ESP) assume a pivotal role within the electricity market [6]. Presently, the pivotal concern for ESP, when contemplating electricity procurement from the trading market, centers on judiciously apportioning the purchase volume in response to market dynamics. This endeavor seeks to curtail the risks posed by fluctuations in price and load, thereby amplifying operational revenues [7].

Now, many scholars have carried out extensive research on the market operation environment of ESP. In [8], researchers harnessed the conditional value-at-risk methodology to quantitatively assess the price fluctuation risks encountered by ESP, culminating in the formulation of an electricity purchase strategy optimization model. Building upon this foundation, Reference [9] delved into the interplay between monthly power deviation assessments and time-of-use electricity pricing, offering insights into the determination of ESP's optimal power purchase ratios across distinct market scenarios. Further expanding the horizon, [10] explored the influence of controllable loads and electricity mutual insurance strategies, culminating in the development of an ESP pricing model tailored to mitigate the divergence in electricity assessments. While significant contributions have been made, challenges persist in the electricity market trading process, notably stemming from the uncertainties surrounding end-user load demand and the volatility of spot market prices. Consequently, the development of optimal trading strategies for ESP that considers multiple uncertainties remains an area with ongoing imperfections. This research endeavors to bridge this crucial gap by conducting a thorough analysis and formulating optimal trading strategies within this context of dual uncertainty.

The pervasive presence of uncertainty within energy systems exerts a profound influence on optimal decision-making, contingent upon problem structure and data accuracy. The energy system encompasses diverse input parameters that harbor uncertainties, thereby posing operational challenges to system operators and ESP. Furthermore, this uncertainty introduces novel complexities to the decision-making processes of pertinent ESP, thereby shaping their planning endeavors [11]. A notable contribution, documented in [12], introduces the scenario analysis method to grapple with electricity price uncertainty, culminating in an energy scheduling model tailored to enhance risk aversion for smart grid retailers. Notably, the accuracy of the scenario-based approach is intricately tied to the number of scenarios employed, with increased accuracy stemming from a greater number, albeit at the cost of computational efficiency. Recognizing these nuances, [13] advocates for a robust optimization approach to tackle uncertainty and conduct risk analysis of market prices. Further enriching the discourse, [14] delineates a transaction optimization model for retailers, reliant on scenarios to delineate the stochastic variables inherent in the electricity market. The conditional value at risk (CVaR) emerges as a salient risk assessment index for electricity purchasing and selling, underpinned by the twin tenets of robust optimization and CVaR. However, conventional approaches often overlook the inherent intricacies related to decision makers' risk preferences. While these methodologies have showcased effectiveness in evaluating uncertainty risk and reducing operational costs for ESP, they are inherently dependent on the availability of precise and abundant data. This reliance presents limitations, particularly in scenarios characterized by restricted information, as observed in China's electricity market [15]. This study comprehensively investigates these methodologies within the context of China's electricity market, aiming to clarify both their suitability and potential constraints.

The Information Gap Decision Theory (IGDT), pioneered by Ben Haim, offers a robust framework for quantifying uncertainty within contexts characterized by unknown probability distributions and fluctuation ranges. Particularly relevant in scenarios where data are sparse, IGDT maximizes parameter perturbations to assess decision outcomes surpassing expected targets. This theory's applicability thrives in contexts of limited information and heightened uncertainty, orchestrating a nuanced integration of decision makers' risk attitudes to derive strategies that encompass risk avoidance and risk seeking, thereby broadening their decision landscape [16]. Leveraging the tenets of IGDT, Ref. [17] formulates a decision-making model catering to power retailers, facilitating electricity procurement and sales across diverse retail contract modalities, whilst navigating the volatility of market-clearing prices. Furthering the discourse, Ref. [18] devises a bidding strategy model for microgrid operators, cognizant of the fluidity of upstream power grid prices. The far-reaching utility of IGDT extends to encompass multifarious aspects of power systems, encompassing the realms of renewable energy generation, spot market pricing, and load dynamics [19]. The integration of IGDT empowers system operators and decision makers with an enhanced grasp of uncertainties, underpinning effective quantification and enabling the formulation of informed risk management strategies. However, prior IGDT optimization-based work on uncertain electricity price mainly focuses on the random characteristic of price, while the fuzzy characteristic is not considered in addressing the human incomplete knowledge caused by the environment, weather, and operational conditions. Therefore, this paper not only expounds upon the conceptual credibility theory into IGDT to describe the random and fuzzy characteristics of electricity price but also explores its pervasive application across diverse domains within electricity trading under multiple uncertainties, accentuating its role in optimizing decision-making processes amidst an uncertain and evolving landscape.

The electricity market landscape is marked by substantial price fluctuations and a notable degree of uncertainty pertaining to both electricity prices and user loads. This may result in high and unnecessary costs for ESP. Consequently, devising apt bidding and scheduling strategies emerges as a formidable challenge for participants within the electricity market. In response to this intricate milieu, this study endeavors to mitigate the multifaceted impact of uncertainties inherent in power market transactions. To this end, we adopt the IGDT framework to model the uncertainty associated with spot market electricity prices and power loads and develop a credibility theory-based IGDT (CTbIGDT) to improve robustness of electricity trading under uncertainties. Our approach is underscored by an earnest commitment to accommodating the risk preference characteristics intrinsic to decision makers. By navigating the terrain of multi-uncertainty, we formulate an innovative power market trading strategy. This strategy is meticulously crafted to harmonize diverse uncertainties while ensuring that operating costs are controlled within the expected range. The overarching objective is to amplify the robustness of ESP within the realm of power trading. This research contributes a comprehensive and adaptable framework that not only embraces the realities of uncertainty within the electricity market but also empowers ESPs to make informed and strategic decisions in the pursuit of optimized power trading outcomes.

2. The Optimization Model of ESP with Multiple Market Transactions

In the pursuit of mitigating the adverse impact of uncertainty, ESP navigates an intricate landscape of electricity market transactions to procure power through a versatile array of approaches. This paper introduces a comprehensive spectrum of electricity market trading methods, encompassing the spot market, option market, and bilateral contract market. To fortify the operational framework, we integrate energy storage units (ESUs) and self-generated power generation equipment, such as gas turbines (GTs), into the system architecture. The ESUs effectively accumulate surplus electrical energy, while synchronized GT output facilitates peak shaving during periods characterized by elevated electricity prices, thereby ameliorating electricity procurement costs. This amalgamation of strategies empowers ESPs with the flexibility to adeptly tailor power purchase approaches to the unique exigencies of distinct scenarios.

Figure 1 graphically illustrates the trading structure of the ESP market, underpinning the multifaceted nature of the adopted strategies. Through this synthesized approach, ESPs stand poised to engender heightened resilience and optimization in power procurement dynamics. The detailed optimization model of ESP with spot market, option market, and bilateral contract market is presented below. It should be mentioned that while our approach indeed utilizes call option contracts as part of a conservative trading strategy, it is important to recognize that this strategy is tailored to address the complexities of spot price uncertainty in the specific environment considered in this study. We acknowledge that in a power system featuring variable renewable energy sources, the assumption of continuously increasing energy prices may not always hold true. However, the use of call options offers a tool for risk management in this dynamic environment, allowing market participants to make informed decisions based on their individual risk tolerance, market expectations, and the need to balance potential costs and benefits. The selection of diverse trading methods offers traders greater flexibility in crafting their trading strategies, enabling them to mitigate the effects of electricity price uncertainties to a certain extent.

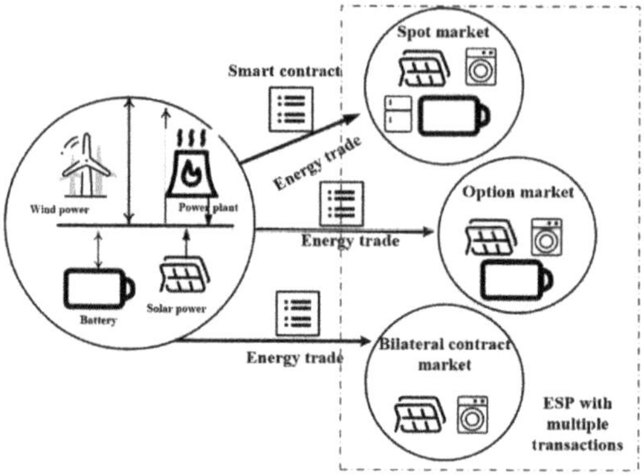

Figure 1. ESP with multiple transaction structure.

2.1. Objective Function

The objective of the model is to minimize the operating cost of an ESP, taking into consideration various cost components. These components include the cost of purchasing electricity from the power market, the operating cost of GTs, and the operating cost of ESUs. The expression for the operating cost function is as follows:

$$C_{\mathrm{ESP}} = C_{\mathrm{m}} + C_{\mathrm{GT}} + C_{\mathrm{ESU}} \tag{1}$$

where C_{ESP} denotes the system operation cost of ESP; C_{m} represents the cost of electricity purchase in the electricity market; C_{GT} is the output cost of GT; and C_{ESU} is the operating cost of ESUs. The explicit mathematical expressions representing each of these costs are provided as follows:

(1) Transaction cost of ESP under multiple markets

In this paper, T periods are considered. T_1 and T_2 are the sets of off-peak and peak hours of power demand, respectively, satisfying $T = T_1 + T_2$. Market transaction costs are the aggregate of transactions at the spot market, option contracts and bilateral contracts. While it is true that buyers of call option contracts may face a disadvantage if the prices fall instead of rise, resulting in the actual price being significantly below the fixed price previously contracted, it is crucial to note that the implementation of these options is typically a strategic decision aimed at mitigating the potential price risk in situations where prices are anticipated to rise. The use of call options involves a trade-off between the potential costs and benefits, and it is essential for market participants to make informed decisions based on their risk tolerance and market expectations. The rationale behind applying the call option exclusively during the peak hours, despite it representing a constant and fixed trading volume, lies in its optimal ability to mitigate price risk within the spot market during periods of heightened electricity consumption [20,21]. Therefore, it is assumed that call option contracts are utilized exclusively during the peak period of electricity consumption, and the trading volume does not change over time. The specific representation of the cost of each transaction mode is given by the following formula:

$$\begin{cases} C_{\mathrm{m}} = C_{\mathrm{s}} + C_{k} + C_{\mathrm{oc}} \\ C_{\mathrm{s}} = \sum\limits_{t \in T} [\lambda_t^{\mathrm{s}} p_t^{\mathrm{s}}] \\ C_{k} = \sum\limits_{t \in T} \sum\limits_{k \in K} \left[\lambda_k^{\mathrm{b}} p_{t,k}^{\mathrm{b}}\right] \\ C_{\mathrm{oc}} = \sum\limits_{t \in T_2} [\min\{\lambda_{\mathrm{ck}}, \lambda_t^{\mathrm{s}}\} p_t^{\mathrm{oc}} + \lambda_0 p_t^{\mathrm{oc}}] \end{cases} \tag{2}$$

where $t \in T$ denotes a collection of periods; $k \in K$ designates a set of optional bilateral contracts; C_S represents the spot market cost of electricity; C_K indicates the transaction cost of a bilateral contract; C_{OC} corresponds to the option transaction cost; λ_t^s characterizes the spot market price at time t; p_t^s denotes the spot market trading volume at time t; λ_k^b corresponds to the electricity price attributed to the kth bilateral contract; $p_{t,k}^b$ represents the trading volume of the kth bilateral contract at time t; and λ_{ck} and λ_0 are the exercise price and option price of the call option, respectively. If $\lambda_{ck} > \lambda_t^s$, ESP executes the option contract to purchase electricity at a fixed price; if $\lambda_{ck} < \lambda_t^s$, ESP does not execute option contracts, and the acquisition of electricity is undertaken through transactions at the spot market price.

(2) Gas turbine output cost

$$C_{GT} = \sum_{t \in T} \sum_{n \in N} \left[a_{GT} (P_{t,n}^{GT})^2 + b_{GT} (P_{t,n}^{GT}) + c_{GT} \right] \quad (3)$$

where $n \in N$ denotes the node set of the power network; $P_{t,n}^{GT}$ represents the GT output of node n at time t; and a_{GT}, b_{GT}, and c_{GT} are the cost coefficients of the GT output.

(3) Operating cost of ESUs

$$C_{ESU} = \sum_{t \in T} [d_{ESU} (P_t^{ei} + P_t^{eo})] \quad (4)$$

where P_t^{ei}, P_t^{eo} are, respectively, the charging and discharging power of the ESU at time t, and d_{ESU} denotes the operating cost coefficient of the ESU.

2.2. Constraint Condition

The parameters governing the secure operation of the ESP encompass an amalgamation of constraints emanating from multiple facets, including the electricity market trading, power flow dynamics within the electrical network, and the inherent limitations of the energy storage system. The formulations detailing these constraints are articulated as follows.

(1) Electricity market transaction constraints

$$\begin{cases} p_{t,\min}^s \le p_t^s \le p_{t,\max}^s \\ p_{t,\min}^{oc} \le p_t^{oc} \le p_{t,\max}^{oc} \\ p_k^{b,\min} s_k \le p_{t,k}^b \le p_k^{b,\max} s_k \\ p_t^{sm} + \sum\limits_{k \in K} p_{t,k}^{bc} = p_t^{grid}, \forall t \in T_1 \\ p_t^s + \sum\limits_{k \in K} p_{t,k}^b + p_t^{oc} = p_t^{grid}, \forall t \in T_2 \end{cases} \quad (5)$$

where s_k is the 0–1 variable, which is used to determine whether to choose k contract at time t; $p_{t,\min}^s$, $p_{t,\min}^{oc}$, $p_k^{b,\min}$ denote the minimum trading volume of the spot market, the option contract, and the kth bilateral contract in the tth period, respectively; $p_{t,\max}^s$, $p_{t,\max}^{oc}$, $p_k^{b,\max}$ represent the maximum trading volume of the spot market, the option contract, and the kth bilateral contract in the t period, respectively; and p_t^{grid} is the total amount of electricity purchased from the electricity market in period t.

(2) Power flow constraints of the electrical network

Preserving the operational integrity and stability of the power network constitutes an imperative objective. To this end, the scheduling of power market transactions necessitates strict adherence to the power flow security constraints inherent within the distribution network. In practical terms, this mandates a comprehensive consideration and fulfillment of the power flow limitations characteristic of the distribution network during the course of power trading and scheduling endeavors. The fundamental rationale behind this approach is to ensure that the transmission of electrical energy within the distribution network remains well within the capacity limits of the network's transmission lines.

By accommodating these constraints, the aim is to avert the occurrence of any detrimental power flow phenomena, such as overloading, voltage deviations, or any other adverse ramifications stemming from power flow imbalances. Such meticulous adherence contributes decisively to upholding the operational safety and stability of the overarching power system.

$$\begin{cases} P_{t,n}^{\text{grid}} + P_{t,n}^{\text{GT}} + f_{t,i(l)}^{\text{p}} = \sum\limits_{l \in L} f_{t,o(l)}^{\text{p}} + D_{t,n}^{\text{p}} \\ Q_{t,n}^{\text{grid}} + Q_{t,n}^{\text{GT}} + f_{t,i(l)}^{\text{q}} = \sum\limits_{l \in L} f_{t,o(l)}^{\text{q}} + D_{t,n}^{\text{q}} \\ \left(R_l f_{t,l}^{\text{p}} + X_l f_{t,l}^{\text{q}} \right) / U_0 = U_{t,i(l)} - U_{t,o(l)} \end{cases} \tag{6}$$

where $l \in L$ denotes the set of power lines; I denotes injection, and o represents release; $P_{t,n}^{\text{grid}}$ and $Q_{t,n}^{\text{grid}}$ are, respectively, the active and reactive power of n nodes injected into the power grid at time t; $Q_{t,n}^{\text{GT}}$ represents the reactive power output of GT to n nodes at time t; $f_{t,i(l)}^{\text{p}}$ and $f_{t,i(l)}^{\text{q}}$ are the active and reactive power of the node n injected by the power line l at time t, respectively; $f_{t,r(l)}^{p}$ and $f_{t,r(l)}^{q}$ are the active and reactive power flowing from node n to line l at time t, respectively; the active and reactive loads of node n at time t are designated as $D_{t,b}^{p}$ and $D_{t,b}^{q}$, respectively; R_l and X_l are the resistance and reactance of power line l, respectively. The reference voltage of the slack bus is signified by U_0. Additionally, $U_{t,i(l)}$ and $U_{t,o(l)}$ correspond to the initial and terminal voltages associated with power line l, respectively.

(3) Energy storage constraints

$$\begin{cases} E_{t+1} = E_t + p_t^{ei} \eta_{+}^{\text{ESU}} - p_t^{eo} / \eta_{-}^{\text{ESU}} \\ 0 \le E_t \le E_t^{\max} \\ S_i(t) + S_o(t) \le 1 \end{cases} \tag{7}$$

where E_t denotes the electricity stored in ESUs at time t; $\eta_{+}^{\text{ESU}}, \eta_{-}^{\text{ESU}}$ are, respectively, the charging and discharging efficiency of the ESU; $E_t^{\max}$ represents the maximum capacity of the ESU; and $S_0(t), S_1(t)$ correspond to the working state of the energy storage device.

In summation, when both spot price and load demand are predicted values, the ESP trading model under the deterministic environment is

$$\begin{cases} \min C_{\text{ESP}} \\ \text{s.t. } (5)\text{–}(7) \end{cases} \tag{8}$$

In this deterministic model, the objective function is to minimize the total cost C_{ESP} of ESP, and Equations (5)–(7) represent power market transaction constraints, power network flow constraints, and energy storage constraints, respectively.

3. CTbIGDT Model for ESP under Uncertain Environment

Subjected to the various influences of a myriad of factors, the operational landscape of power market transactions is inherently rife with a plethora of uncertainties. A prominent illustration of this uncertainty stems from the involvement of intermittent renewable energy sources with variable predictability, culminating in pronounced undulations in electricity prices within the market. Furthermore, the unpredictability inherent in user-side load variations exerts a substantial sway over the deliberations guiding the ESP's engagement in electricity market transactions [22].

It is noteworthy that the ESP trading model, as expounded in the preceding section, regrettably overlooks the risks engendered by this intricate nexus of uncertainties [23]. Consequently, a pivotal imperative arises to construct an ESP transactional decision model within the purview of this section, one that adeptly accommodates the ramifications of electricity market transactional risk and the intricate fabric of multiple uncertainties. This endeavor is underpinned by the theoretical framework of credibility theory and IGDT, ushering forth a comprehensive and robust model calibrated to navigate the intricate terrains of uncertainty-laden contexts.

3.1. Credibility Theory

The theory of credibility measures based on measure theory address a limitation in possibility measures by introducing self-duality [24]. This advancement furnishes a novel tool for investigating fuzzy decision problems, circumventing the potential pitfalls in decision-making arising from conventional membership degree calculations.

Credibility measures can be represented using the least upper bound of variables in a fuzzy event set. For any given set $A \in \Re$, the credibility measure of a fuzzy variable is defined as follows [25]:

$$\mathrm{Cr}\{\xi \in A\} = \frac{1}{2}\left(\sup_{x \in A}\mu(x) + 1 - \sup_{x \in A^c}\mu(x)\right) \tag{9}$$

In this context, $\sup f(x)$ and $1 - \sup f(x)$ respectively denote the possibility measure and necessity measure of set A. The complement of set A is represented by A^c and represents the membership function of the fuzzy variable A, and sup denotes the upper bound. Notably, the coefficient $1/2$ in the equation ensures the establishment of self-duality. Furthermore, for any non-empty set Θ, Equation (9) follows four axioms. The four axioms and a brief introduction are given by Appendix A Axiom A1. The detailed proof process can be found in Reference [25].

3.2. Credibility Theory-Based Information Gap Decision Theory

In light of the intricate web of uncertainties encompassing electricity prices and user loads, the ESP can adeptly embrace a risk-averse strategy in orchestrating its electricity market trading blueprint. In this paper, we establish credibility theory as a foundational risk assessment methodology for uncertain price and load, incorporating both necessity and randomness measures. Subsequently, we advance the concept by developing the CTbIGDT optimization model, grounded in the consideration of expected costs, with the primary aim of fortifying the robustness of electricity trading practices. Through the manipulation of diverse anticipated cost targets, CTbIGDT emerges as a potent instrument, capably quantifying distinct risk inclinations. This methodology effectively engenders power market trading strategies that are uniquely aligned with the diverse risk preferences exhibited by ESPs.

Distinguished from alternative methods of uncertainty modeling, CTbIGDT demonstrates a distinctive attribute: it circumvents the need for an extensive corpus of data to enact uncertainty modeling. Rather than demanding an exorbitant volume of data, CTbIGDT harnesses existing information on uncertainty parameters to optimize within the permissible bounds of uncertainty, all while adhering to predefined objectives. By doing so, CTbIGDT furnishes operators with insights into both potential favorable and adverse outcomes that may transpire due to uncertainty. These insights enable operators to make judicious and reasoned choices, grounded in safety or calculated risk. Unlike stochastic programming and chance-constrained methodologies, CTbIGDT-driven approaches sidestep the necessity of comprehending the probability distribution function of uncertain parameters. Notably, CTbIGDT distinguishes itself from the realm of fuzzy uncertainty modeling by obviating the requirement to ascertain or assign membership functions to uncertain parameters.

At the crux of CTbIGDT lies the exploration of plausible ramifications attributable to uncertain parameters while upholding preset objectives. In this context, "information" pertains to the gamut of uncertain factors that impact the objective function, while the term "gap" serves to demarcate the differential expanse between what is known and what remains undisclosed. Notably, this approach eschews the need for an exhaustive comprehension of intricate probability distributions tied to uncertain parameters, operating rather within the confines of upper and lower bounds. In essence, the application of CTbIGDT furnishes ESPs with a robust and adaptable methodology that navigates the intricate landscape of uncertainty, underpinning the decision-making process and facilitating optimal outcomes while respecting predetermined parameters.

CTbIGDT stands as a judicious decision-making approach tailored to address intricate uncertainties. Within this framework, two discernible strategies come to the fore: the Risk Averse Strategy (RAS) and the Risk Seeker Strategy (RSS). RAS embarks on the construction of a robust function by accentuating the amplification of uncertainty's influence on the resultant solutions. On the contrary, RSS endeavors to extract the utmost potential gains from uncertain risks, thereby constructing an opportunity function of maximal returns. Considering the focal objective of this paper—to mitigate the sway of uncertainty emanating from electricity prices and load demand on ESP operational outcomes—RAS is judiciously adopted to architect a CTbIGDT-based robust optimization model [26]. The overarching design of the CTbIGDT robust model assumes the form as elucidated below:

$$\left\{\begin{array}{l} \max \mathrm{Cr}(w) \\ \text{s.t.} \left\{\begin{array}{l} \max F(s,w) \leq F_C \\ F_C = (1+\beta)F_0 \\ G(s,w) = 0 \\ K(s,w) \leq 0 \\ U(\mathrm{Cr}(w), w_{\mathrm{p}}) = \left\{\mathrm{Cr}(w) : \left|\frac{w-w_{\mathrm{p}}}{w_{\mathrm{p}}}\right| \leq \mathrm{Cr}(w)\right\} \end{array}\right. \end{array}\right. \tag{10}$$

where $\mathrm{Cr}(w)$ symbolizes the parameter of uncertainty, referred to as the "uncertainty radius." This parameter serves as an indicator of the oscillation scope inherent inthe uncertain variable under consideration. F denotes the objective function, while s represents the decision variable. The variable w embodies the actual value attributed to the uncertain parameter, whereas w_{p} signifies the anticipated value of said uncertain parameter. The term encapsulates the optimal solution yielded by the deterministic model, designated as F_0. Furthermore, F_C assumes the role of the maximum anticipated value deemed acceptable by the decision-making entity. β is the deviation factor, indicating the degree of deviation between the expected robust optimization target value and the deterministic model. $G(s,w) = 0$ indicates equality constraints, including power balance constraints shown by (5), energy storage constraints at adjacent moments given by (6), and energy balance constraints in market transactions illustrated by (7). Conversely, $K(s,w) \leq 0$ represents inequality constraints, consisting of the upper and lower limits of energy transaction, energy storage charging and discharging, node voltage, and GT outputs. Importantly, $U(\mathrm{Cr}(w), \widetilde{w})$ represents the amplitude of fluctuations attributed to the uncertain variables, contributing a pivotal layer to the formulation's nuanced framework.

3.3. CTbIGDT Model for ESP under Price and Load Uncertainty

In order to assess the likelihood of uncertain events, Zadeh introduced the concept of possibility measures in [27] and necessity measures in [28]. While both measures have been demonstrated to satisfy the properties of normality, non-negativity, and monotonicity, they are not dual, meaning that the sum of possibilities or necessities for an uncertain event and its complement does not equal one, as discussed in [29]. To address this limitation, the authors in [29] introduced the use of credibility measures to capture the uncertainty arising from the linguistic vagueness of possible values associated with uncertain variables. To this end, we extend this theory by developing a credibility theory-based IGDT framework to enhance the robustness of electricity trading under conditions of uncertainty, particularly with respect to uncertain spot prices and load demands. Drawing upon the preceding discourse, this study designates the electricity price and load demand within the electricity market as pivotal uncertain variables. The scope of oscillation inherent to these variables can be succinctly articulated as follows:

$$U(\mathrm{Cr}(\lambda_t^{\mathrm{s}}), \lambda_t^{\mathrm{s,p}}) = \left\{ \mathrm{Cr}(\lambda_t^{\mathrm{s}}) : \left| \frac{\lambda_t^{\mathrm{s}} - \lambda_t^{\mathrm{s,p}}}{\lambda_t^{\mathrm{s,p}}} \right| \leq \mathrm{Cr}(\lambda_t^{\mathrm{s}}) \right\} \tag{11}$$

$$U(\mathrm{Cr}(D_t), D_t^{\mathrm{p}}) = \left\{ \mathrm{Cr}(D_t) : \left| \frac{D_t - D_t^{\mathrm{p}}}{D_t^{\mathrm{p}}} \right| \leq \mathrm{Cr}(D_t) \right\} \tag{12}$$

where $\lambda_t^{\mathrm{s,p}}$ and D_t^{p} are, respectively, the predicted values of spot market electricity price and load. $\mathrm{Cr}(\lambda_t^{\mathrm{s}})$ and $\mathrm{Cr}(D_t)$ denote the credibility measures associated with electricity price and load calculated by Equation (9) to study the risk brought by uncertain price and load, respectively.

As a result, the CTbIGDT model for optimizing ESP decisions within an uncertain context is derived as presented below:

$$\begin{cases} \max[\mathrm{Cr}(\lambda_t^{\mathrm{s}}), \mathrm{Cr}(D_t)] \\ \mathrm{s.t.} \begin{cases} \max C_{\mathrm{ESP}} \leq (1+\beta)C_0 \\ \lambda_t^{\mathrm{s}} \in U\left(\mathrm{Cr}(\lambda_t^{\mathrm{s}}), \lambda_t^{\mathrm{s,p}}\right) \\ D_t \in U\left(\mathrm{Cr}(D_t), D_t^{\mathrm{p}}\right) \\ \mathrm{s.t.}(5) \sim (7) \end{cases} \end{cases} \tag{13}$$

where C_0 denotes the objective function value derived from Equation (8) within a deterministic environment and symbolizes the factor of robust deviation.

Considering the intricate nature of the model described above, which operates within the framework of a bi-level multi-objective optimization, solving it directly presents a formidable challenge. To address this complexity, we initiate the optimization process by transforming the objective function into $\max\min([\mathrm{Cr}(\lambda_t^{\mathrm{s}}), \mathrm{Cr}(D_t)])$ using robust optimization principles. Additionally, it is important to highlight that in typical scenarios, the simultaneous presence of peak spot prices and increased electrical load demand in the electricity market results in the highest aggregate operating costs for ESP. Therefore, the term $\max C_{\mathrm{ESP}} \leq (1+\beta)C_0$ within the robust model can be equivalently expressed as $C_{\mathrm{ESP}} \leq (1+\beta)C_0$.

In summation, the robust optimization framework governing ESP trading within an uncertain context can be succinctly expressed as follows:

$$\begin{cases} \max\min([\mathrm{Cr}(\lambda_t^{\mathrm{s}}), \mathrm{Cr}(D_t)]) \\ \mathrm{s.t.} \begin{cases} C_{\mathrm{ESP}} \leq (1+\beta)C_0 \\ \lambda_t^{\mathrm{s}} = (1+\mathrm{Cr}(\lambda_t^{\mathrm{s}}))\lambda_t^{\mathrm{s,p}} \\ D_t = (1+\mathrm{Cr}(D_t))D_t^{\mathrm{p}} \\ \mathrm{s.t.}(5) \sim (7) \end{cases} \end{cases} \tag{14}$$

4. Case Study

This paper undertakes a comprehensive numerical investigation within the context of a modified IEEE 33-node distribution network system. The distinct topology of the network is visually depicted in Figure 2. Notably, two GTs are interconnected with nodes 5 and 10, respectively, serving as sources of power for end users. The ESU is integrated into the system at node 2. This paper assumes the presence of ten bilateral contracts available for selection. Contracts 1–5 are applicable during off-peak periods, while contracts 6–10 are suitable for selection during peak periods. The strike price and option price of call options are 52.1$/MWh and 3.2$/MWh, respectively. Moreover, for a comprehensive view of the detailed data, please refer to the specifics outlined in Reference [30]. Comprehensive information regarding bilateral contracts and specifications for GT units can be found in Appendix A Tables A1 and A2, respectively. Furthermore, Figure 3 delineates the anticipated spot price trends and electric load projections. In this paper, a mathematical model is formulated using the MATLAB platform, and we utilize the SCIP solver within the YALMIP toolbox to solve the optimization problem.

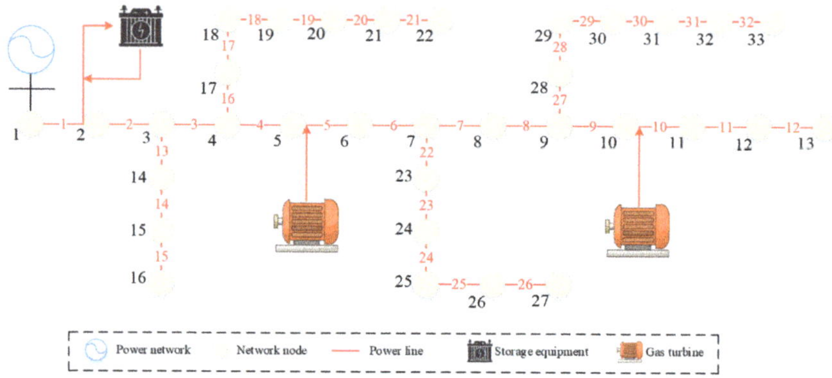

Figure 2. Topology of modified IEEE 33-node distribution network.

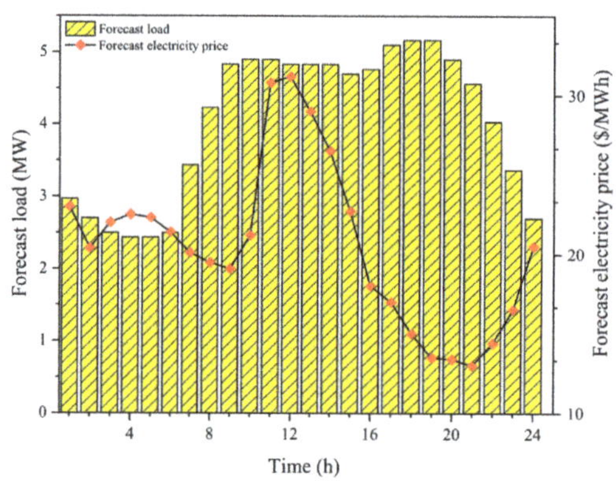

Figure 3. Forecast spot electricity price and electricity load.

4.1. Analysis of ESP Electriticy Trading Strategy under Different Markets

Initially, we assume that the deterministic environment mirrors the actual electricity price, and the predicted load serves as the baseline. The uncertain environment is represented when the robust deviation factor β is 0.1. The costs under both the deterministic and uncertain environments have been organized systematically in Table 1 for reference within the narrative. From the table, we can see that compared with the deterministic environment cost, the main increasing cost in an uncertain environment happens in electricity trading, which increases by 10%. Under the proposed optimization framework, ESP enhances the robustness of system operations by accepting certain costs considering the uncertain electricity price.

Table 1. Cost comparison under different environments.

	Deterministic Environment	Uncertainty Environment
C_{ESP}	$1804.3	$1984.8.
C_m	$1273.5	$1429.6
C_{GT}	$530.40	$554.80
C_{ESU}	$0.4636	$0.4053

In the deterministic environment, the corresponding trading strategies for electricity purchases in different electricity markets are intuitively represented in Figure 4. As gleaned from Figure 4, a discernible pattern emerges wherein the electricity market activities of ESP display distinct preferences across different time periods. During off-peak intervals, the prevailing modus operandi of ESP's electricity market interactions predominantly centers on spot market involvement coupled with bilateral contract engagements. In contrast, as peak hours ensue, the primary focus shifts to a blend of spot market participation and option contract negotiations. Notably, during specific time slots, namely at 11:00 and 12:00, the conspicuous escalation in spot prices instigates a strategic response from ESP. Driven by the objective of curbing electricity procurement expenses, ESP adeptly adjusts its transactional approach by tempering the volume of spot market transactions while concurrently augmenting the volume of transactions executed through the more economically advantageous bilateral contracts.

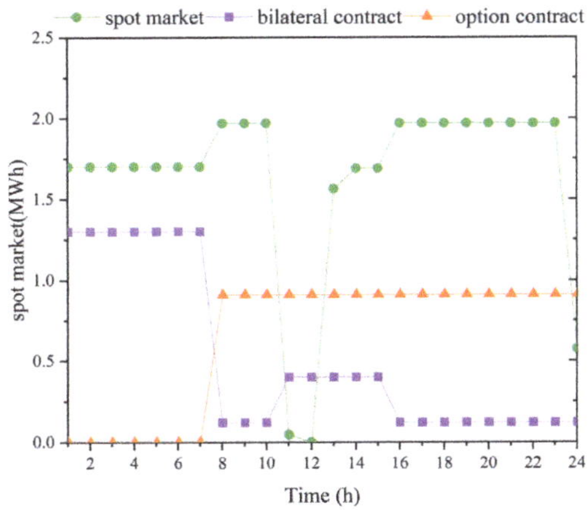

Figure 4. Trading strategy of ESP with different markets in deterministic environment.

In the broader spectrum, the comprehensive breakdown of transactions reveals that electricity procurement from the spot market totals 39.13 MWh. Simultaneously, the volume of bilateral contract transactions amounts to 12.54 MWh, whereas option contracts contribute substantially, amounting to 15.47 MWh. This strategic utilization profile, discernible from the insights gleaned from Figure 4, underscores the dynamic and astute decision-making employed by ESP in navigating the intricate landscape of electricity market dynamics. The nuanced blend of spot market, bilateral contracts, and option contracts attests to ESP's strategic acumen in optimizing cost efficiencies aligned with real-time market fluctuations.

The strategic blueprint governing ESP's electricity market transactions within an uncertain milieu is illustrated in Figure 5. For further insight into the specifics of distinct bilateral contract procurements within both certain and uncertain environments, readers are directed to Appendix A Table A3.

According to Figure 5, distinct patterns emerge in ESP's electricity market transactions across varying time periods. During off-peak intervals, the predominant mode of transaction centers around both the spot market and the bilateral contract arena. In contrast, as peak hours ensue, a shift occurs wherein electricity market dealings pivot towards a combination of option contracts and bilateral contracts. A conspicuous observation emerges when comparing ESP's electricity market purchasing strategy in a certain environment with that within an uncertain environment. Evidently, ESP displays an inclination to elevate the trading volume of option contracts and bilateral contracts in the face of uncertainty. This strategic shift can be attributed to ESP's deliberate efforts in mitigating the ramifications of multiple uncertainties. By opting for comparatively more stable bilateral contracts and option contracts, ESP effectively mitigates exposure to uncertain spot market dynamics, thus leading to a reduction in transaction volume within this realm.

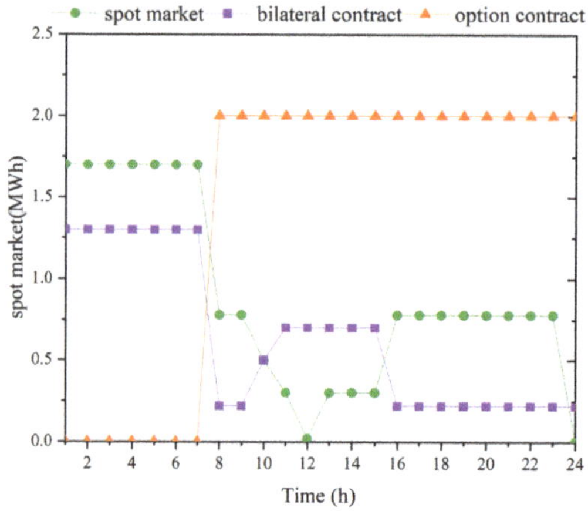

Figure 5. Trading strategy of ESP with different markets under uncertainty environment.

In broader terms, the delineation of transactions reveals a procurement of 21.42 MWh from the spot market, while the bilateral contract transactions contribute a volume of 15.52 MWh. Significantly, the option contract transactions emerge as a prominent component, amounting to 34 MWh. This strategic recalibration of transactional volumes, evident from Figure 5, underscores ESP's proactive approach in confronting uncertainty, strategically leveraging bilateral contracts and option contracts to curtail exposure to volatile spot market conditions.

4.2. ESP Trading Strategy Analysis under Uncertainties

The trading strategies adopted by ESP within deterministic and uncertain environments are visually depicted in Figure 6.

The insights gleaned from Figure 6a reveal ESP's strategic maneuvers within the off-peak period, whereby a strategic decision is made to accumulate surplus electricity within the ESU. A distinct contrast emerges during peak hours, particularly within the 11:00–12:00 timeframe, where the pronounced escalation in spot prices compels ESP to significantly curtail its spot market electricity procurement. Driven by the imperative of cost minimization, ESP strategically opts to truncate electricity purchases from the market while concurrently amplifying GT output and tapping into the stored energy reserves within the ESU, especially during off-peak intervals.

Turning attention to Figure 6b, a conspicuous departure is discernible in comparison to Figure 6a. Here, ESP's modus operandi maintains a more pronounced emphasis on electricity market transactions during peak periods. When contextualized with Figure 5, this strategic shift can be attributed to ESP's heightened engagement in bilateral contracts and option contracts. The rationale underpinning this shift is grounded in the fixed pricing nature of bilateral contracts and option contracts, serving as a strategic hedge against the volatility inherent in spot prices. In navigating the uncertainty-laden environment, ESP adeptly leverages these contracts to secure load demand while meticulously managing operational risks arising from multifaceted uncertainties. This nuanced adaptation, elucidated through Figure 6a,b, underscores ESP's dynamic responsiveness to the prevailing market dynamics and its strategic acumen in mitigating risks within an intricate and uncertain energy landscape.

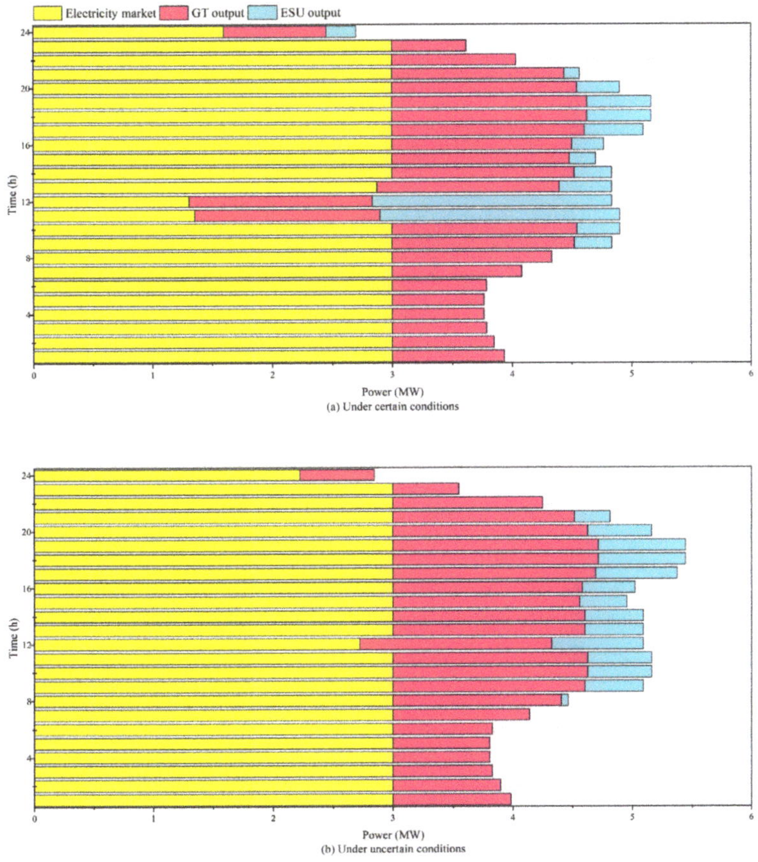

Figure 6. ESP trading strategies under deterministic and uncertain environments.

4.3. The Influence of Robust Deviation Factor and Uncertainty on ESP Trading Strategy

Under different robust deviation factors, the operating cost of ESP and the corresponding change of uncertainty radius are shown in Figure 7.

Figure 7 unveils a discernible trend wherein the escalation of the deviation factor corresponds to an augmentation in the uncertainty radius, consequently driving an increase in ESP's operational cost. This shift can be attributed to decision makers harboring a pessimistic outlook, apprehending the adverse repercussions of uncertainty in spot electricity prices and load dynamics on cost optimization. Remarkably, the application of CTbIGDT ensures that the system's scheduling cost remains below the anticipated value even in the face of fluctuating uncertainties. This, in turn, guarantees the secure and stable operation of the system. Hence, in real-world operation, ESP must adeptly strike a balance between safety and economy, devising a power market transaction scheduling strategy tailored to the prevailing circumstances to optimally fulfill operational requisites.

To validate the efficacy of the ESP optimization model grounded in IGDT in bolstering the robustness of ESP's power market transactions, this study embarks upon the following scenarios. Scenario 1: Only addressing the uncertainty of spot price. Scenario 2: Only considering the uncertainty of electrical load. Scenario 3: Simultaneously accounting for the uncertainty of spot electricity price and electrical load.

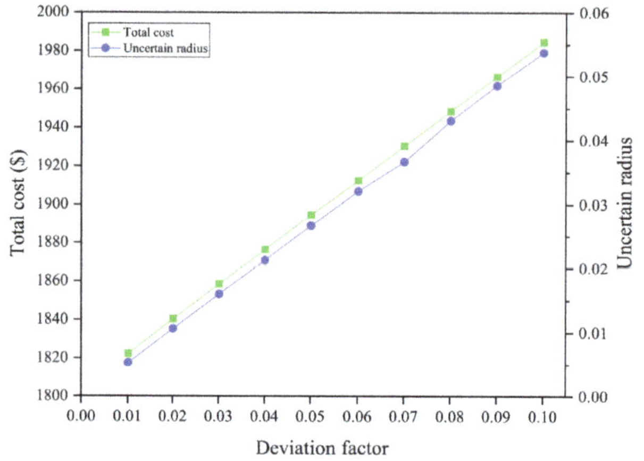

Figure 7. The operating cost and uncertainty radius of ESP under different robust deviation factors.

Across these diverse scenarios, Table 2 meticulously delineates the altering cost and uncertainty radius for distinct facets of ESP, contingent on the robust deviation factor. The strategic scheduling strategies for scenarios 1 and 2, where the robust deviation factor is set at 0.1, are visually portrayed in Figure 8a,b, respectively. The scheduling strategy of Scenario 3 is portrayed in Figure 6b above, and to avoid repetition, further analysis pertaining to this scenario has been omitted.

Table 2. The influence of robust deviation factor on cost in different scenarios.

	Scenario 1			Scenario 2			Scenario 3		
	$\beta = 0.01$	$\beta = 0.05$	$\beta = 0.1$	$\beta = 0.01$	$\beta = 0.05$	$\beta = 0.1$	$\beta = 0.01$	$\beta = 0.05$	$\beta = 0.1$
C_{ESP}	$1822.4	$1894.6	$1984.8	$1822.4	$1894.6	$1984.8	$1822.4	$1894.6	$1984.8
C_m	$1277.8	$1364.9	$1454.4	$1287.2	$1348.2	$1392.1	$1278.5	$1345.6	$1429.6
C_{GT}	$544.08	$529.33	$530.13	$534.74	$545398	$592.29	$543.39	$548.50	$554.80
C_{ESU}	$0.5014	$0.4607	$0.2685	$0.4560	$0.4180	$0.4019	$0.4853	$0.4425	$0.4053
$Cr(w)$	0.0193	0.1048	0.2329	0.0074	0.0.364	0.0586	0.0053	0.0267	0.0538

A comprehensive analysis of Table 1 underscores a recurring pattern: within the same scenario, an escalation in the robust deviation factor invariably precipitates an expansion of the uncertainty radius. This compelling observation serves as a testament to the capacity of CTbIGDT to not only mitigate risks but also enhance its resistance to uncertainties. In essence, this amplifies the system's robustness, exemplifying the potency of CTbIGDT in bolstering resilience. Turning attention to Figure 8, a nuanced panorama comes into focus, revealing distinct deviation radius magnitudes engendered by diverse uncertain factors under identical deviation factors. Additionally, these variances in uncertain factors concurrently yield disparate power trading strategies.

Scenario 1 exclusively contemplates the uncertainty associated with electricity prices. Under this circumstance, ESP strategically opts to pare down electricity market transactions, thereby curtailing costs prior to satisfying load demand. In contrast, Scenario 2 solely addresses the uncertainty linked to load demand. Here, ESP opts to amplify electricity market transactions, ensuring alignment with load demand imperatives. Furthermore, during off-peak intervals, ESP judiciously stockpiles energy within the ESU to act as a safeguard against the potential impact of uncertain load demand during peak periods. In summation, ESP must meticulously harmonize its risk aversion proclivity with the enhancement of transactional robustness, all while steadfastly upholding economic considerations. This orchestration ensures a judicious equilibrium, custom-tailored to ESP's unique operational context.

In a context where generation availability is uncertain and price fluctuations can be volatile, it is essential to emphasize that adopting a conservative approach does not always guarantee lower costs, as is often the case in financial trading. Our optimization suggestion extends beyond the realms of academic exercise. It is specifically designed to navigate the intricate landscape of uncertainty, delivering tangible benefits not only to the system's stability but also to the economic considerations of market participants. By embracing a balanced approach that carefully considers risk and return, our methodology serves as a practical tool to assist both system operators and market participants in making well-informed decisions that can yield both cost-efficiency and system reliability.

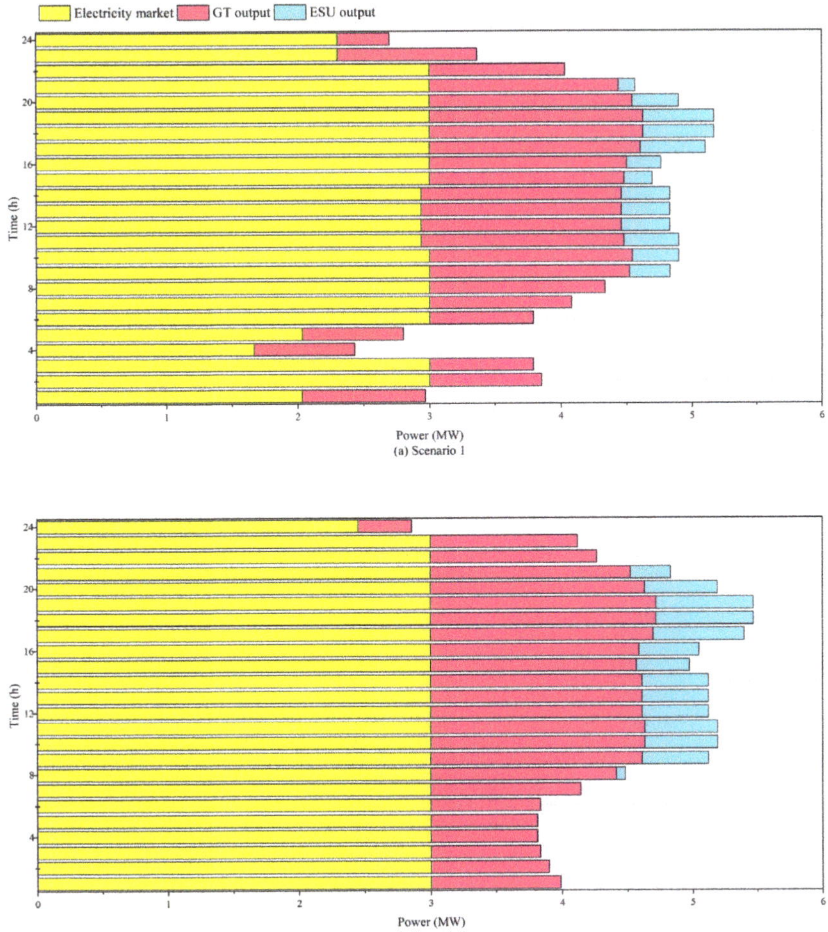

Figure 8. Comparison of electricity transactions of ESP under Scenarios 1 and 2.

5. Conclusions

Aligned with the multifaceted uncertainties arising from spot electricity prices and user load demand, which pose significant challenges to power service providers during operational activities in power trading, this study introduces an ESP power trading optimization framework rooted in CTbIGDT. Through empirical illustration, the findings accentuate a discernible trend: a rise in the robust deviation factor corresponds to an elevation in the anticipated cost, paralleled by an expanded uncertainty radius. Consequently, with a heightened focus on mitigating the negative impact of multiple uncertain risks, ESP enhances the robustness of system operations by accepting certain costs. Notably, when confronted with these uncertain risks, ESP opts to amplify the trading volume associated with bilateral contracts and option contracts within its market trading strategy while concurrently reducing involvement in the spot market. This strategic adjustment effectively enhances the resilience of system optimization operations.

The ESP optimization model, forged on the foundations of CTbIGDT, scrupulously embraces the risk inclination of decision makers. This comprehensive approach culminates in the derivation of transactional and systemic scheduling strategies adeptly calibrated to attain predefined cost targets. This pivotal step invariably enhances ESP's robustness within the realm of power trading. By amalgamating these nuanced elements, the model effectively equips power service providers with innovative tools and conceptual paradigms, thereby furnishing valuable insights to guide risk-informed decision-making within the intricate domain of power trading participation. This pioneering approach provides a new approach and pragmatic solution for power service providers facing complex risk-related dilemmas, reducing the high cost risks that uncertainty factors may bring while navigating the multi-faceted environment of power transactions.

Author Contributions: X.Z.: investigation, writing—original draft. P.W.: writing—original draft, supervision. Q.L.: writing—original draft, conceptualization. Y.L.: writing—original draft, validation. Z.L.: writing—original draft, investigation; L.F.: writing—original draft, investigation, supervision. J.C.: writing—original draft, validation. All authors have read and agreed to the published version of the manuscript.

Funding: This work was supported by a grant from the Science and Technology Program of State Grid Shandong Electric Power Company (520625230011).

Informed Consent Statement: Not applicable.

Data Availability Statement: Images used can be shared by contacting the corresponding author.

Conflicts of Interest: The authors declare that they have no known competing financial interest or personal relationships that could have appeared to influence the work reported in this paper.

Appendix A

Axiom A1:

$$\begin{aligned}&\text{Axioms 1}: \ \mathrm{Cr}\{\Theta\} = 1\\&\text{Axioms 2}: \mathrm{Cr}\{A_1\} \leqslant \mathrm{Cr}\{A_2\}, A_1 \subseteq A_2 \subseteq \Theta\\&\text{Axioms 3}: \mathrm{Cr}\{A\} + \mathrm{Cr}\{A^c\} = 1, A \subseteq \Theta\\&\text{Axioms 4}: \mathrm{Cr}\{\cup_i A_i\} = \sup_i \mathrm{Cr}\{A_i\}, \sup_i \mathrm{Cr}\{A_i\} < 0.5\end{aligned}$$

Axiom 1 says that the trustworthiness measure of the entire nonempty set Θ is equal to 1; Axiom 2 states that if $A_1 \subseteq A_2 \subseteq \Theta$, then their trustworthiness measure satisfies $\mathrm{Cr}\{A_1\} \leqslant \mathrm{Cr}\{A_2\}$. Axiom 3 indicates that the trustworthiness measure of set A and the trustworthiness measure of its complement A^c add to 1. Axiom 4 says that if $\sup_i \mathrm{Cr}\{A_i\} < 0.5$, then $\mathrm{Cr}\{\cup_i A_i\} = \sup_i \mathrm{Cr}\{A_i\}$ is satisfied.

Table A1. GT unit parameters.

GT	$P_{t,n}^{\mathrm{GT}}$ (MW)	$Q_{t,n}^{\mathrm{GT}}$ (MVar)	a_{GT} (\$/MW^2)	b_{GT} (\$/MW)	c_{GT} (\$)
GT1	[0, 1.5]	[0, 0.5]	0.12	20.0	0
GT2	[0, 2.0]	[0, 1.0]	0.09	15.0	0

Table A2. Bilateral contract parameters.

Contract Number	Time (h)	Minimum Value (MW)	Maximum Value (MW)	Contract Price ($/MWh)
1	peak absence	0.06	0.15	17
2	peak absence	0.08	0.2	16.5
3	peak absence	0.1	0.25	15
4	peak absence	0.1	0.3	14
5	peak absence	0.12	0.4	13
6	peak period	0.06	0.15	25
7	peak period	0.08	0.2	24.5
8	peak period	0.1	0.25	23.5
9	peak period	0.1	0.3	23
10	peak period	0.12	0.4	22

Table A3. Comparison of bilateral contract transaction volume in different environments.

Contract Number	1	2	3	4	5	6	7	8	9	10
Deterministic environment	0	0	1.15 MW	1.50 MW	2.80 MW	0	0	0	0	4.56 MW
Uncertain environment	0.78MW	1.04 MW	1.75 MW	2.10 MW	2.80 MW	0	0	0	0	5.96 MW

References

1. Wang, L.; Li, K. Research on renewable energy consumption and emission reduction in power market based on bi-level decision making in China. *Energy* **2022**, *260*, 125119. [CrossRef]
2. Li, G.; Li, G.; Zhou, M. Market transaction model design applicable for both plan and market environment of China's renewable energy. *Front. Energy Res.* **2022**, *10*, 862653. [CrossRef]
3. Sreenivasulu, G.; Sahoo, N.C.; Balakrishna, P. A coordinated stochastic dispatch model for hybrid energy markets with renewable energy uncertainties using moth flame optimization. *Energy Syst.* **2022**, *12667*, 1–41. [CrossRef]
4. Goodarzi, S.; Perera, H.N.; Bunn, D. The impact of renewable energy forecast errors on imbalance volumes and electricity spot prices. *Energy Policy* **2019**, *134*, 110827. [CrossRef]
5. Rai, A.; Nunn, O. On the impact of increasing penetration of variable renewables on electricity spot price extremes in Australia. *Econ. Anal. Policy* **2020**, *67*, 67–86. [CrossRef]
6. Yang, H.; Yang, S.; Xu, Y.; Cao, E.; Lai, M.; Dong, Z. Electric vehicle route optimization considering time-of-use electricity price by learnable partheno-genetic algorithm. *IEEE Trans. Smart Grid* **2015**, *6*, 657–666. [CrossRef]
7. Zhang, Z.; Wang, X.; Li, H.; Peng, D.; Ren, B. Ability model of the influence of demand response on electricity demand. *Proc. CSU-EPSA* **2017**, *29*, 115–121.
8. Yan, G.J.; Chen, J.J.; Liu, J.Y.; Chen, W.; Xu, B. Random clustering and dynamic recognition-based operation strategy for energy storage system in industrial park. *J. Energy Storage* **2023**, *73*, 109192. [CrossRef]
9. Feng, L.; Zhang, I.; Li, G.; Zhang, B. Cost reduction of a hybrid energy storage system considering correlation between wind and PV power. *Prot. Control Mod. Power Syst.* **2016**, *1*, 11. [CrossRef]
10. Furkan, A.; Mohammand, S.A.; Mohammand, A. Developments in xEVs charging infastructure and energy management system for smart microgrids including xEVs. *Sustain. Cities Soc.* **2017**, *35*, 552–564.
11. Majidi, M.; Mohammadi-Ivatloo, B.; Soroudi, A. Application of information gap decision theory in practical energy problems: A comprehensive review. *Appl. Energy* **2019**, *249*, 157–165. [CrossRef]
12. Wang, X.; Gong, Y. Air conditioner fast dispatching model based on load aggregator and direct load control. *J. Eng.* **2017**, *13*, 2535–2538. [CrossRef]
13. Nojavan, S.; Nourollahi, R.; Pashaei-Didani, H.; Zare, K. Uncertainty-based electricity procurement by retailer using robust optimization approach in the presence of demand response exchange. *Int. J. Electr. Power Energy Syst.* **2019**, *105*, 237–248. [CrossRef]
14. Yang, S.; Tan, Z.; Tan, Z.; Liu, Z.; Lin, H.; Ju, L.; Zhou, F.; Li, J. A multi-objective stochastic optimization model for electricity retailers with energy storage system considering uncertainty and demand response. *J. Clean Prod.* **2020**, *277*, 124017. [CrossRef]
15. Jiang, X.; Li, Y.; Yu, S.; Lin, Z.; Ji, C.; Zhang, Z. Power purchasing optimization of electricity retailers considering load uncertainties based on information gap decision theory. *Energy Rep.* **2022**, *8*, 693–700. [CrossRef]
16. Sun, B.; Wu, X.; Xie, J.; Sun, X. Information gap decision theory-based electricity purchasing optimization strategy for load aggregator considering demand response. *Energy Sci. Eng.* **2021**, *9*, 200–210. [CrossRef]
17. Najafi-Ghalelou, A.; Nojavan, S.; Zare, K. Robust thermal and electrical management of smart home using information gap decision theory. *Appl. Therm. Eng.* **2018**, *132*, 221–232. [CrossRef]

18. Moradi-Dalvand, M.; Mohammadi-Ivatloo, B.; Amjady, N.; Zareipour, H.; Mazhab-Jafari, A. Self-scheduling of a wind producer based on Information Gap Decision Theory. *Energy* **2015**, *81*, 588–600. [CrossRef]
19. Mohammadi-Ivatloo, B.; Zareipour, H.; Amjady, N.; Ehsan, M. Application of information-gap decision theory to risk-constrained self-scheduling of GenCos. *IEEE Trans. Power Syst.* **2012**, *28*, 1093–1102. [CrossRef]
20. Deng, S.J.; Oren, S.S. Electricity derivatives and risk management. *Energy* **2006**, *31*, 940–953. [CrossRef]
21. Azevedo, F.; Vale, Z.A.; de Moura Oliveira, P.B. A decision-support system based on particle swarm optimization for multiperiod hedging in electricity markets. *IEEE Trans. Power Syst.* **2007**, *22*, 995–1003. [CrossRef]
22. Liu, J.; Chen, J.; Yan, G.; Chen, W.; Xu, B. Clustering and dynamic recognition based auto-reservoir neural network: A wait-and-see approach for short-term park power load forecasting. *iScience* **2023**, *26*, 107456. [CrossRef] [PubMed]
23. Zheng, J.H.; Guo, J.C.; Li, Z.; Wu, Q.H.; Zhou, X.X. Optimal design for a multi-level energy exploitation unit based on hydrogen storage combining methane reactor and carbon capture, utilization and storage. *J. Energy Storage* **2023**, *62*, 106929. [CrossRef]
24. Shao, L.P.; Chen, J.J.; Pan, L.W.; Yang, Z.J. A Credibility Theory-Based Robust Optimization Model to Hedge Price Uncertainty of DSO with Multiple Transactions. *Mathematics* **2022**, *10*, 4420. [CrossRef]
25. Liu, B. A survey of credibility theory. *Fuzzy Optim. Decis. Mak.* **2006**, *5*, 387–408. [CrossRef]
26. Peng, C.; Chen, L.; Zhang, J. Multi-objective optimal allocation of energy storage in distribution network based on classified probability chance constraint information gap decision theory. *Proc. CSEE* **2020**, *40*, 2809–2819.
27. Zadeh, L.A. Fuzzy sets as a basis for a theory of possibility. *Fuzzy Sets Syst.* **1978**, *1*, 3–28. [CrossRef]
28. Zhou, J.; Li, X.; Pedrycz, W. Mean-semi-entropy models of fuzzy portfolio selection. *IEEE Trans. Fuzzy Syst.* **2016**, *24*, 1627–1636. [CrossRef]
29. Huang, X. Mean-entropy models for fuzzy portfolio selection. *IEEE Trans. Fuzzy Syst.* **2008**, *16*, 1096–1101. [CrossRef]
30. Li, R.; Wei, W.; Mei, S.; Hu, Q.; Wu, Q. Participation of an energy hub in electricity and heat distribution markets: An MPEC approach. *IEEE Trans. Smart Grid* **2018**, *10*, 3641–3653. [CrossRef]

energies

MDPI

Article

A Voltage-Aware P2P Power Trading System Aimed at Eliminating Unfairness Due to the Interconnection Location

Satoshi Takayama * and Atsushi Ishigame

Department of Electrical and Electronic Systems Engineering, Osaka Metropolitan University, Sakai City 599-8531, Japan; ishigame@omu.ac.jp
* Correspondence: takayama@omu.ac.jp

Abstract: P2P power trading is necessary for efficiently using consumer electricity not subject to FIT. However, the execution rules for P2P power trading do not include restrictions on voltage, and there is a trade-off between the activation of the P2P power trading market through the mass introduction of PV and the optimization of the voltage of the power distribution system. In addition, there is a tendency for output curtailment to be biased toward consumers connected to the end of the grid. Since consumers cannot choose the interconnection location, there are concerns about unfairness. In this study, we investigate a new P2P model that includes voltage constraints for the execution rules of P2P power trading to avoid voltage deviation while ensuring benefits and fairness for the participants. In the proposed model, to increase the incentive to participate in the P2P power trading market, we consider compensating consumers who receive output curtailment signals due to voltage constraints. In addition, the profit is secured by differentiating the compensation cost unit price depending on the contract's availability. A case study was conducted on this model using the IEEE 33 bus system. The results show that the proposed model is superior.

Keywords: P2P power trading; distribution network; voltage regulation; photovoltaic generation

1. Introduction

In August 2009, the Japanese government introduced a "new purchase program for photovoltaic power generation" to transition to a low-carbon society. The "new purchase program for photovoltaic power generation" is a system that obliges general electric utilities to purchase surplus electricity from qualifying photovoltaic power generation facilities at a fixed price for a certain period. It is called the surplus power purchase program. The purchase price per kWh for the first year was 48 yen for residential use and 24 yen for non-residential use, about twice the voluntary purchase price of electric utilities. In August 2011, the "Act on Special Measures Concerning the Procurement of Renewable Electricity by Electric Utilities" was implemented on a trial basis. In July 2012, a feed-in tariff system was introduced, which obliges general electric utilities to purchase surplus electricity from renewable energy generation facilities that meet the requirements at a fixed price for a certain period [1]. The power supply subject to the feed-in tariff program was the power supply subject to the surplus power purchase program.

On the other hand, residential solar power generation with an output of less than 10 kW, for which the ten-year purchase period, gradually expires in November 2019. The cumulative number of residential photovoltaic power generation systems the purchase period of which ended by 2021 is said to total 1 million, and the cumulative amount of power generated is said to total 3.96 million kW [2]. Households whose purchase period has expired must have the option of increasing consumption or selling excess electricity at a lower price. However, due to the uneven output of solar power, the price of electricity sold is below 10 JPY/kWh for many businesses, which means the price will drop significantly. As a new trading venue for such surplus power, P2P power trading, which utilizes blockchain to conduct power transactions, is being considered [3–6]. P2P is an abbreviation for Peer to Peer, meaning transactions between equals, and the P2P power trading market is a market where consumers who sell surplus power (hereinafter referred to as "prosumers") and consumers who purchase power (hereinafter referred to as "consumers") exchange power. In P2P power trading, a blockchain (BC)-based power trading system has been proposed and tested [7]. In these studies, we evaluate automatic bidding, matching, profit in P2P power trading use, and intelligent contract processing capability in power trading built on BC. Because the unit price for purchasing PV power at the end of the purchase period has become very low, the prosumers will benefit from participating if they can sell power at a slightly higher unit price in P2P power trading. Similarly, if consumers can purchase electricity at a unit price lower than the retail electricity rate, they can reduce their electricity bills. In other words, P2P power trading has potential benefits for participants.

On the other hand, with the expansion of PV installation, the surplus power generated by PV exceeds the electricity demand during the hours when PV generation is high. As a result, the voltage rises that cause the end-of-consumer voltage to deviate from the specified range have become apparent. For this reason, installing power conditioners for PV systems (PV-PCS) with the output curtailment method is mandatory. However, in the conventional output curtailment method using a PV-PCS [8], the increase in the own-end voltage is suppressed by suppressing the active power output, resulting in inequity in the surplus power sold depending on the installation location in the power distribution system. Therefore, P2P power trading is recommended for systems that tend to be low-voltage with large interconnected loads [9].

However, for grids that tend to be high-voltage with a low interconnected load or when considering the future increase in the surplus power due to the spread of PV, the problem of voltage rise may occur. To solve the voltage rise problem, P2P power trading that takes the grid voltage into account has been studied [10–13]. These papers solve the over-voltage problem in addition to increasing profits. However, since these papers estimate the effects of voltage and power flow on a per-matching, unfairness occurs depending on the interconnection location of the distribution system, such that trading fails when voltage problems arise. In addition, these P2P rules lost trading opportunities because the tradable power is equalized and limited in advance to deal with grid voltage problems. Furthermore, there was a significant profit difference depending on the interconnection location within the distribution network.

In this study, we propose a P2P power trading system that considers the unfairness caused by voltage constraints. Inequity here refers to the fact that the amount of PV output curtailment varies depending on the location of the grid connection because the output curtailment of PV is required by considering voltage constraints. We contribute to developing P2P power trading systems by constructing a P2P power trading system that eliminates this unfairness and increases the benefits to the participants. Specifically, we propose a P2P power market system that introduces financial compensation for PV output curtailment, which is necessary to manage grid voltage to consider the unfairness of interconnection locations due to voltage constraints. In this system, we introduce a contract price as a variable cost to compensate for the source of financial compensation for PV output curtailment. The proposed method sets different unit prices of output curtailment for the availability of P2P transactions by prosumers. Then, the amount of PV output curtailment for each consumer is determined to minimize the total compensation cost for output curtailment. The amount of output curtailment is selected to reduce the total compensation within the voltage constraints.

The remainder of this paper is organized as follows. Section 2 describes the P2P power trading model proposed in this paper. Section 3 describes the kernel density estimation methodology to create PV output scenarios. In Section 4, we simulate using the IEEE 33 bus system and evaluate the effectiveness of the proposed P2P model. Finally, in Section 5, a brief conclusion of this paper is given.

2. Proposed P2P Power Trading System

2.1. Outline

A schematic diagram of the proposed P2P electricity trading system in this study is shown in Figure 1. This paper has four players in this market system: consumers, prosumers, system operators, and retail electricity providers. The consumers are the residences that consume electricity and do not have PV systems. They procure electricity from the P2P electricity market. If they cannot procure electricity from the P2P electricity market, they procure electricity from retailers.

Prosumers supply their demand with PV power and earn profits by bidding their surplus power to the P2P power market. If they cannot sell their surplus power in the P2P power market, they can sell it to the retailers at the graduated FIT unit price.

The system operator needs to manage the grid voltage and can request the prosumers to curtail the PV surplus output if the grid voltage is expected to deviate from the regulated range. This study supposes that the system operator introduces proposed P2P power trading. In other words, if the introduction of P2P power trading rules eliminates voltage deviations, there will be no need to install new equipment and a cost benefit for the system operator.

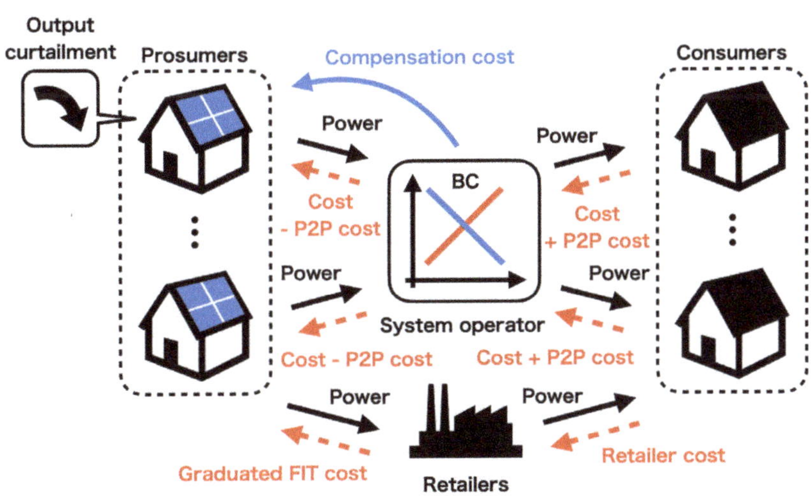

Figure 1. Schematic diagram of the proposed P2P electricity trading system in this study.

On the other hand, the voltage distribution in the distribution system during times of heavy reverse power flow tends to increase at the end of the distribution system. Suppose the grid operator requests consumers to curtail their output without specific measures. In that case, consumers connected to the end of the distribution system receive many curtailment signals, leading to a cost disadvantage for the consumers connected to the end. Therefore, this paper proposes that the grid operator pays a compensation fee to consumers who receive output curtailment signals for output curtailment. This compensation fee is calculated by multiplying the indicated amount of output curtailment by the unit price. Consumers who have established P2P power trading contracts receive a compensation fee equal to the clearing price if targeted for output curtailment, which means that the grid operator fully compensates the prosumer for the cost of output curtailment. Conversely, prosumers who fail to make a contract receive compensation lower than the FIT unit price. Initially, if P2P power trading is not implemented, there is no financial compensation for output curtailment. However, if they participate in P2P power trading, they will be financially compensated by the grid operator even if they are subject to output curtailment. This will encourage further prosumer participation, thus stimulating P2P power trading. However, if the P2P market compensates more than the FIT unit price, the advantage of selling electricity in the P2P market will be lost. So, the unit compensation price for output curtailment will be 8 JPY/kWh or less in subsequent studies.

The daily flow of a P2P power market is shown in Figure 2. First, the system operator opens the P2P electricity market. Then, prosumers and consumers bid on the P2P market. The system operator makes a pending contract based on their bidding. The contracting scheme is described in detail in Section 2.2, which refers to the reference in [14]. The system operator then checks the voltage profile based on the calculation result using the pending contract result and the system information. When the voltage at each node is controlled within the regulated range, the system operator decides to execute the contract with the pending contract. When there is a voltage deviation, the system operator determines the prosumer's output curtailment and updates the contract result. The method of determining the output curtailment is described in Section 2.3. Finally, the system operator sends the updated contract result to the participant.

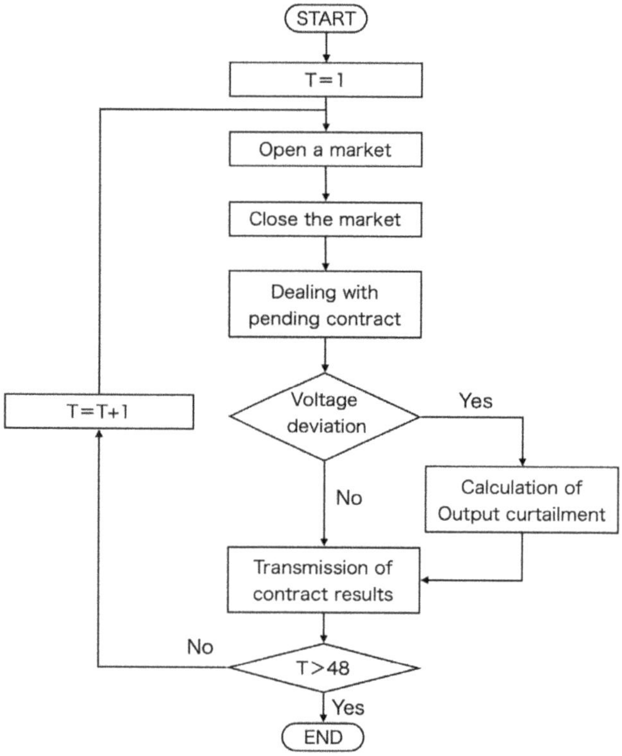

Figure 2. Daily flow of proposed P2P power market.

2.2. Contract Scheme

The flowchart of the matching process applied in this paper is shown in Figure 3, where p represents the bid price, e represents the bid quantity, i represents the consumer number, j represents the prosumer number, k represents the total number of consumers at time t, and n represents the total number of prosumers at time t. At the beginning of the auction period, market participants submit their trade prices and energy quantities to the market. The system operator sorts the prosumers' and consumers' bidding information based on the price to create supply and demand curves. Then, the market clearing price between the matched buy and sell prices is calculated using the matching method shown in Figure 3. The matching method used in this paper is a double auction. The system operator determines the clearing price by half the price from the consumer with the highest bid and the prosumer with the lowest bid. When there is a difference in the contracted power between the consumer with the highest bid and the prosumer with the lowest bid, the system operator matches the remaining bidding power with the next participant. The transaction volume equals the minimum volume between the matched orders. Finally, at the end of the auction period, orders that do not match the energy residuals are balanced by the retail electric utility at the ToU and FIT prices. Note that the pricing strategies of all the market participants will be bound between the FiT and the ToU to ensure economic benefits.

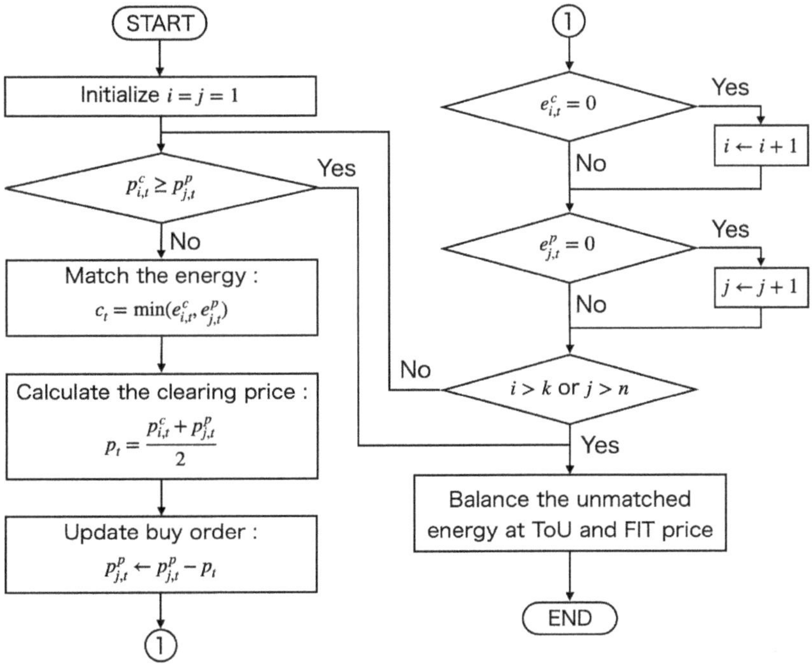

Figure 3. Flowchart of the matching process.

2.3. Calculation of Output Curtailment

As mentioned above, the system operator pays a compensation fee for the output curtailment to the consumer who receives the output curtailment signal. However, the unit price of the compensation fee varies depending on the contract's execution. This paper formulates the problem of determining the amount of PV output curtailment as an optimization problem. In the proposed P2P power market transaction, the system operator compensates the prosumers who suppress their output to maintain the system voltage. By reducing the total amount of compensation, the participation fee and transaction fee for P2P power market transactions paid by prosumers and consumers, which are the source of funds, are reduced, and the system operator reduces the compensation risk, creating a win–win relationship for all players. Therefore, this paper sets the objective function of the optimization problem to minimize the financial compensation to the prosumers who have implemented output curtailment to maintain the system voltage. The optimization formulation is shown below.

$$\min \sum_{i=1}^{n} P_i^{cur} \cdot C_t^{P2P} + \sum_{j=1}^{m} P_j^{cur} \cdot C^{com} \tag{1}$$

$$P_i - P_i^{cur} = V_i \sum_{k=1}^{n+m} V_k \{G_{ik} \cos(\theta_i - \theta_k) + B_{ik} \sin(\theta_i - \theta_k)\} \tag{2}$$

$$P_j - P_j^{cur} = V_j \sum_{k=1}^{n+m} V_k\{G_{jk}\cos(\theta_j - \theta_k) + B_{jk}\sin(\theta_j - \theta_k)\} \quad (3)$$

$$Q_i = V_i \sum_{k=1}^{n+m} V_k\{G_{ik}\sin(\theta_i - \theta_k) - B_{ik}\cos(\theta_i - \theta_k)\} \quad (4)$$

$$Q_j = V_j \sum_{k=1}^{n+m} V_k\{G_{jk}\sin(\theta_j - \theta_k) - B_{jk}\cos(\theta_j - \theta_k)\} \quad (5)$$

$$V^{\min} \leq V_k \leq V^{\max} \quad (k = 1 \ldots n+m) \quad (6)$$

$$-\pi \leq \theta_k \leq \pi \quad (k = 1 \ldots n+m) \quad (7)$$

$$0 \leq P_k^{cur} \leq P_k \quad (k = 1 \ldots n+m) \quad (8)$$

where i denotes the number of prosumers contracted, j denotes the number of prosumers not contracted, P^{cur} denotes the amount of output curtailment, C^{P2P} denotes the clearing price for the P2P power transaction, C^{com} denotes the compensation fee for the PV output curtailment, P denotes the amount of active power bid, Q denotes the reactive power, V denotes the node voltage, θ denotes the phase angle of the node, and P^{sur} denotes the surplus power. G and B denote the conductance and susceptance components between nodes, and $V^{\min}$ and $V^{\max}$ denote the minimum and maximum voltage in the specified range of the node voltage, respectively.

Equation (1) shows the objective function formulated to minimize the sum of compensation charges for consumers who have contracted in P2P power trading and those who have not. This paper considers that the prosumers who could contract in P2P power trading set the contracted price to the compensation unit price to prevent opportunity losses due to output curtailment, and the compensation unit price for prosumers who could not contract is considered a parameter. Equations (2)–(5) show the tidal equations for active and reactive power, respectively, and by adding a term for PV output curtailment P^{cur} to Equations (2) and (3), the voltage upper and lower limit constraints are satisfied by substituting a value for P^{cur} when the voltage upper and lower limit constraints shown in Equation (6) are not satisfied. Equation (7) shows the phase angle constraints for each node, and Equation (8) shows the upper bound constraints on the amount of output curtailment so that no more curtailment than surplus power is allowed.

2.4. P2P Cost Determination Method

When the voltage conditions are considered in determining the contract quantity and price for P2P power trading, the frequency of output curtailment varies depending on the interconnection location, causing inequity among consumers. To reduce this inequity, we implement monetary compensation for output curtailment by prosumers. The system operator needs to compensate prosumers financially for the opportunity loss to curtail electricity sales due to output curtailment. However, since consumers can receive supply from retail electricity suppliers even if the transaction cannot be executed, this study supposes that the system operator does not provide monetary compensation to consumers. In addition, for the system operator to provide monetary compensation, it must have financial resources. We consider that the system operator collects P2P costs from P2P market participants and uses them as the resource; the P2P costs consist of participation fees from the market participants, shown in Equation (9), and the transaction fees (meaning consignment fees) for contracts, shown in Equation (10).

$$F_b = \sum_{k=1}^{n+m} C^{pf} \tag{9}$$

$$F_{con} = \sum_{k=1}^{n+m} \omega \cdot C^{P2P} \cdot (P_K - P_k^{cur}) \tag{10}$$

where F_b denotes the total participation fee from all the participants, which is determined based on the profit that the system operator wishes to secure in this study; the participation fee is considered necessary for the operation of the P2P power trading system. We considered it to correspond to the profit of the system operator in this study. However, it is necessary to determine the participation fee while considering the profit of the participants in P2P power trading because if the participants in P2P power trading are disadvantaged, it will lead to the decline of the market; C^{pf} denotes the participation fee per consumer. In this study, the system operator settles and collects a uniform amount of C^{pf} from the participants based on their profit performance so that the profit of the least profitable participant equals the profit of the system operator. Note that the participation fee is collected once a month, like a basic fee. F^{con} is the total compensation for PV output curtailment; ω is the percentage of the transaction fee; and C^{P2P} is the clearing price of a contract. Since $P_k - P_k^{cur}$ indicates the final contracted quantity, we set the market participant to pay the system operator ω% of the traded amount in the P2P power transaction. The transaction fee is necessary for grid maintenance and is considered equivalent to the consignment fee in the current system. In Japan, the demand side bears 100% of the transmission charges. However, to expand the introduction of renewable energy in the future, a study is underway to make cost-sharing more equitable by requiring power generators and grid users to bear a portion of the charges along with consumers. In this study, the generation side is the prosumer, and the retail side is the consumer, referring to the previous study. Hence, the prosumers are supposed to bear 10% of the transaction fee and the consumers 90%. This study reduces the burden of grid usage fees for prosumers to increase PV installation and stimulate P2P power trading.

3. Generation of Scenario of PV Output

This paper evaluates the proposed method's usefulness using PV data over a long period and in various weather conditions. It also creates multiple PV scenarios based on actual data. In particular, we applied kernel density estimation to the probability distribution of the PV outputs from measurement data. This chapter aims to estimate the probability density of the PV output. Figure 4 shows the normalized statistical results of the PV output. As shown in Figure 4, two peaks are caused by sunny and rainy days. Using the parametric model for the probability density estimation of the PV output is difficult. The authors employed kernel density estimation, a nonparametric method. In particular, reference [15] shows that the Epanechnikov kernel can be used as the kernel function to transition the probability density with reasonable accuracy, and this paper uses the Epanechnikov kernel to estimate the PV output distribution in this study.

When the data $X_1, X_2 \cdots$, and X_n are observed according to an unknown distribution q, the kernel density estimator for that distribution is defined as

$$\hat{f}_x(x,h) = \frac{1}{nh}\sum_{i=1}^{n} K\left(\frac{x - X_i}{h}\right) \tag{11}$$

where K is the kernel function, and h is the bandwidth. The kernel function has the following properties.

$$\int K(u)\mathrm{d}u = 1 \tag{12}$$

$$\int uK(u)\mathrm{d}u = 0 \tag{13}$$

$$\int u^2K(u)\mathrm{d}u = \sigma_K^2 < \infty \tag{14}$$

The Gaussian and the Epanechnikov kernel, expressed in equation (15), have been used as kernel functions in various studies. In this study, the Epanechnikov kernel, which provides a probability density function with high accuracy, was used [16].

$$K_{EPA}(u) = \begin{cases} \frac{3}{4}(1-u^2) & (|u| \leq 1) \\ 0 & (otherwise) \end{cases} \tag{15}$$

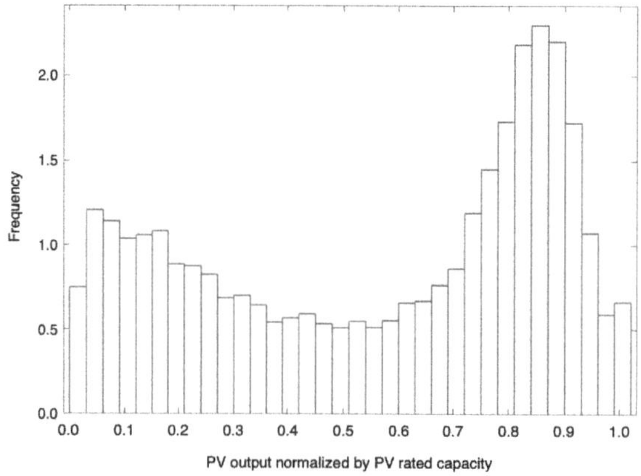

Figure 4. Frequency distribution of PV output.

Then, we consider the integration of the mean square error of the density function for the bandwidth, and by transforming and approximating, we obtain Equation (16).

$$ISME\left(\hat{f}_x(x,h)\right) \approx \frac{1}{nh}\int K(t)^2\mathrm{d}t + \frac{1}{4}h^4k_2^2\int f''(x)^2\mathrm{d}x \tag{16}$$

where

$$k_2 = \int t^2K(t)\mathrm{d}t \tag{17}$$

Let $\hat{h}$ be the h for which Equation (16) is minimal, and we can derive Equation (18).

$$\hat{h} = k_2^{-2/5}\left(\int K(t)^2\mathrm{d}t\right)^{1/5}\left(\int f''(x)^2\mathrm{d}x\right)^{-1/5}n^{-1/5} \tag{18}$$

Then, we created PV scenarios using the probability density function obtained from the kernel density estimation, using PV data measured at the Osaka Metropolitan University in April and May 2011. Due to the limited number of data, we normalized the PV-rated capacity at each period during the two months to 1.0 p.u. We created one probability distribution for each period when the PV generated power. The probability density distribution for the period between 11:30 and 12:00, obtained by the kernel density estimation, is shown in Figure 5. The vertical axis shows the probability density, and the horizontal axis shows the normalized PV output. As in Figure 5, the probability density distribution has two peaks representing sunny and cloudy weather; we can model the PV output appropriately.

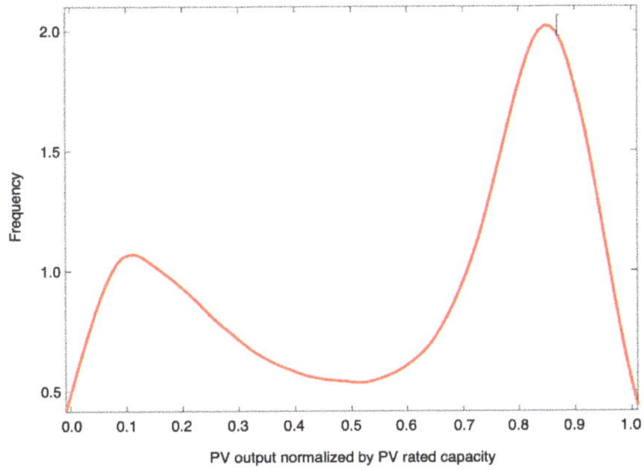

Figure 5. The probability density function of PV output obtained from the kernel density estimation.

Finally, Figure 6 shows an example of the PV output data, which was converted from p.u. to kW using the values used for normalization at each period, with upper and lower limit constraints placed on the PV output variation to prevent abrupt PV output fluctuations from occurring. Note that different colors indicate different scenarios.

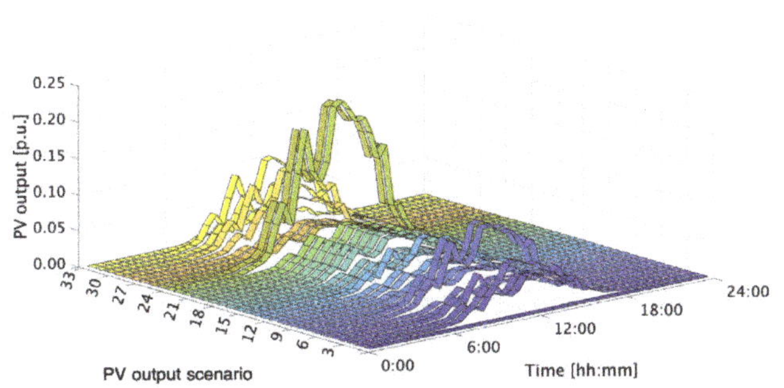

Figure 6. The PV output scenario from the kernel density estimation.

4. Numerical Verification

In this chapter, we use the PV scenario to verify the P2P electricity market transactions numerically. We describe the simulation's content and assumptions, followed by its results and discussion. We used "fmincon" in MATLAB2021a's Optimization toolbox, precisely the sequential quadratic programming method within "fmincon".

4.1. Simulation Conditions for the Distribution Network

Figure 7 shows the middle-voltage grid model of the IEEE 33 bus system [17] used in this study. There are 30 consumers installed at each node of the simulation model, and an aggregator manages each node. The aggregator determines and notifies the next day's bid volume and price on the previous day. The specific bid amounts and bid prices are described below in Section 4.2. In addition, we consider a pattern for the PV installation: to install the PV at even-numbered nodes to alternate the PVs' locations.

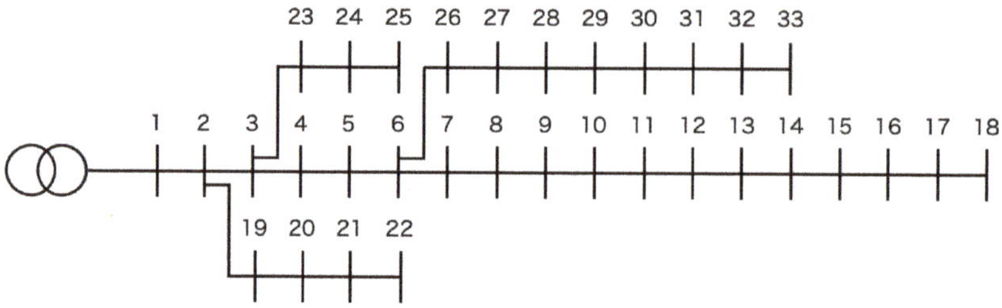

Figure 7. The IEEE 33-bus system model.

We used the power demand data of consumers in Ota City, Gunma Prefecture, Japan, measured within the NEDO demonstration project. We modified the number of consumers so that their MW values matched those of the demand data listed in the IEEE 33-bus system (Figure 8). Note that different colors indicate different nodes of the IEEE 33-bus system.

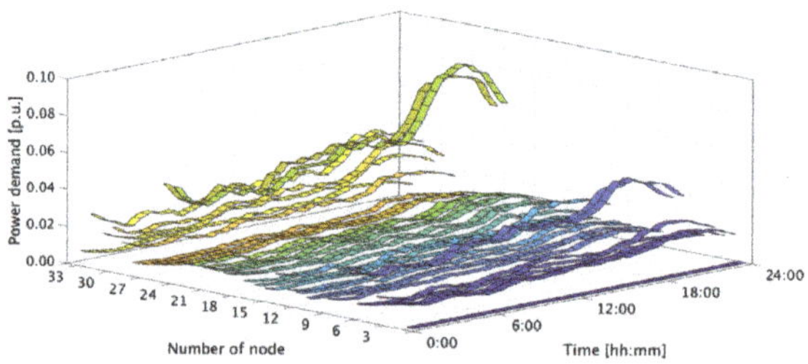

Figure 8. Demand load data in a certain day.

4.2. Simulation Conditions for the P2P Power Trading

This section describes a model of the bidding behavior of market participants. The prosumers bid the amount of surplus PV output, and the consumers bid the required amount of demand for P2P electricity trading. The market participants also set their bid prices between 8 and 20 yen using uniform random numbers. We set these conditions concerning the electricity prices and the purchase price of the PV surplus power of the Kansai Electric Power Co., Inc. Multiple simulations under these conditions simulate various bidding behaviors and evaluate the effectiveness of the proposed model based on statistical indices.

As mentioned above, C_{P2P} sets the unit price equal to the clearing price to compensate for the profit to the prosumer who could execute the transaction. Conversely, the compensation unit price C_{com} for prosumers who could not perform the transaction is between 1 and 8 yen, referring to the unit price of the surplus PV output purchase. Note that we consider the compensation unit price by varying it as a parameter.

4.3. Simulation Results

Figure 9 shows the voltage transition for each method when setting either output curtailment minimization or compensation cost minimization as the objective function. In addition, we show the voltage transition when all the reverse power is supplied to the grid for comparison.

As shown in Figure 9, when the PV output is not curtailed, the voltage exceeds the upper bound of 1.05 p.u. In contrast, the proposed optimization method can keep the voltage within the regulation range by curtailing the PV output. Note that there is almost no difference in the amount of output curtailment due to the difference in the objective function, and the voltage transition is nearly the same.

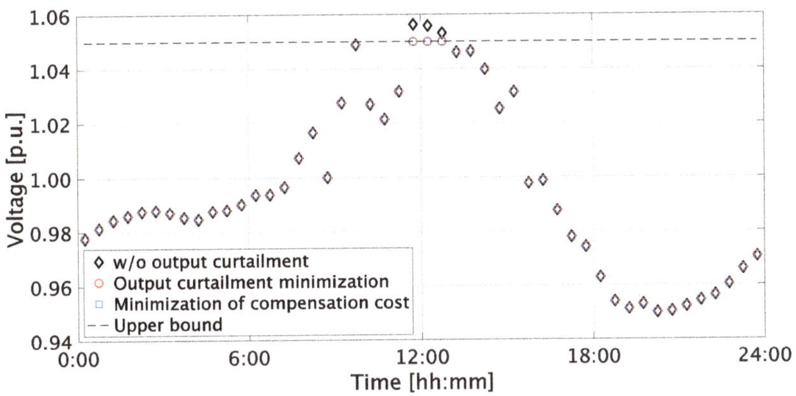

Figure 9. Voltage transition in each method.

Then, we evaluate the PV output curtailment for different objective functions. Figure 10 shows each objective function's PV output curtailment results. Figure 10a shows the result of minimizing the PV output curtailment, and Figure 10b shows the result of reducing the compensation cost.

As shown in Figure 10a, the PV output of node 18, the end node, is curtailed because each node's output curtailment power is determined to minimize the PV output curtailment. On the other hand, from Figure 10b, since node 18 was able to contract, the compensation cost is higher, and as a result, node 18 avoided the PV output curtailment. Therefore, the profit of node 18 from the P2P power transaction is not impaired. Conversely, the PV output of nodes 14 and 16, which could not contract, is curtailed. Thus, both objective functions can keep the voltage within the specified range, as shown in Figure 9. However, the PV output curtailment of each node varies depending on the difference in the objective function. Therefore, the results suggest that the proposed P2P model can achieve voltage management while eliminating the unfairness of output curtailment.

Figure 11 shows the improvement in electricity prices with and without P2P electricity trading. As shown in Figure 11, the electricity price of node 18, which is the prosumer at the end of the grid, is more significant than that of node 2, which is the prosumer at the beginning of the grid. We considered that these are losses of the end prosumers' electricity sales associations due to the impact of the output curtailment. However, after the introduction of P2P power trading, the difference in electricity prices becomes smaller. In other words, a P2P power trading system that introduces financial compensation for PV output curtailment to maintain voltage can contribute to eliminating the unfairness caused by grid interconnection locations. Financial compensation for prosumers at the end of the grid, where output curtailment is relatively common, will help to eliminate inequity due to grid interconnection location. Compared to the situation before the introduction of the P2P power trading system, the electricity prices of P2P power trading participants have decreased regardless of the unit price of compensation. Therefore, the prosumers at the beginning of the grid with less output curtailment also benefit from the system. Moreover, the study sets the same compensation cost unit price at a discrete value for all consumers, which results in the terminal consumers benefiting unfairly. Thus, this problem can be solved by formulating and solving the setting of compensation unit costs as an optimization problem.

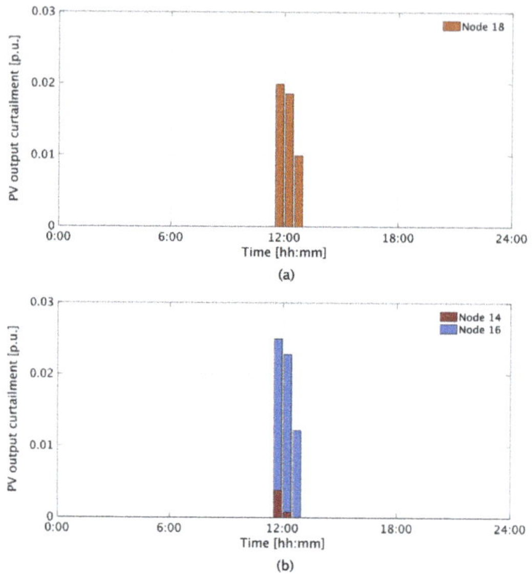

Figure 10. PV output curtailment for each objective function; (**a**) PV output curtailment minimization; (**b**) compensation cost minimization.

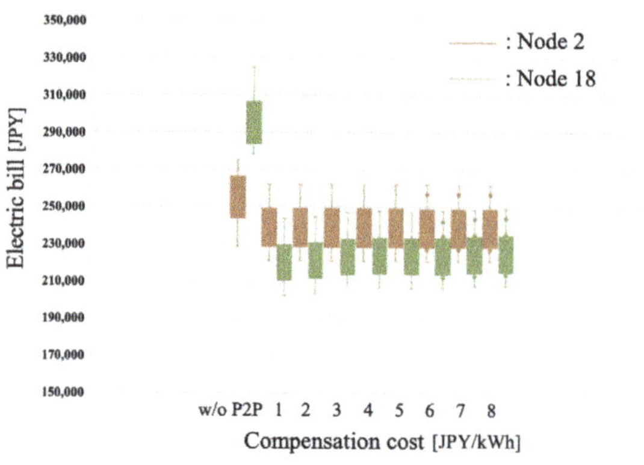

Figure 11. Electricity bills with introduction of P2P power trading system.

Figure 12 shows the total compensation amount and total output curtailment at each compensation unit price. The horizontal axis of Figure 12 shows the unit price of compensation for monetary compensation for the PV output curtailment of prosumers who were unable to commit. Figure 12a shows that increasing the compensation's unit price decreases the total compensation amount. Figure 12b shows that increasing the compensation's unit price decreases the total output curtailment amount. These results suggest that when determining the amount of output curtailment, the total compensation amount is minimized by appropriately implementing the necessary amount of output curtailment on the required portion as the compensation unit price increases.

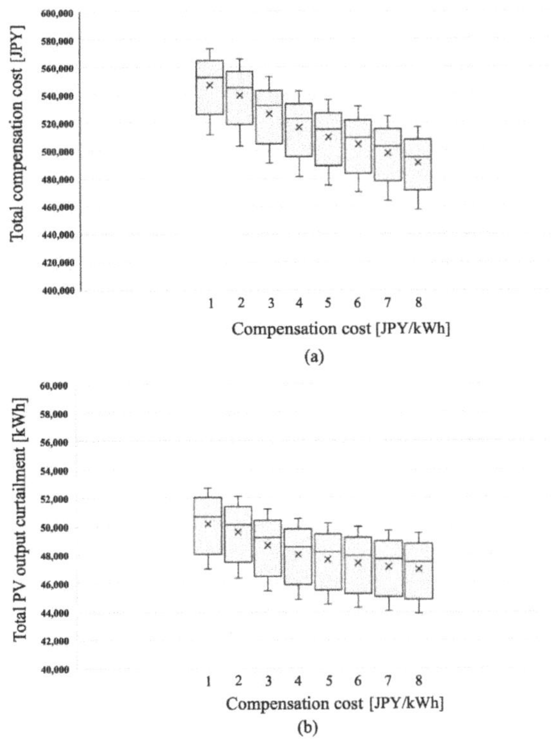

Figure 12. Total compensation cost and total PV output curtailment for each compensation cost; (**a**) total compensation cost; (**b**) total PV output curtailment.

Table 1 shows the transaction fees determined based on the above-mentioned simulation results and Equations (9) and (10) and Table 2 show the P2P participation fees to secure the source of the compensation costs. As shown in Table 1, since expanding the market participants increases the overall market's profit, the system operator determines transaction fees to ensure no participants lose money when applying the proposed rule. Moreover, the consumer fee was about 40% of the transaction profit, and the prosumer fee was about 4% of the transaction profit. In addition, we set the participation fee to ensure that the profit of the system operator is equal to the profit of the participants before and after the introduction of P2P power trading, resulting in a fee of approximately 350 JPY per participant. Figure 13 shows the reduction in electricity costs with and without P2P power trading for node 1 (consumer), node 2 (prosumer), node 18 (prosumer), and the system operator. The horizontal axis in Figure 13 corresponds to the first row of Table 1. For example, the system operator collects a transaction fee of 4.42 [%] of the contract profit from the prosumers and 39.8 [%] of the contract profit from the consumers to compensate 1 [yen/kWh] to the prosumers, subject to output curtailment, respectively. We can see from Figure 13 that the participation fee is charged so that the profit of the system operator is appropriately secured. Thus, the profit of the participant and the system operator are comparable. These results suggest that the P2P electricity trading participants' profits can be increased while maintaining the grid voltage and ensuring the profits of the system operator.

Table 1. Transaction cost for compensation cost based on the simulation result.

[%]	1	2	3	4	5	6	7	8
Prosumer	4.42	4.36	4.25	4.17	4.11	4.08	4.03	3.97
Consumer	39.8	39.2	38.3	37.6	37.1	36.7	36.2	35.7

Table 2. Participation cost for compensation cost based on the simulation result.

[JPY]	1	2	3	4	5	6	7	8
Participants	340	343	352	360	365	368	373	376

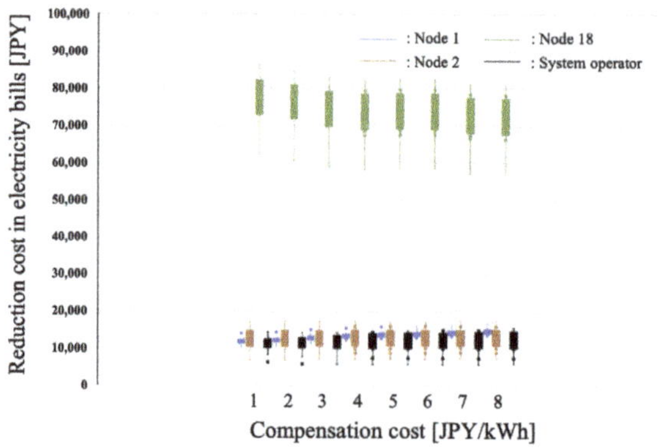

Figure 13. Reduction cost in electricity bills in each node.

Finally, the case where PV is installed at nodes 8, 9, 10, 11, 12, 13, 14, 15, 16, 17, 18, 29, 30, 31, 32, and 33, which are the terminal nodes of the system with severe voltage increases, is examined and discussed. Figure 14 shows the reduction in the power costs with and without P2P power trading for each node at the terminal connection. Figure 14 shows that when PV is concentrated at the end node, the electricity cost reduction is always negative for node 1, the consumer. This is because more PV is suppressed than before due to P2P power trading, including voltage constraints, which reduces the amount of electricity consumers can purchase from the market. It is also because we set the ratio of commission rates at 9:1 (consumers: prosumers), suggesting the need to re-set commission rates according to the PV interconnection status. On the other hand, for other participants and grid operators, it can be confirmed that the introduction of P2P power trading generates cost benefits.

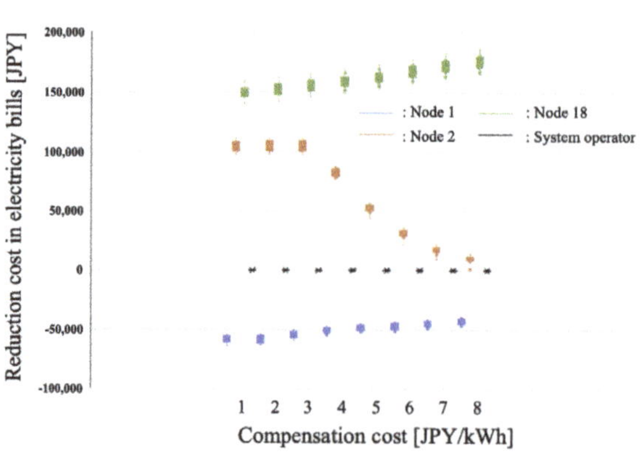

Figure 14. Reduction cost in electricity bills in each node when PV is interconnected to the end node.

5. Conclusions

This paper studies the introduction of monetary compensation for PV output curtailment in a P2P power trading system, which is necessary for grid voltage management. The study aims to increase the benefits for participants and the system operator. First, simulations were performed using PV scenarios. In each scenario, we determined the transaction fee and participation fee that would be the source of compensation and showed that increasing the profits of the participants and the system operator is possible. Introducing a P2P power trading system could eliminate the inequity in electricity prices. In addition, by increasing the unit price of compensation, the total amount of compensation and the total amount of output curtailment can be reduced, which will benefit participants in the P2P power trading system and those using clean energy. In addition, since the overall profit of all the P2P power trading system participants will increase, even if PV is unevenly distributed in the grid, all the participants can benefit if they cooperate in sharing the profits. Next, we determined the transaction fees and participation rates from the simulation of the previous scenario. The results showed that transaction fees and participation rates can be defined in advance while ensuring that the P2P electricity trading system benefits participants and the system operator. Future work includes examining the distribution of profits between market participants and system operators and when market participants have bidding strategies. Furthermore, we plan to study and evaluate the stability of real-time control and P2P power markets in relation to the [18]. The method of determining the percentage of prosumer and consumer transaction fees is also an issue for future work.

Author Contributions: S.T.; Conceptualization, methodology, software, validation, formal analysis, investigation, resources, data curation, writing—original draft preparation, visualization, project administration, supervision, funding acquisition, A.I.; writing—review and editing. All authors have read and agreed to the published version of the manuscript.

Funding: This work was supported by JSPS KAKENHI Grant Number JP20K04420.

Data Availability Statement: Data are contained within the article.

Conflicts of Interest: The authors declare that they have no known competing financial interests or personal relationships that could have appeared to influence the work reported in this paper.

References

1. Kimura, K. *Feed-in Tariffs in Japan: Five Years of Achievements and Future Challenges*; Renewable Energy Institute: Minato, Japan, 2017.
2. IEA. World Energy Outlook 2022. Available online: https://www.iea.org/reports/world-energy-outlook-2022 (accessed on 28 December 2023).
3. Khubrani, M.M.; Alam, S. Blockchain-Based Microgrid for Safe and Reliable Power Generation and Distribution: A Case Study of Saudi Arabia. *Energies* **2023**, *16*, 5963. [CrossRef]
4. Eltamaly, A.M.; Ahmed, M.A. Performance Evaluation of Communication Infrastructure for Peer-to-Peer Energy Trading in Community Microgrids. *Energies* **2023**, *16*, 5116. [CrossRef]
5. Damisa, U.; Nwulu, N.I.; Siano, P. Towards Blockchain-Based Energy Trading: A Smart Contract Implementation of Energy Double Auction and Spinning Reserve Trading. *Energies* **2022**, *15*, 4084. [CrossRef]
6. Matsuda, Y.; Yamazaki, Y.; Oki, H.; Takeda, Y.; Sagawa, D.; Tanaka, K. Demonstration of Blockchain Based Peer to Peer Energy Trading System with Real-Life Used PHEV and HEMS Charge Control. *Energies* **2021**, *14*, 7484. [CrossRef]
7. Nagatsuka, T.; Sano, M.; Yamaguchi, N. Experiment of Execution Time in Decentralized Transaction System of Surplus PV Output using Bloackchain. *IEEJ Trans. Power Energy* **2019**, *3*, 176–185. (In Japanese) [CrossRef]
8. Yokoyama, A. Toward Deregulated, Smart and Resilient Power Systems with Massive Integration of Renewable Energy in Japan. *IEEJ Trans. Elec. Electron. Eng.* **2022**, *17*, 1242–1254. [CrossRef]
9. Paudel, A.; Khorasany, M.; Gooi, H.B. Decentralized local energy trading in microgrids with voltage management. *IEEE Trans. Ind. Inform.* **2021**, *17*, 1111–1121. [CrossRef]
10. Azim, M.I.; Tushar, W.; Saha, T.K. Regulated P2P energy trading: A typical Australian distribution network case study. In Proceedings of the IEEE Power and Energy Society General Meeting, Montreal, QC, Canada, 2–6 August 2020; pp. 1–5.
11. Guerrero, J.; Chapman, A.C.; Gregor, V. Decentralized P2P Energy Trading Under Network Constraints in a Low-Voltage Network. *IEEE Trans. Smart Grid* **2019**, *10*, 5163–5173. [CrossRef]
12. Park, S.W.; Cho, K.S.; Son, S.Y. Voltage management method of distribution system in P2P energy transaction environment. *IFAC-PapersOnLine* **2019**, *52*, 324–329. [CrossRef]

13. Azim, M.I.; Wayes, T.; Saha, T.K. Coalition Graph Game-based P2P Energy Trading with Local Voltage Management. *IEEE Trans. Smart Grid* **2021**, *12*, 4389–4402. [CrossRef]
14. Qiu, D.; Wang, J.; Wang, J.; Strbac, G. Multi-Agent Reinforcement Learning for Automated Peer-to-Peer Energy Trading in Double-Side Auction Market. In Proceedings of the Thirtieth International Joint Conference on Artificial Intelligence, Montreal, QC, Canada, 19–27 August 2021; pp. 2913–2920.
15. Wand, M.P.; Jones, M.C. *Kernel Smoothing*; Taylor and Francis: Abingdon, UK, 1994.
16. Trashchenkov, S.; Pimentel, S.P.; Astapov, V.; Annuk, A.; Marra, E.G. Kernel Density Estimation for Stochastic Modeling of PV Power Output. In Proceedings of the 2018 7th International Conference on Renewable Energy Research and Applications (ICRERA), Paris, France, 14–17 October 2018; pp. 1179–1183. [CrossRef]
17. Dolatabadi, S.H.; Ghorbanian, M.; Siano, P.; Hatziargyriou, N.D. An Enhanced IEEE 33 Bus Benchmark Test System for Distribution System Studies. *IEEE Trans. Power Syst.* **2021**, *36*, 2565–2572. [CrossRef]
18. Ma, D.; Liu, M.; Zhang, H.; Wang, R.; Xie, X. Accurate Power Sharing and Voltage Regulation for AC Microgrids: An Event-Triggered Coordinated Control Approach. *IEEE Trans. Cybern.* **2022**, *52*, 13001–13011. [CrossRef] [PubMed]

energies

MDPI

Article

On the Dynamics of Spot Power Prices across Western Europe in Pandemic Times

Luis María Abadie [1] and José Manuel Chamorro [2,*]

[1] Metroeconomica, Av. Zugazarte 8, 3rd Floor (Modules 5 & 6), 48930 Getxo, Spain; lm.abadie@outlook.es
[2] Department of Financial Economics II, Institute of Public Economics, University of the Basque Country UPV/EHU, Av. Lehendakari Aguirre 83, 48015 Bilbao, Spain
* Correspondence: jm.chamorro@ehu.eus

Abstract: Learning the dynamics of power prices in a given market is important for a number of players (e.g., producers, consumers, and policy makers) at both macro- and microeconomic levels. This paper analyzes the recent behavior of spot prices in eight Western European countries. The sample period coincides with the COVID-19 pandemic for the most part: it starts in April 2020 and runs until May 2023; it includes the start of the Russia–Ukraine war. We introduce a new model for the hourly spot price of electricity. The deterministic component includes yearly, weekly, and daily seasonalities; the stochastic component accounts for volatility, mean reversion, and discrete jumps. We estimate the model with publicly available hourly data. Regarding the development of the internal market for electricity, we find that core mainland power markets now move closer in step with one another than before, but the integration process of the Iberian Peninsula seems to have kicked into reverse. As for the dynamics of power prices, in the last part of the sample period the speed of reversion falls everywhere, and price volatility increases noticeably; the expected number of jumps per hour decreases, but their average size turns to positive and they become more volatile.

Keywords: spot power price; COVID-19 pandemic; seasonality; mean reversion; volatility; discrete jumps

1. Introduction

It has been a long time since the adoption of the Single European Act (28 February 1986) [1], which aimed to add new momentum to European integration and to complete the 'internal market' by 1 January 1993. As far as the power sector is concerned, there have been several milestones along the way [2–5].

A number of researchers have assessed the extent to which the goal of completing 'an area with no internal borders and in which there is free movement' of electricity in particular has been accomplished. The assessment draws on the notion of market integration via price convergence; see for instance [6–8], among others. These two concepts are not to be confused. According to [9], the level of market integration at a particular time shows the (static) degree to which the single European market is attained. Instead, price convergence is the (dynamic) measure for the development of prices toward a single European price.

Other researchers look at power prices from a different perspective. They focus on the peculiar dynamics of electricity prices in day-ahead wholesale markets. Understanding price dynamics is necessary for the proper valuation and risk management of both real assets and financial contracts on electricity; see [10–13], to name a few. Ref. [14] provide a taxonomy of electricity price models applied to these purposes. As pointed out, the proposal of a particular model responds to the user's objectives, e.g., practical use, good price representation (of the main features of prices), price consistency (no arbitrage opportunities), or identifiability (liable to estimation from data).

No doubt, the impact of the COVID-19 pandemic has rippled way beyond global health concerns. Thus, a number of papers have addressed its impact on power systems, most of them from a physical or technical perspective, e.g., power demand and supply [15–17]. They usually adopt a top-down viewpoint (i.e., policy makers and system operators) and use metrics related to security of supply, system operation, frequency deviations, and the like. The Russia–Ukraine war has further compounded the global scenario. Ref. [18] adopts an economic perspective; in particular, the authors assess the ability of purchase power agreements to protect energy consumers from energy price spikes.

Arguably, the aftermath of the pandemic and the war may well run deeper than price levels. Specifically, have these events altered the inner dynamics of spot power prices? If so, is there any common pattern in the impact(s) across countries? From another perspective, have these events brought the internal market a bit closer or rather the opposite? Ref. [19] examines how the market anomalies caused by these events have affected European electricity markets. Following a novel approach, the authors convert time series into networks and then compare their degree distributions using the overlap coefficient. This transformation allows them to expand the set of properties analyzed simultaneously.

To our knowledge, no other paper explicitly addresses the above questions. Yet, our approach is quite different. Following previous research, we introduce a model for the hourly 'spot' price of electricity that encompasses four characteristics simultaneously: volatility, seasonality, mean reversion, and jumps. It comprises a deterministic component along with a stochastic component. The former accounts for seasonalities with different time frames. The latter combines mean reversion and discrete jumps. Next, we estimate our model, drawing on publicly available hourly data from eight Western European countries: Belgium, France, Germany, Italy, The Netherlands, Portugal, Spain, and the UK. The time horizon stretches from April 2020 through to May 2023, i.e., practically the whole COVID-19 pandemic and the outbreak of war. We are interested in the differential impact of these events across space and time. To this end, we analyze and then compare the behavior of the above markets before them and during them. See Scheme 1 below.

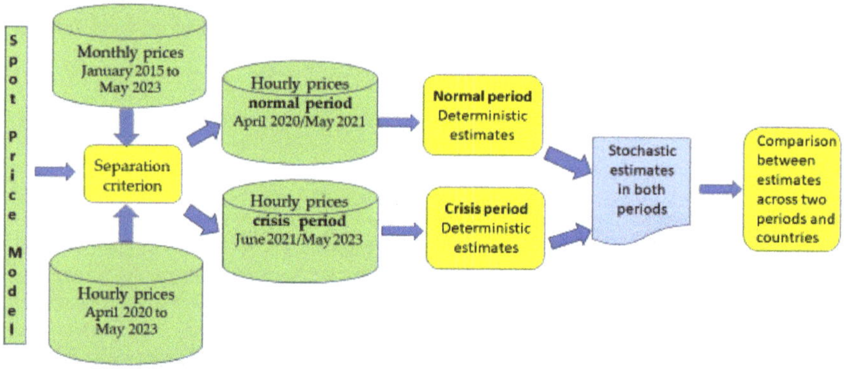

Scheme 1. Graphical representation of the process followed in developing the analysis.

Drawing on our data, we identify two sub-periods: April 2020–May 2021 and June 2021–May 2023 (end of the pandemic according to the WHO [20]); the first one covers the initial lockdowns and subsequent recovery toward the 'new normal', while the second one corresponds to the energy (price) crisis. Thus, Tables 4 and 5 show the deterministic estimates in the normal period and the crisis period, respectively; similarly, Tables 7 and 8 show the stochastic estimates in these periods. Relative to the first period, in the second one prices in the mainland markets show greater co-movement, whereas Iberian prices deviate from them. In addition, the speed of reversion falls everywhere, and price volatility increases noticeably; the expected number of jumps per hour decreases, but their average size turns to positive and they become more volatile.

The structure of the paper is as follows. Section 2 proposes the theoretical model for the spot price. Section 3 introduces two data sets; they include both hourly prices and monthly average prices, with each data set serving a different purpose. Section 4 provides numerical estimates of all the underlying parameters. The first ones correspond to the deterministic part in the two sub-periods. Next, the stochastic part is derived (as a residual, by subtracting the deterministic series from the price series). It is subsequently broken down into two series (corresponding to the mean-reverting process and the discrete-jumps process). Each series is used to estimate the parameters in the corresponding process in the two sub-periods. Section 5 discusses the main results, and Section 6 concludes.

2. Methodology (Stochastic Model for the Electricity Price)

Ref. [10] proposes a model for the natural logarithm of the day-ahead hourly price. Ref. [21] uses the logarithm of weekday medians (i.e., medians of 120 hourly observations). Logarithmic transformation has some advantages. Thus, log prices potentially mitigate the heteroscedastic properties of prices by minimizing the effects of high volatility and the outlier effects [22]. Nonetheless, a drawback is the inability to account for negative power prices, which some wholesale power markets allow (e.g., the Central Western Europe (CWE) market and the Irish Single Electricity Market (SEM)) and can be observed ever more frequently. Ref. [11] uses average daily prices (consequently, they only consider weekly and monthly seasonalities). Unlike these papers, we propose a model for the (absolute) spot price in levels and use hourly data. Absolute prices lead to straightforward interpretations, for instance when it comes to jumps' frequency and size. They allow for positive and negative values. In addition, hourly prices in particular allow analyze daily seasonality.

Specifically, we propose a process, P, for the (absolute) spot price in the physical world, which comprises a deterministic part, D, and a stochastic part, S:

$$P_t^i = D^i(t) + S_t^i \tag{1}$$

with i ={Belgium, France, Germany, Italy, The Netherlands, Portugal, Spain, UK}. Time t appears in two different formats: as an argument, $Z(t)$ denotes a deterministic function of time; instead, as a subscript, Z_t stands for the value of a stochastic process at time t. Time (t) is measured in years on an hourly basis. For instance, the first hour in April 2020 is $t = (1/366/24)$; this is the first hour for which there are data from every country. Similarly to previous papers, this model includes the main features in wholesale power markets: seasonality, mean reversion, volatility, and jumps.

By assumption, the mathematical expectation of the stochastic part S^i is zero. Ref. [23] put it in a different way: "Based on the rather deterministic demand for electricity (which in turn is due to highly seasonal temperature patterns influencing the demand to a large extent), we assume that deviations from a deterministic seasonal function are temporary". Therefore, when it comes to forecasting mean or expected values, only D^i is relevant (for point values, instead, both parts–deterministic and stochastic–play a role):

$$D^i(t) = \beta_1^i + \beta_2^i t + \beta_3^i D_0(t) t^2 + YC^i(t) + WC^i(t) + DC^i(t) \tag{2}$$

Thus, the model for D^i includes an intercept, β_1^i, and a time trend, $\beta_2^i t + \beta_3^i D_0(t) t^2$ (where $D_0(t) = 0$ in the normal period and $D_0(t) = 1$ in the crisis period); we set both a linear and a quadratic time trend because, as displayed later on in Figures 1 and 2, there is a first period when prices are more or less stable, which is then followed by another one with an inverse U-shape. The model also includes the following:

(a) A yearly cycle, $YC^i(t)$: it encompasses different seasonal components (annual, semi-annual, quarterly, semi-quarterly, monthly, ...), as many as determined by statistical significance:

$$YC^i(t) = \sum_{j=1}^{5} [\beta_{2+2j}^i \sin(2j\pi t) + \beta_{3+2j}^i \cos(2j\pi t)] \tag{3}$$

Equation (3) represents the yearly cycle with up to five sine and cosine components. In principle, there are 10 beta coefficients, numbered from β_4 to β_{13}. If a specific coefficient turns out to be statistically non-significant (at standard confidence levels) in the estimation process, then it will be dropped (alternative model specifications and elimination procedures are available, e.g., [24]).

(b) A weekly cycle, $WC^i(t)$, according to the day of the week:

$$WC^i(t) = \beta^i_{14}D_1(t) + \beta^i_{15}D_2(t) + \beta^i_{16}D_3(t) + \beta^i_{17}D_4(t) + \beta^i_{18}D_5(t) + \beta^i_{19}D_6(t) \quad (4)$$

In Equation (4), there are six dummy variables. $D_1(t)$ equals 1 if it is Monday, and zero otherwise. $D_2(t) = 1$ on Tuesdays, and zero otherwise, and so on through to the sixth dummy variable. On the seventh day, Sunday, the six dummies are zero.

(c) A daily cycle, $DC^i(t)$, based on the particular hours, with their own seasonalities:

$$DC^i(t) = \sum_{j=1}^{5} [\beta^i_{18+2j} \sin(2j\pi\tau)/24 + \beta^i_{19+2j} \cos(2j\pi\tau)/24] \quad (5)$$

Equation (5) represents the daily cycle in hours. The index τ indicates the hour: τ = 1, 2, . . ., 24. There are 10 beta coefficients, from β_{20} to β_{29}.

Summing up, there are 29 beta parameters in the deterministic part $D^i(t)$: the constant, 2 time trends, 10 betas in the yearly cycle, 6 in the weekly one, and 10 in the daily one. As mentioned above, if any of them are not statistically significant, they will be suppressed, in which case the model is re-estimated again with fewer parameters.

On the other hand, the stochastic part, S^i_t, follows a continuous mean-reverting process with discrete jumps:

$$dS^i_t = \left(\alpha^i - \kappa^i S^i_t\right)dt + \sigma^i dW^i_t + J^i\left(\mu^i_j, \sigma^i_j\right)dq^i_j \quad (6)$$

Specifically, Equation (6) comprises three terms on the right hand. The first two of them constitute a so-called Ornstein–Uhlenbeck (OU) process; the third one is a Poisson process. Now, proceeding step by step in Equation (6), the first term is a function of S^i_t, while the other two are stochastic. Leaving the latter aside for a moment, the equation can be rewritten as $dS^i_t = (\alpha^i - \kappa^i S^i_t)dt = \kappa^i(\frac{\alpha^i}{\kappa^i} - S^i_t)dt$. Thus, the stochastic part of the electricity price in each country tends toward α^i/κ^i in the long term, with a reversion speed κ^i. If S^i_t falls below its long-run quilibrium level the parenthesis will be positive, which induces an increase in its value ($dS^i_t > 0$); conversely, if S^i_t rises above α^i/κ^i the parenthesis will be negative, pushing S^i_t downwards ($dS^i_t < 0$). In sum, when S^i_t departs from its long-term equilibrium (due to the impact of stochastic shocks, namely OU and jumps), the first term tends to restore the equilibrium (always subject to shocks). In addition, the higher the speed of reversion κ^i, the sooner S^i_t approaches its equlibrium value. Now, the second term generates a continuous random behaviour (without jumps): the volatility of the mean-reverting process is σ^i; dW^i_t is the increment of a standard Wiener process. The third term accounts for jumps in the electricity price with intensity λ^i (the mean rate of event occurrence); thus, if the time unit is an hour then λ^i jumps are expected per hour. The jump size is normally distributed with mean μ^i_j and volatility σ^i_j. Here, dq^i_j is a Poisson process such that $dq^i_j = 1$ with probability $\lambda^i dt$, and $dq^i_j = 0$ with probability $1 - \lambda^i dt$. We assume that dW^i_t and dq^i_j are independent.

Note that the earlier assumption $E(S^i) = 0$ implies that, in terms of average or expected values, the mean-reverting component and the discrete-jump component offset each other: if any one of them has a positive expected value then the other must have a negative one.

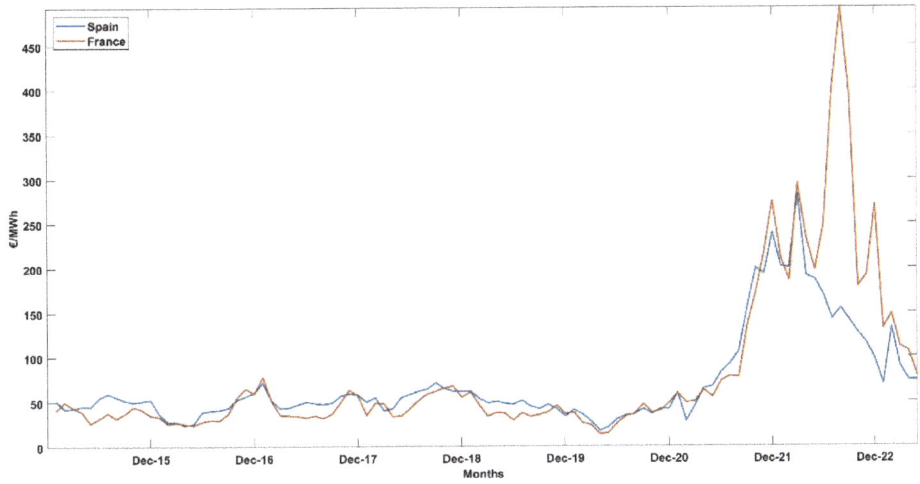

Figure 1. Electricity monthly spot prices in France and Spain during the 'normal' period (April 2020–May 2021) and the 'crisis' period (June 2021–May 2023).

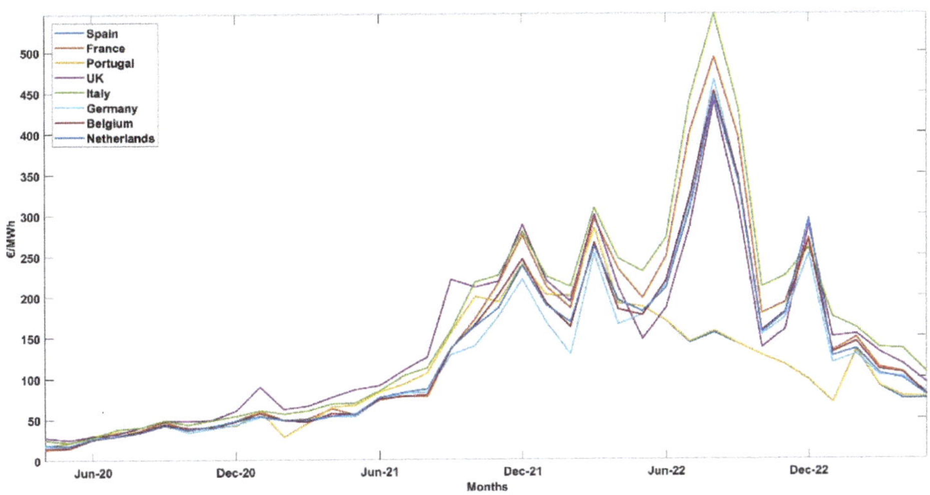

Figure 2. Monthly spot prices in eight European countries during the sample period.

3. Data

There are different definitions or notions of convergence. For example, in [25] price convergence means the reduction in international price level dispersion over time. According to [26], convergence is defined by convention as the percentage of hours during which absolute price differences are below 0.1 EUR/MWh. Ref. [7] distinguishes between partial convergence (prices in two markets approximate each other) and full convergence (law of one price holds). There are also definitions tailored to the particular econometric/theoretical model at hand, among them [7,9,22]. The empirical approaches include correlation analysis, regression analysis, cointegration analysis, Kalman filter analysis, principal components analysis, etc. Ref. [27] provides a list of them.

In our case, we keep the analysis simple and consider two different data sets. The first one includes three countries: France, Portugal and Spain. For each of them we have average monthly prices from January 2015 to May 2023 (i.e., 101 average prices for each country). A new inter-connector between Baixas (France) and Santa Llogaia (Spain) started operating in October 2015; its 2000 MW transmission capacity was ramped up progressively (Ref. [28] considers that on 1 January 2016 it was fully available). As a matter of fact, prices in Spain and Portugal are identical in many hours owing to the integration of both markets. Indeed, there is a joint Iberian power market operator, OMIE, for both Portugal and Spain.

The second set accounts for all the eight countries: the former three plus Belgium, Germany, Italy, The Netherlands, and the UK. It comprises hourly prices from April 2020 to May 2023 (i.e., 27,000 prices for each country). Thus, the two data sets overlap from April 2020 through to May 2023.

3.1. Monthly Overview

Monthly prices provide a first glimpse of the two sub-periods. Figure 1 shows their time path in France and Spain. Until early 2021, they fluctuate around a more or less stable level but then prices increase steeply.

Our first aim is to break down the time horizon into two parts: the normal period and the crisis period. Both hourly and daily electricity prices usually show high volatility. At this point, we stick to monthly prices; they are less volatile than hourly prices, which in turn makes it easier to reliably identify a structural break. The separation criterion draws on 24-month moving windows. We compute the price average and volatility for each month (starting from January 2017). Considering a standard normal distribution, the value that leaves 99.5% of the cumulative probability to its left (or, alternatively, 0.5% to the right) is 2.57583. We set the start of the crisis period when the monthly price is greater than the average plus 2.57583 times the volatility over 7 successive months; this happens in June 2021 (for both France and Spain). This result is in line with reports from the Spanish transmission system operator (TSO). According to Red Eléctrica de España [29]: "The average daily electricity market price in 2021 has been 111.93 €/MWh, the highest in history... more than triple that of last year (itself very low due to the pandemic)... with prices already above 90 €/MWh since the end of May".

The above result for France and Spain is assumed to apply to the eight countries. Thus, we split the spot price series into a normal period (running from April 2020 through to May 2021) and a crisis period (June 2021 to May 2023). Note that the normal period includes the initial months of the COVID-19 pandemic.

Figure 2 displays power prices in the eight countries over the whole sample period. Monthly prices are highest in Italy and France. They are lowest in Spain and Portugal; the latter are hard to identify because they are the same as the former very frequently.

3.2. Hourly Overview

Now, Table 1 shows the summary statistics of hourly prices in both sub-periods. Average levels and volatilities increase significantly from the first one to the second. The upward jumps are comparatively weaker in the Iberian Peninsula than in the other countries. Thus, percentage changes in Portugal and Spain fall short of 300%, but they surpass 400% and even 500% elsewhere.

Regarding skewness, the UK leads the sample in both the normal period (17.10) and the crisis period (3.81); second comes Portugal in the former (0.61) and France in the latter (2.08). Germany is the only place with negative skewness (−0.23, normal period). Positive skewness suggests that the probability mass is relatively more concentrated on the left of the distribution and the tail turns up on the right. This phenomenon is reinforced during the crisis period: skewness increases everywhere (except in the UK); the lowest rate is a non-negligible 31% (in Portugal).

When it comes to kurtosis, again the UK stands apart in the normal period (519.38); Germany (3.13) comes second, far behind. Italy is the only country with negative kurtosis (−0.24, normal period). As before, in the crisis period this statistic increases everywhere (except in the UK), with all the values above 3.0; this suggests that the probability distribution generates more outliers and/or more extreme outliers than the normal distribution. Kurtosis is minimum in Italy (3.68) and maximum in The Netherlands (25.12).

Table 1. Hourly electricity prices (EUR/MWh), April 2020–May 2023: summary statistics.

Country	Normal Period (10,224 h)		Crisis Period (16,776 h)		% Change	
	Mean	**Std. Dev.**	**Mean**	**Std. Dev.**	**Mean**	**Std. Dev.**
Spain	40.8	20	150	72.5	267.6%	262.5%
France	41.1	21.1	210	137	410.9%	549.3%
Portugal	40.8	20	150	72	267.6%	260.0%
UK	53.0	43.7	204	128	284.9%	192.9%
Italy	47.5	19.2	237	128	398.9%	566.7%
Germany	38.6	21.5	182	127	371.5%	490.7%
Belgium	39.8	21.7	192	124	382.4%	471.4%
The Netherlands	39.8	19.8	189	121	374.9%	511.1%
Country	Normal period (10,224 h)		Crisis period (16,776 h)		% Change	
	Skewness	Kurtosis	Skewness	Kurtosis	Skewness	Kurtosis
Spain	0.60	0.41	0.78	8.05	31.1%	1847.7%
France	0.31	1.26	2.08	6.97	571.8%	452.6%
Portugal	0.61	0.44	0.80	5.92	31.0%	1237.9%
UK	17.10	519.38	3.81	6.15	−77.7%	−98.8%
Italy	0.26	−0.24	1.35	3.68	412.9%	−1622.0%
Germany	−0.23	3.13	1.42	6.40	−724.7%	104.3%
Belgium	0.00	2.68	1.25	10.61	62,656.9%	296.6%
The Netherlands	0.41	2.14	1.33	25.12	226.6%	1073.4%

Tables 2 and 3 show the correlation coefficients between national hourly prices over both sub-periods; in a sense, the correlation coefficient provides a (static) measure of the level of market integration. Some markets show high correlations, among them those pertaining to the CWE electricity market. It is possible to distinguish two blocks. The first one comprises Portugal and Spain, where the correlation is almost perfect (99.8%) over both time horizons. In the other block, the correlations are not so extreme. Up to May 2021, they are higher than 75% between continental countries (France, Italy, Germany, Belgium, and The Netherlands), with the UK lagging behind (around 50%).

Table 2. Pearson correlation between hourly prices in normal period (April 2020–May 2021).

Country	Spain	France	Portugal	UK	Italy	Germany	Belgium	The Netherlands
Spain	1.000							
France	0.743	1.000						
Portugal	0.998	0.738	1.000					
UK	0.465	0.526	0.463	1.000				
Italy	0.749	0.858	0.745	0.532	1.000			
Germany	0.623	0.892	0.618	0.474	0.776	1.000		
Belgium	0.679	0.930	0.675	0.536	0.807	0.927	1.000	
The Netherlands	0.639	0.896	0.634	0.518	0.816	0.91	0.939	1.000

Table 3. Pearson correlation between hourly prices in energy crisis period (June 2021–May 2023).

Country	Spain	France	Portugal	UK	Italy	Germany	Belgium	The Netherlands
Spain	1.000							
France	0.454	1.000						
Portugal	0.998	0.454	1.000					
UK	0.437	0.755	0.439	1.000				
Italy	0.404	0.925	0.406	0.737	1.000			
Germany	0.380	0.889	0.380	0.751	0.869	1.000		
Belgium	0.443	0.912	0.443	0.782	0.882	0.954	1.000	
The Netherlands	0.445	0.896	0.444	0.778	0.880	0.953	0.972	1.000

Nonetheless, from June 2021 onwards correlations get stronger (Table 3): above 85% between continental countries, and higher than 70% for the UK. Thus, the energy crisis has contributed to further integrating these power markets. Instead, the evidence for the Iberian Peninsula is the opposite: hourly prices have de-coupled from those beyond the Pyrenees, as shown by correlations falling from over 60% in the normal period to under 40% in the crisis period.

4. Numerical Estimates of National Price Processes

Regarding empirical analyses, Ref. [10] considers three European wholesale power markets: the APX (The Netherlands), EEX (Germany), and PPX (France). The same number of European markets applies to [11]: Amsterdam Power Exchange (APX, The Netherlands), NordPool (Scandinavia) and Spain. In [21], the number grows to six: APX (The Netherlands), EEX (Germany), EXAA (Austria), NordPool (Scandinavia), Omel (Spain), and Powernext (France). The authors of [22] use data from seven markets: European Energy Exchange (EEX, Germany), Belgian Power Exchange (BELPEX, Belgium), Energy Exchange Austria (EXAA, Austria), Amsterdam Power Exchange (APX, The Netherlands), Nord Pool Power Exchange (ELSPOT, Scandinavia), Single Electricity Market (SEM, Northern Ireland and Republic of Ireland), and APX Power UK (former UKPX, Great Britain). We estimate our model drawing on hourly data from eight Western European countries.

4.1. Deterministic Parts

Tables 4 and 5 display the estimation results during the 'normal' and 'crisis' periods, respectively. Full details for each country in each period appear in Supplementary Materials. We run an OLS linear regression analysis with heteroskedasticity-consistent (HAC) robust standard errors.

Table 4. Deterministic parameters in the normal period (April 2020–May 2021).

Parameters	Spain	France	Portugal	UK	Italy	Germany	Belgium	The Netherlands
β_1	3.19	4.00	3.56	14.00	11.40	5.20	5.17	9.74
β_2	47.49	44.02	47.39	55.37	46.81	36.29	41.43	36.94
β_3	n.s.	n.s.	n.s.	n.s.	n.s.	n.s.	n.s.	n.s.
β_4	−5.25	−4.85	−5.24	n.s.	−2.86	−3.18	−4.93	−3.07
β_5	−6.72	n.s.	−6.63	3.78	−0.87	n.s.	n.s.	1.18
β_6	−3.34	2.72	−3.40	n.s.	0.88	2.22	1.71	1.17
β_7	4.40	2.46	4.52	4.01	0.87	1.85	2.33	1.92
β_8	n.s.	n.s.	n.s.	3.68	−1.06	n.s.	1.20	0.90
β_9	5.65	3.37	5.73	7.15	2.34	2.29	3.09	2.28
β_{10}	1.59	n.s.	1.66	2.90	n.s.	n.s.	n.s.	n.s.
β_{11}	4.32	2.69	4.37	3.01	1.00	n.s.	1.90	0.95
β_{12}	4.25	n.s.	4.34	2.52	n.s.	n.s.	n.s.	n.s.
β_{13}	n.s.	−3.67	n.s.	n.s.	−2.51	−3.42	−3.12	−2.75
β_{14}	10.96	13.85	10.53	5.54	10.03	13.91	11.75	9.90
β_{15}	13.00	16.62	12.66	7.94	11.89	17.01	15.30	12.19
β_{16}	13.44	17.54	13.07	12.43	12.80	17.27	16.38	12.30
β_{17}	14.15	16.63	13.82	9.49	12.30	17.34	15.11	12.77
β_{18}	11.57	14.15	11.23	8.45	11.28	16.22	13.45	11.02
β_{19}	4.45	6.29	4.27	2.77	4.75	7.71	6.02	4.71
β_{20}	−2.76	−4.11	−2.80	−12.44	−4.71	−3.58	−3.78	−3.84
β_{21}	0.54	−1.05	0.55	n.s.	−0.50	n.s.	0.76	n.s.
β_{22}	−5.83	−7.84	−5.69	−9.87	−7.50	−8.69	−8.71	−8.40
β_{23}	1.24	−0.66	1.25	n.s.	−1.01	−1.53	n.s.	−1.18
β_{24}	−0.49	1.53	−0.50	3.30	1.51	1.72	1.97	2.05
β_{25}	−0.67	−0.55	−0.70	−6.42	−0.20	−0.78	−0.72	−0.62
β_{26}	1.00	2.03	0.94	5.52	1.40	1.71	1.84	1.72
β_{27}	−0.66	0.35	−0.63	−0.70	−1.00	−0.48	−0.39	−0.57
β_{28}	n.s.	−0.69	n.s.	n.s.	n.s.	−0.58	−0.45	−0.58
β_{29}	n.s.	0.26	n.s.	3.69	n.s.	0.16	0.63	0.38

n.s. = not significant at the 10% level.

Table 5. Deterministic parameters in the energy crisis period (June 2021–May 2023).

Parameters	Spain	France	Portugal	UK	Italy	Germany	Belgium	The Netherlands
β_1	−347.52	−1229.5	−345.44	−660.4	−1214.4	−949.4	−953.53	−933.10
β_2	550.37	1328.36	549.31	828.20	1342.57	1017.06	1047.56	1035.03
β_3	−143.14	−297.21	−142.77	−190.7	−297.86	−223.53	−232.80	−231.27
β_4	−23.82	26.72	−23.35	n.s.	35.33	31.75	17.80	19.40
β_5	23.13	−43.88	23.25	−35.71	−51.82	−46.54	−42.30	−39.47
β_6	n.s.	−34.12	n.s.	−39.40	−37.03	−27.27	−27.01	−28.28
β_7	19.94	−12.49	20.16	0.00	−13.68	−12.98	−11.65	−8.67
β_8	−4.78	n.s.	−4.61	19.83	10.00	14.14	9.86	13.35
β_9	2.99	36.62	3.22	33.83	33.55	30.33	30.10	28.73
β_{10}	−13.28	−33.51	−13.10	−37.46	−26.06	−33.79	−34.53	−32.41
β_{11}	−3.84	7.56	−3.88	7.03	8.69	n.s.	n.s.	n.s.
β_{12}	n.s.	24.09	n.s.	14.41	12.31	15.15	20.55	15.73
β_{13}	−11.69	−8.76	−11.72	−6.78	−9.24	−8.88	−8.49	−7.44
β_{14}	23.08	52.75	21.77	28.26	33.10	53.14	47.12	42.84
β_{15}	24.94	66.76	23.51	33.61	42.06	70.89	57.76	51.98
β_{16}	22.91	65.03	21.31	25.32	41.22	66.41	53.62	47.87
β_{17}	24.76	63.19	23.54	30.67	44.69	60.15	53.13	47.78
β_{18}	21.38	56.24	19.92	27.88	41.69	49.00	45.95	42.52
β_{19}	9.43	22.08	8.07	n.s.	15.55	17.72	14.23	13.25
β_{20}	−3.88	−18.65	−4.23	−34.23	−19.44	−16.60	−16.60	−15.72
β_{21}	10.43	−7.05	10.02	−4.39	−2.40	n.s.	3.17	5.57
β_{22}	−21.12	−31.69	−20.78	−32.99	−27.81	−31.65	−33.13	−33.77
β_{23}	1.22	−6.27	1.30	n.s.	−12.71	−17.75	−14.71	−16.91
β_{24}	−1.49	5.85	−1.46	13.95	7.18	7.73	8.41	9.71
β_{25}	−1.51	n.s.	−1.48	−11.19	n.s.	n.s.	n.s.	1.72
β_{26}	4.25	8.65	4.12	11.09	6.96	6.97	7.42	6.10
β_{27}	−3.68	n.s.	−3.56	−2.34	−1.98	0.89	n.s.	n.s.
β_{28}	n.s.	−2.12	n.s.	−1.49	n.s.	−2.37	−1.93	−1.93
β_{29}	n.s.	n.s.	n.s.	5.69	0.79	n.s.	0.66	n.s.

n.s. = not significant at the 10% level.

Table 4. The regression intercept (β_1), which is statistically significant everywhere, varies markedly across markets. It is highest in the UK (14.00 EUR/MWh) and Italy (11.40), while the lowest values correspond to Portugal (3.56) and Spain (3.19). The linear time trend (β_2) is positive everywhere and shows less dispersion. Again, the UK (55.37 EUR/MWh over a year) stands out, followed by Spain (47.49) and Portugal (47.39). The minimum value corresponds to Germany (36.29). As for the quadratic time trend, β_3 is not statistically significant anywhere at the 10% level in this period (which is not surprising in view of the left part in Figures 1 and 2).

Regarding the yearly cycle (β_4 through β_{13}), some coincidences arise. For example, there is at least one non-significant parameter in every country (though it is never the same for every one). Yet, some groupings show up. For instance, β_8 is non-significant in France–Germany–Spain–Portugal. Instead, β_{10} and β_{12} are non-significant in Belgium–France–Germany–Italy–The Netherlands. There are also parameters that are significant in every market and even show the same sign, e.g., β_7 and β_9, both positive. Conversely, β_4 is consistently negative (except in the UK, where it is not significant).

Unlike the former, the weekly cycle (β_{14} through β_{19}) is statistically significant every day in all of the markets. Across space, the lowest estimates correspond to the UK, and the highest ones to Germany and France. Across time, the estimates are highest on Wednesdays and Thursdays, and lowest on Saturdays.

When it comes to the daily cycle (β_{20} through β_{29}), France is the only country where all of the estimates are statistically significant (at the 10% level). Similarly to the yearly cycle, some parameters are significant in every market and even show the same sign, either negative (β_{20}, β_{22}, and β_{25}) or positive (β_{26}). On the other hand, a few groups turn up. For instance, both β_{28} and β_{29} are non-significant in Portugal–Spain–Italy.

Table 5. In the energy crisis period, some results change dramatically. Thus, the numerical estimates of the regression intercept (β_1) bear no resemblance to the earlier ones. They turn negative and quite sizeable: around (−1200) in UK–Italy, and (−350) in Portugal–Spain. The linear time trend (β_2) remains positive but jumps above 1000 in Belgium–France–Germany–Italy–The Netherlands, while reaching 500 in Portugal–Spain. The quadratic time trend, β_3, which was not significant anywhere in the normal period, now becomes statistically significant and shows a negative sign, with Belgium–France–Germany–Italy–The Netherlands at one end (between −223.53 and −297.86) and Portugal–Spain at the other (around −143).

The estimates of the yearly cycle are very different too (relative to the normal period). The number of non-significant estimates drops from 23 before to 9 now. Some estimates switch from negative to positive or the other way round. The absolute values increase noticeably, sometimes by a factor of 10 or more. Still, other findings remain, e.g., β_9 is significant and positive in all of the markets. Conversely, now β_{10} is significant everywhere and negative (no longer positive).

Concerning the weekly cycle (β_{14} through β_{19}), the main difference is the size of the estimates, which increases across both space and time, sometimes by a factor of five or more. Again, the highest values correspond to Germany and France. The lowest ones arise in Portugal and Spain (not the UK). Across time, now the estimates are highest mostly on Tuesdays; again, the lowest are on Saturdays.

As for the daily cycle (β_{20} through β_{29}), now the number of non-significant coefficients grows from 12 to 17. As before, β_{20} and β_{22} remain significant and negative everywhere, while β_{26} continues to be positive in every market. β_{25} ceases to be significant in Belgium–France–Germany–Italy. Similarly to the above parameters, the absolute values jump upward: in Table 4 just one of them reaches 10; now, values above 20 and even 30 are common. Again, β_{28} is non-significant in Portugal–Spain–Italy. β_{29} continues non-significant in Portugal–Spain but now France–Germany–The Netherlands join the list.

To gain additional insights, the following figures display some of the above results for the three types of seasonalities. Regarding yearly patterns, Figures 3 and 4 show $YC^i(t)$ over the two periods in Germany (core) and Spain (periphery), respectively. A cursory look allows the observation that the energy crisis has led to much wilder swings in the former than in the latter (as suggested in Tables 4 and 5 by the changes in β_4 through β_{13}). An absence of pattern changes in the yearly cycle would imply perfect positive correlation (+1.00) across the two periods. In Spain it is 0.3783. In Germany it is −0.30643.

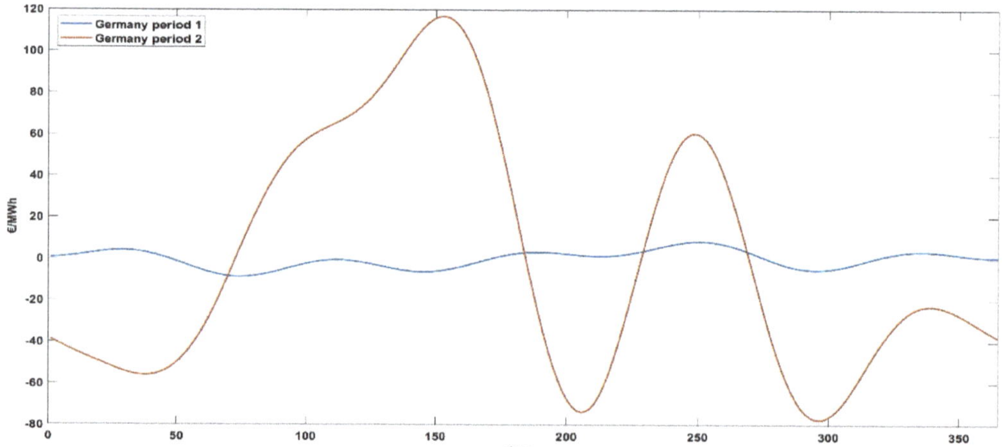

Figure 3. Germany: yearly seasonal effect, $YC(t)$, during the two periods.

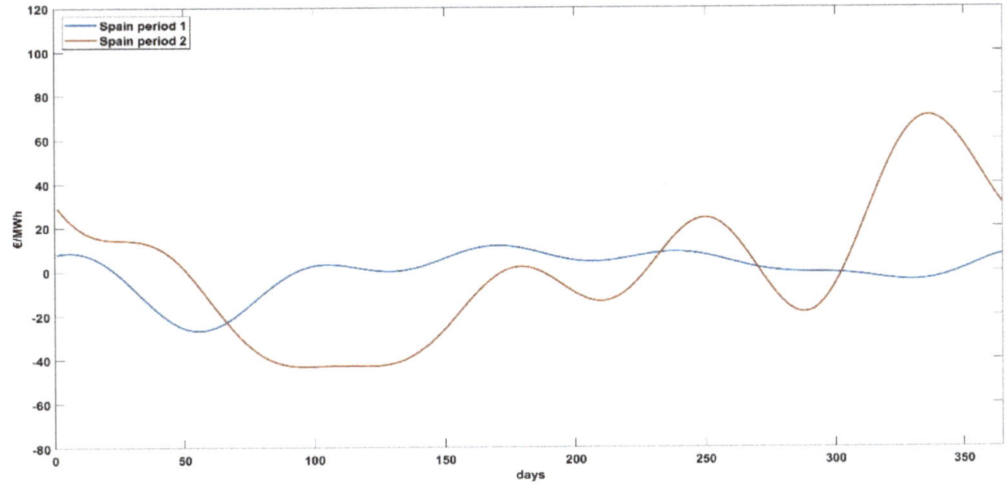

Figure 4. Spain: yearly seasonal effect, $YC(t)$, during the two periods.

Figure 5 focuses on the energy crisis period specifically. The two national cycles are very different during the first part of the year; in the second, instead, they describe similar paths. At any time, the German yearly cycle displays wider amplitude than the Spanish one.

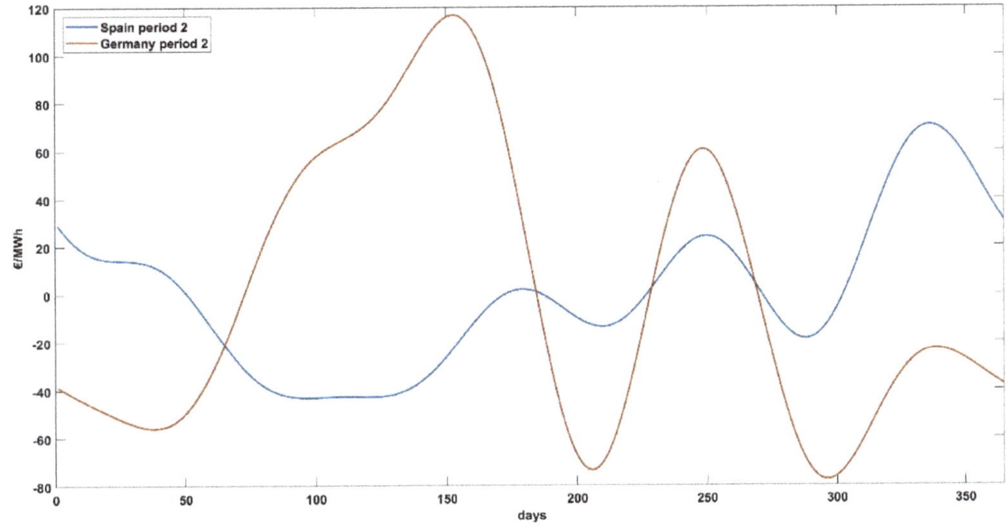

Figure 5. Germany and Spain: yearly seasonal effect, $YC(t)$, during the crisis period.

As for the weekly effects, Figures 6 and 7 show that $WC^i(t)$ get more prominent in both Germany and Spain during the crisis period. As before, the changes are much bigger in the former than in the latter. On the other hand, Saturday stands apart from the working days.

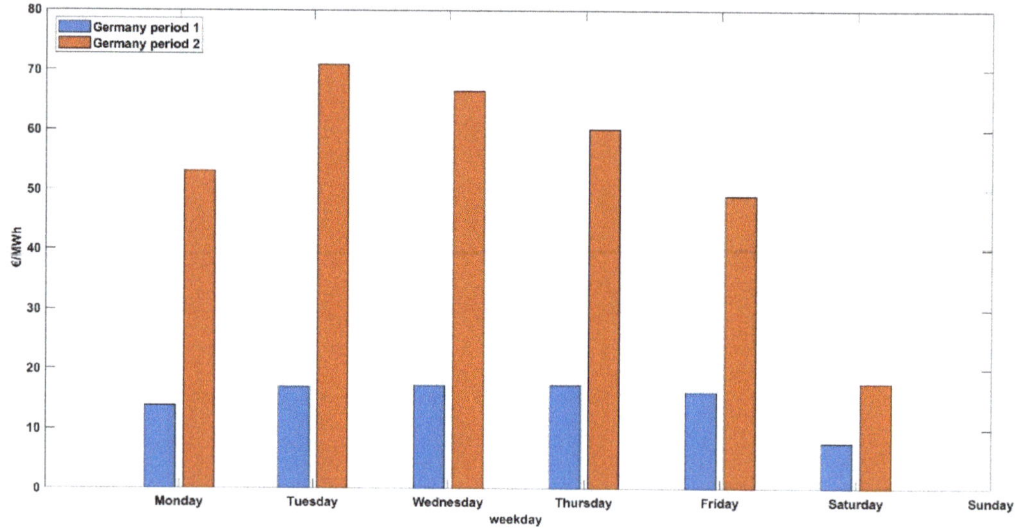

Figure 6. Germany weekly effects, $WC(t)$, during the two periods.

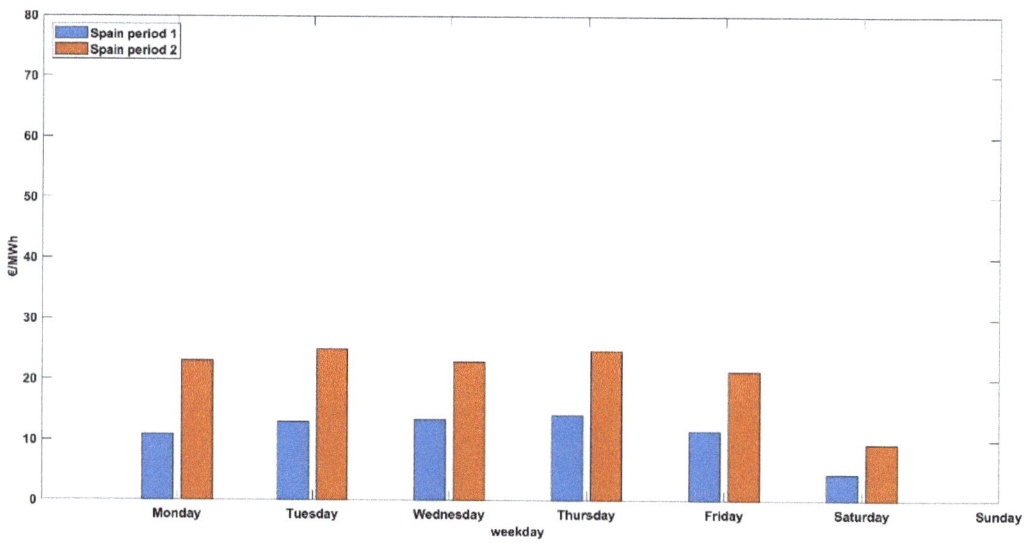

Figure 7. Spain weekly effects, $WC(t)$, during the two periods.

Concerning hourly effects over the day, $DC^i(t)$, here the differences between the two countries are smaller than before. In Germany, the correlation of the hourly cycles during the two periods is 0.9355; see Figure 8. In Spain it is slightly lower, 0.9087; see Figure 9. In both countries the swings become wider during the crisis period.

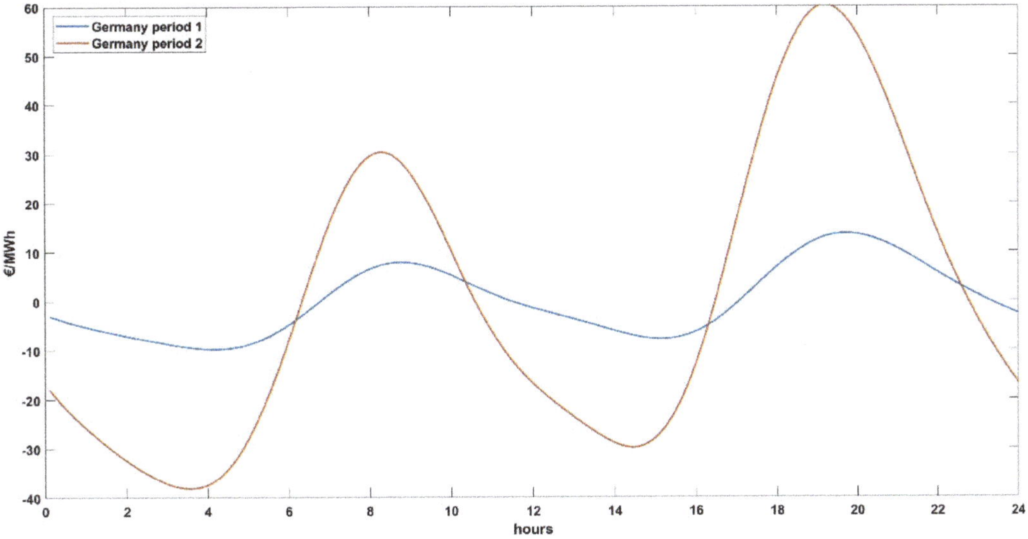

Figure 8. Germany, daily effects, $DC(t)$ over the 24 h, during the two periods.

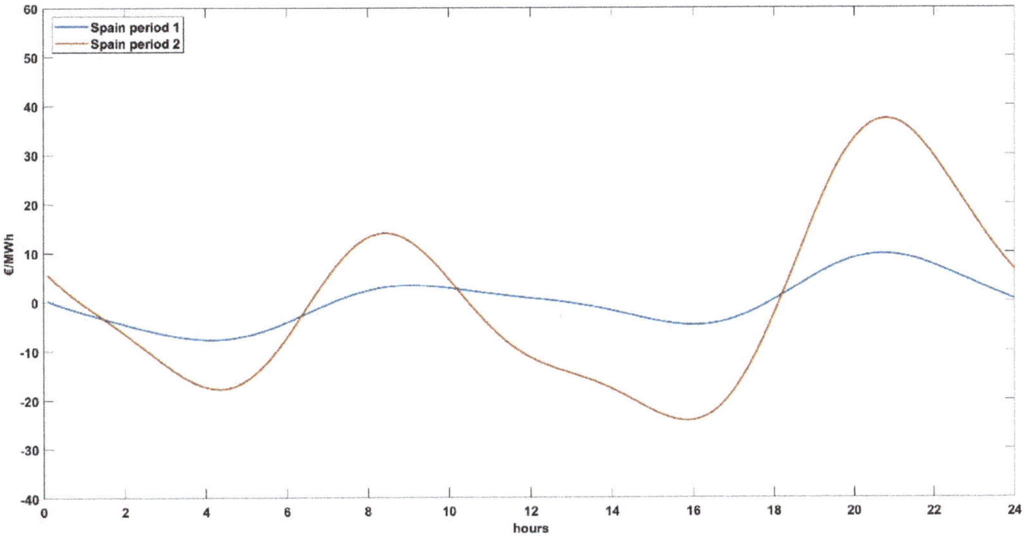

Figure 9. Spain, daily effects, $DC(t)$ over the 24 h, during the two periods.

Figure 10 displays the national hourly cycles during the energy crisis period. The paths are similar, with Spanish patterns following German ones with a delay of 1 to 2 h in the second part of the day.

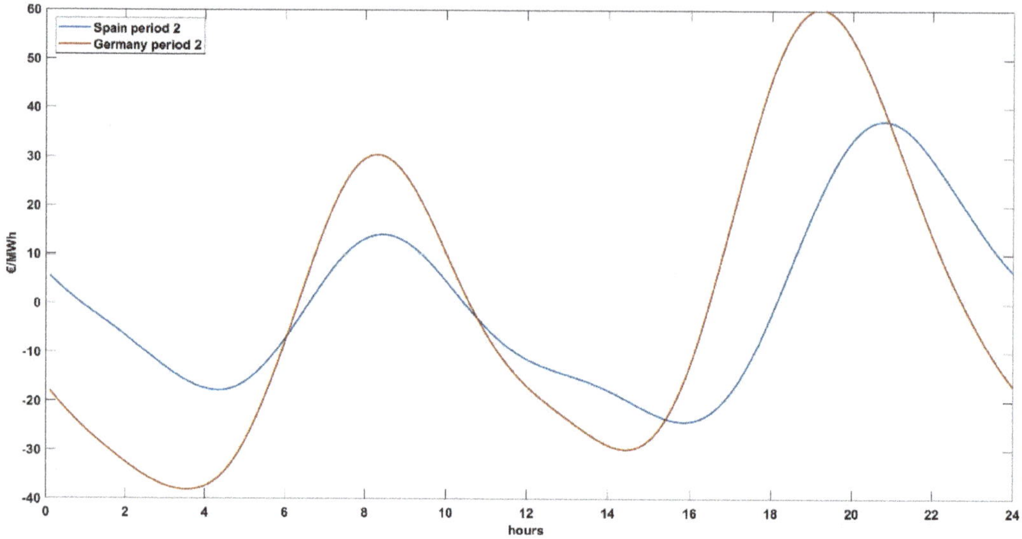

Figure 10. Germany and Spain, daily effects, $DC(t)$ over the 24 h, during the crisis period.

For further comparison purposes, Figures 11 and 12 show hourly spot prices in Germany and Spain over the energy crisis period along with the respective deterministic components. There is evidence of (yearly) seasonality, volatility, jumps, and mean reversion, especially in the particular case of Germany.

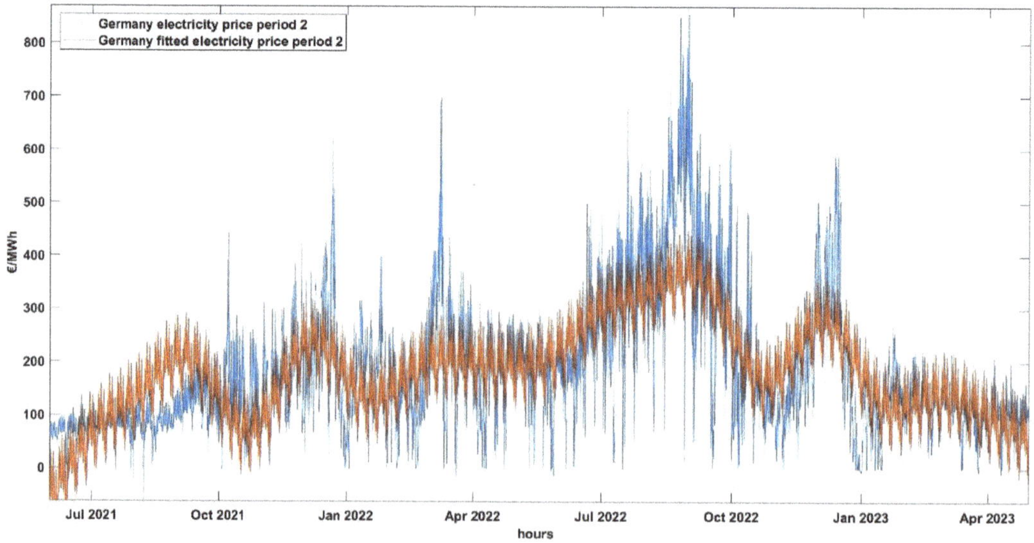

Figure 11. Germany, electricity prices (blue) and deterministic part (red) in crisis period.

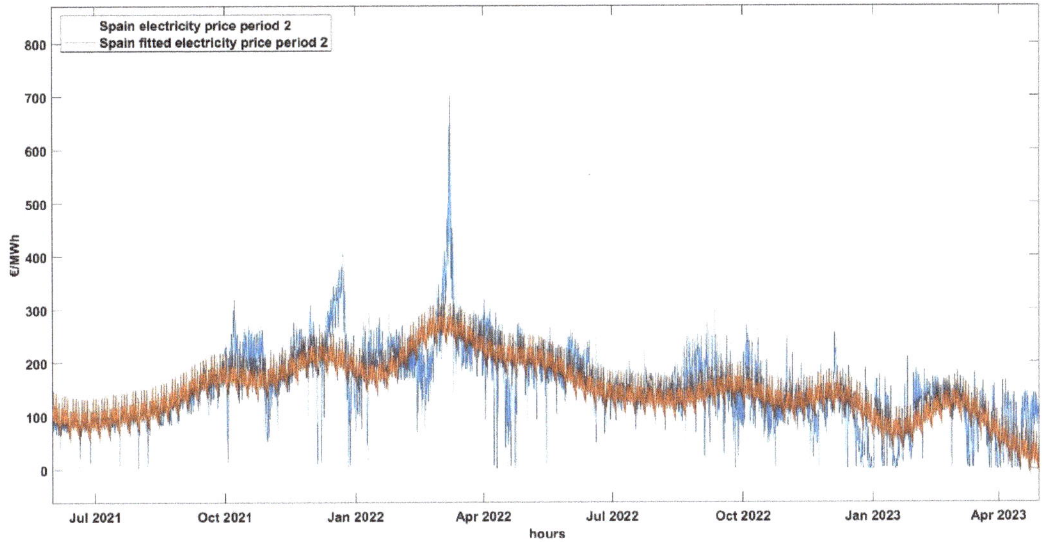

Figure 12. Spain, electricity prices (blue) and deterministic part (red) in crisis period.

4.2. Stochastic Parts

Now, moving on to the stochastic part of the hourly spot price entails getting the series of S_t^i residuals ($S_t^i = P_t^i - D^i(t)$, i.e., the residuals of the OLS regression). As expected, during the energy crisis (June 2021–May 2023) its volatility grows significantly (see Table 6). This fact renders future prices less predictable than in normal times. The biggest percentage increases take place in Italy and France; the opposite is true in the Iberian Peninsula and the UK, where it is smallest.

Table 6. Hourly volatility (EUR/MWh) of the S_t^i series in both periods for the sample countries.

Volatility	Spain	France	Portugal	UK	Italy	Germany	Belgium	The Netherlands
Normal	12.17	12.32	12.14	36.16	8.83	15.12	14.35	12.93
Crisis	45.52	88.19	45.34	96.52	72.61	86.85	84.09	81.89
% Change	274.1%	616.1%	273.6%	167.0%	722.3%	474.5%	485.9%	533.5%

According to Equation (6), the stochastic component S_t^i comprises two parts: an OU process and a Poisson process. We separate the mean-reverting part and the discrete-jumps part following a recursive approach. Starting from the initial series of S_t^i residuals, we consider that there is jump at a particular time when (the absolute value of) the residual at this time exceeds three times the standard deviation ('volatility') of that series; the same metrics is used by [6,14], among others. After this first residual is filtered out, the volatility of the initially considered mean-reverting part will be lower; thus, it is possible that new values turn up as jumps (in which case they are treated accordingly). The process finishes when the volatility of the mean-reverting part does not change and therefore no new jump arises. The number of iterations changes across countries and periods. The minimum number is six, and the maximum is nine; Ref. [9] performs a filter and smoother algorithm up to five times, the same number as [6]. In the end, starting from the S_t^i series the recursive procedure leads to two series: one corresponds to the OU process and the other to the Poisson process. Each series allows the derivation of numerical estimates of the parameters underlying it.

Figure 13 displays the decomposition of the S_t^i series for Germany during the energy crisis period. The upper graph shows the original series, i.e., the sum of the OU process and the Poisson process. The (discrete) series of jumps, whether negative or positive, is represented in the middle. The bottom graph shows the (continuous) series of mean-reverting changes. Clearly, whenever the S_t^i residual approaches 400 EUR/MWh in the upper graph the reason is a jump, not mean reversion; just look at the units along the vertical axes of the middle and bottom graphs.

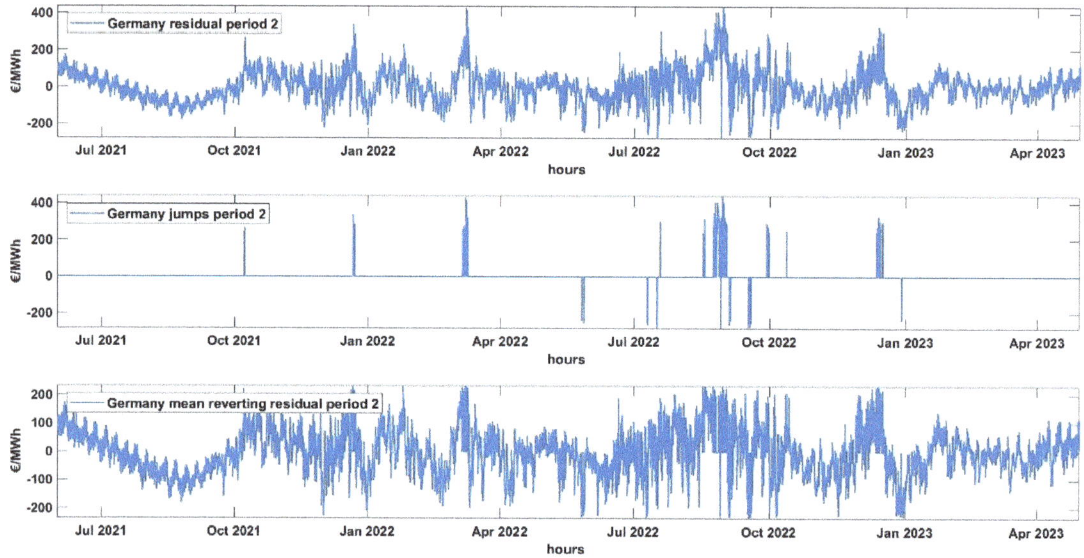

Figure 13. Germany's stochastic part over the crisis period: jumps and mean reversion.

At this point, it is worth remembering that, by assumption, $E(S_t^i)$ is zero. This applies to the upper series in Figure 13. Therefore, the averages of the other two series below must sum to zero. The graph in the middle shows more positive jumps than negative ones (i.e., the mean is positive). This in turn implies that the mean-reverting series in the lower graph must have a negative mean.

Similarly, Figure 14 displays the decomposition of the S_t^i series for Spain during the energy crisis period. The upper graph shows the original series. It is more stable than the German one (as suggested by Figures 11 and 12). The jumps in the middle graph are more abundant; nonetheless, their size is smaller. Interestingly, positive jumps in Spain synchronize well with the German ones, but this is not true for negative jumps, which are more frequent in Spain. The bottom graph of mean-reverting changes shows a similar size reduction: the range in Germany is [−200 EUR/MWh; 200 EUR/MWh], while in Spain it is [−100 EUR/MWh; 100 EUR/MWh].

The next step is to derive numerical estimates of the parameters underlying S_t^i in both periods. We use the residuals of the OLS regression (their average is zero). Upon identification of the jump series, parameter estimation is straightforward. Table 7 shows the results; note that $\Delta t = 1/(365 \times 24)$.

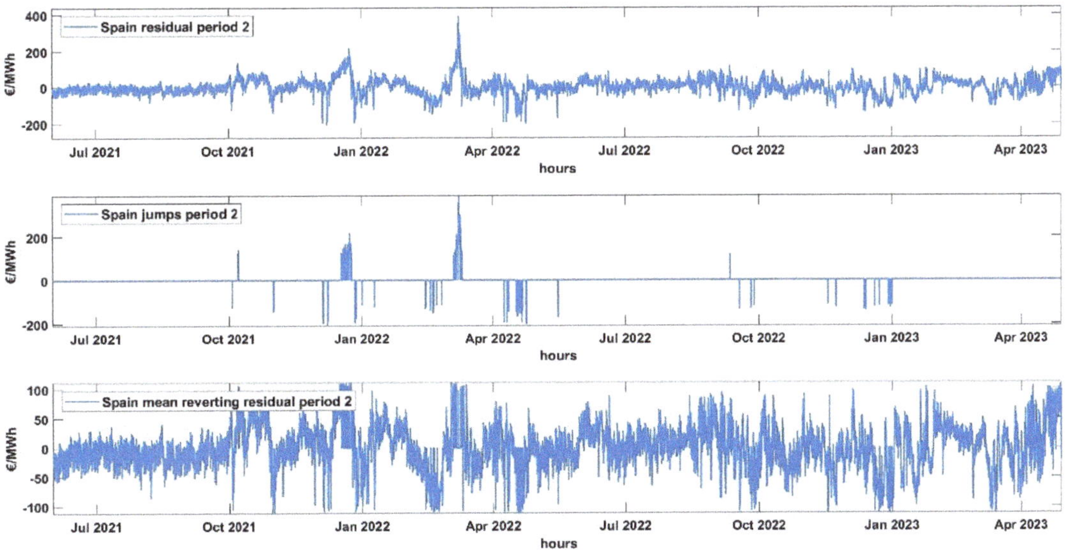

Figure 14. Spain's stochastic part over the crisis period: jumps and mean reversion.

Table 7. Stochastic part S_t^i: parameter estimates of the jump process.

Country	Parameter	Period 1		Period 2	
		Value	95% Confidence Int.	Value	95% Confidence Int.
Spain	λ^i	0.0441	(0.0318, 0.0564)	0.0264	(0.01836, 0.0344)
	μ_j^i	−14.4427	(−17.7146, −11.1708)	18.5654	(3.6591, 33.4717)
	σ_j^i	35.3564	(33.1898, 37.828)	159.637	(149.772, 170.904)
France	λ^i	0.0443	(0.0342, 0.0544)	0.0169	(0.0109, 0.0230)
	μ_j^i	−12.2928	(−15.9278, −8.6578)	278.116	(251.293, 304.938)
	σ_j^i	39.3675	(36.96, 42.1129)	229.641	(212.18, 250.257)
Portugal	λ^i	0.0442	(0.0317, 0.0567)	0.0260	(0.0180, 0.0341)
	μ_j^i	−14.3919	(−17.6526, −11.1313)	21.7014	(6.7194, 36.6833]
	σ_j^i	35.2744	(33.115, 37.7373)	159.351	(149.44, 170.68)
UK	λ^i	0.0247	(0.0180, 0.0315)	0.0255	(0.0183, 0.0326)
	μ_j^i	60.4514	(34.7452, 86.1576)	302.743	(272.567, 332.919)
	σ_j^i	207.616	(190.964, 227.473)	317.244	(297.298, 340.081)
Italy	λ^i	0.0246	(0.0179, 0.0313)	0.0157	(0.0093, 0.0221)
	μ_j^i	1.37807	(−2.4449, 5.2010)	219.776	(202.397, 237.156)
	σ_j^i	30.8143	(28.3383, 33.768)	143.414	(132.135, 156.814)
Germany	λ^i	0.0506	(0.03959, 0.0615)	0.0165	(0.0106, 0.0224)
	μ_j^i	−22.9209	(−26.7685, −19.0734)	199.309	(174.505, 224.114)
	σ_j^i	44.5307	(41.9718, 47.4244)	209.709	(193.579, 228.795)
Belgium	λ^i	0.0512	(0.0403, 0.0620)	0.0183	(0.0124, 0.0242)
	μ_j^i	−14.7873	(−18.6988, −10.8759)	162.406	(136.437, 188.375)
	σ_j^i	45.5337	(42.9313, 48.4744)	231.236	(214.277, 251.132)
The Netherlands	λ^i	0.0432	(0.0336, 0.0528)	0.0212	(0.0149, 0.0274)
	μ_j^i	−5.56887	(−9.6148, −1.5229)	135.724	(110.714, 160.734)
	σ_j^i	43.2802	(40.6028, 46.3386)	239.605	(223.181, 258.658)

Regarding Table 7, we focus specially on Germany and Spain (the analysis can be extended easily to the other markets in the sample). Starting with the former in the two periods, the expected number of jumps per hour (λ^i) drops by two-thirds from the normal period to the crisis period (namely, from 0.0506 to 0.0165). This may be a consequence of more volatile power prices in the crisis period: the higher the volatility, the higher the threshold to overcome (three times) in order to qualify as a jump. The jumps become more acute. The average size (μ_j^i) switches from (−22.9209) to 199.309 EUR/MWh, and volatility (σ_j^i) jumps from 44.5307 to 209.709 EUR/MWh, a factor of 4.7. In the case of Spain, the jumps undergo similar changes, but these are milder. The expected jumps per year (λ^i) decrease by less than one-half (from 0.0441 to 0.0264). Their average size (μ_j^i) reverses from (−14.4427) to 18.5654, and volatility (σ_j^i) increases by a factor of 4.5 (from 35.3564 to 159.637). These increases in jump sizes and volatilities are consistent with Table 6: S_t^i is more volatile in Germany than in Spain and becomes more so (see also Table 1).

The above patterns are broadly similar across all the sample markets. The expected number of jumps per year drops everywhere (except in the UK, where it is almost constant). In the normal period, the average jump size is negative in six (out of eight) markets (the exceptions being the UK and Italy). However, it is positive in all of them during the crisis period; for one, in Italy the average jump rises from 1.37807 to 219.776 EUR/MWh. As for jump volatility, in the normal period the UK is at the top (207.616) and Italy at the bottom (30.8143). They both keep their positions in the crisis period, but the gap compresses noticeably (317.244 and 143.414, respectively).

When it comes to the OU process, the parameters are estimated by OLS with HAC robust standard errors; see Table 8. Again, we look in particular to Germany and Spain; regarding the absolute parameter estimates, note that in Equation (6) the first parenthesis is multiplied by dt, which equals (1/8760) here. As explained in Section 2, the joint parameter α^i/κ^i is the level toward which the mean-reverting part of the electricity price in country i tends in the long term; further, it does so at a reversion speed κ^i. During the normal period, the long-term level of S_t^i is positive in both Germany (1.1574) and Spain (0.6410), both measured in EUR/MWh. Nonetheles, it switches to negative in the crisis period (−3.2956 and −0.4108, respectively). On the other hand, remember that, by assumption, $E\left(S_t^i\right) = 0$: the averages of the two underlying processes must sum to zero. In this regard, whenever the value of μ_j^i in Table 7 is negative, the corresponding α^i/κ^i on Table 8 is positive, and the opposite is also true. Overall, the results are consistent with those in Table 6, namely the higher volatility levels in Germany (whatever the period considered) and also the bigger increase in volatility.

The reversion speed κ^i falls in both countries: by around two-thirds in Germany (from 1431.94 to 507.31) and less than one-fifth in Spain (from 900.05 to 737.42). Note that $k = ln2/t_{1/2}$, where $t_{1/2}$ is the expected half-life of the (deseasonalized) stochastic part, i.e., the time required for the gap between S_0^i and the long-term level α^i/κ^i to halve. A lower speed of reversion means that, when a shock to S_t^i strikes, the impact takes longer to disappear (or S_t^i takes longer to stabilize). In other words, the anchoring effect of long-run levels α^i/κ^i weakens. In turn, intuition suggests that more observations far from the average (because of the slower reversion) will lead to higher volatility in the series. Table 8 shows that the volatility of the mean-reverting process (σ^i) doubles in the crisis period: in Germany from 219.36 to 480.02, and in Spain from 187.94 to 362.41. Again, this is consistent with the results in Table 6.

Now at the sample level, in the normal period, the long-term level α^i/κ^i is negative in just two countries, namely the UK (−1.5009) and Italy (−0.0354). All other countries are somewhere between The Netherlands (0.2383) and Germany (1.1574). Perhaps a possible interpretation is that even in this period, the UK was already hard-pressed in terms of power prices (in Table 1 it tops the rank with the highest average price, 53.0 EUR/MWh, followed by Italy, so the 'natural' path forward is downward). Yet, in the second period all of them display negative values of α^i/κ^i. The lowest ones correspond to the UK (−7.6989) and France (−4.7273). In the opposite extreme, we find Spain (−0.4108) and Portugal (−0.4869). The biggest drops take place in the UK and France, and the smallest ones in Spain and Portugal.

Table 8. Stochastic part S_t^i: parameter estimates of the mean-reverting process.

Country	Parameter	Period 1		Period 2	
		Value	95% Confidence Int.	Value	95% Confidence Int.
Spain	α^i	576.92	(−82.96, 1236.79)	−302.94	(−2145.33, 1539.46)
	k^i	900.05	(789.28, 1010.81)	737.42	(653.18, 821.65)
	σ^i	187.94	(186.67, 189.25)	362.41	(360.48, 364.36)
France	α^i	722.81	(−57.06, 1502.68)	−2668.79	(−5643.07, 305.49)
	k^i	1336.16	(1216.69, 1455.64)	564.55	(485.44, 643.64)
	σ^i	201.44	(200.07, 202.83)	480.17	(477.63, 482.76)
Portugal	α^i	551.88	(−87.67, 1205.88)	−354.12	(−2177.74, 1469.51)
	k^i	872.89	(762.38, 983.4)	727.29	(645.87, 808.71)
	σ^i	186.31	(185.04, 187.6)	360.71	(358.8, 362.66)
UK	α^i	−2916.67	(−4275.61, −1557.73)	−6064.53	(−9442.32, −2686.73)
	k^i	1943.31	(1818.22, 2068.39)	787.72	(706.02, 869.42)
	σ^i	260.43	(258.66, 262.23)	489.85	(487.25, 492.49)
Italy	α^i	−55.61	(−736.76, 625.54)	−1910.29	(−4408.36, 587.77)
	k^i	1570.10	(1457.43, 1682.8)	542.41	(469.32, 615.5)
	σ^i	192.30	(190.99, 193.63)	442.67	(440.32, 445.04)
Germany	α^i	1657.33	(712.19, 2602.47)	−1671.86	(−4724.35, 1380.61)
	k^i	1431.94	(1302.63, 1561.25)	507.31	(436.22, 578.39)
	σ^i	219.36	(217.88, 220.88)	480.02	(477.48, 482.61)
Belgium	α^i	1230.48	(281.95, 2179.01)	−2221.93	(−5894.06, 1450.2)
	k^i	1636.34	(1499.6, 1773.09)	745.24	(661.05, 829.44)
	σ^i	241.23	(239.6, 242.91)	514.47	(511.74, 517.25)
The Netherlands	α^i	431.50	(−528.79, 1391.79)	−2455.92	(−6258.76, 1346.96)
	k^i	1810.56	(1669.2, 1951.93)	853.33	(769.38, 937.28)
	σ^i	221.68	(220.18, 223.22)	518.87	(516.12, 521.67)

Regarding the reversion speed, κ^i falls the most in Italy (65.45%) and Germany (64.57%). It falls the least in Portugal (16.68%) and Spain (18.07%). As before, the volatility of the mean-reverting process (σ^i) increases significantly in the crisis period. It rises by 138.37% in France and 134.06% in The Netherlands. Instead, it does so only by 88.09% in the UK and 92.83% in Spain.

5. Discussion

Looking at hourly prices in both sub-periods, we found that price averages and volatilities rise noticeably from the first one to the second (with the increases comparatively weaker in the Iberian Peninsula). Skewness is also reinforced in the crisis period (except in the UK, where it reaches the highest levels). The same applies to kurtosis, which increases everywhere (except in the UK), well above 3.0 in every market. On the other hand, correlation analysis shows a fragmented market. In the first period, correlation between the five mainland countries is higher than 75%. It falls to around 50% between them and the UK, and somewhere between 62% and 74% between them and Portugal and Spain (which display an almost perfect match, 99.8%). In the second period, we identify two distinct dynamics: the correlation between core countries increases above 85%, and above 70% between them and the UK; instead, between core countries and the two Iberian countries, it falls to somewhere between 38% and 45% (which suggests de-coupling with the mainland). To some extent, these results resemble those in [19]: they too find strong similarities among most markets, while the Spanish and Portuguese electricity exchanges exhibit the weakest similarity to the broader European market.

Then, we estimated the model with hourly data from the above countries over the two periods. Regarding the deterministic component, we ran a linear OLS regression. In the normal period, the intercept is statistically significant everywhere, ranging between 14.00 EUR/MWh in the UK and 3.19 EUR/MWh in Spain. The linear time trend is also significant everywhere and positive (around 50). The quadratic time trend, instead, is not significant anywhere. Not surprisingly, some regularities arise in terms of seasonalities across countries. In the crisis period, there are several dramatic changes. Thus, the regression intercept remains significant but becomes negative and huge (between −1229.5 EUR/MWh in France and −345.44 EUR/MWh in Portugal). The linear time trend remains significant and positive, but jumps above 1000 in some countries. The quadratic time trend now becomes significant and negative. There are also changes in the seasonalities parameters. Common to all of them is a sizeable increase in their absolute values.

When it comes to the stochastic component, the starting point is the series of residuals of the earlier OLS regression. Its volatility grows significantly in the crisis period. As for estimation, the first step is to break the single series down into two, namely the continuous mean-reverting series and the discrete-jumps series. For this purpose we follow a recursive approach (it filters out the residuals whose absolute value exceeds three times the series volatility). The number of iterations ranges between six and nine; it changes across countries and periods.

The series of jumps lends itself to straightforward estimation of the underlying parameters. Overall, in the second period the expected number of jumps per hour decreases across all the markets (apart from the UK, where it is essentially constant). The reason may be higher price volatilities in it, which raise the threshold to qualify as a jump (namely three times volatility). In addition, in the normal period the average jump is negative in most countries, but turns to positive everywhere in the crisis period. Jump volatility also rises in all of the markets, with the UK at the top and Italy at the bottom.

Unlike the former, the mean-reverting process is estimated from the corresponding series by OLS regression. In the normal period, the long-term level of the stochastic component is positive in most of the countries. However, it switches to negative in all of them in the crisis period; the extreme values are found in the UK (−7.6989 EUR/MWh) and Spain (−0.4108 EUR/MWh). The reversion speed falls everywhere in the second period, with the highest drop taking place in Italy (65.45%) and the lowest in Portugal (16.68%). The opposite happens with volatility, which increases significantly in the second period (the most in France and The Netherlands, and the least in the UK and Spain).

In short, the road toward an integrated EU internal market for electricity is far from over. As the impacts of the COVID-19 pandemic have faded away and the economies return to 'normal', core power markets have gotten more correlated; in a sense, it has brought the EU goal closer to fulfillment. Nonetheless, regarding the Iberian Peninsula, the pandemic seems to have put it farther apart from the mainland. This unexpected setback jeopardises ongoing efforts to bridge structural isolation from the main markets.

Our findings can be adressed from different perspectives, e.g., empirical and theoretical. Thus, Ref. [7] mentions several reasons for deviations from uniform spot prices, among them limited inter-connection capacities and differences in market design (e.g., auction design, pricing rules, closing hours...). Because of these 'imperfections', market coupling can play an important role in fostering market integration. But constraints loom. For instance, Ref. [26] finds a convergence of power prices in France and Germany in 2010 and 2011 (in the wake of market coupling). Nonetheless, since 2012 they have diverged; renewable generation in Germany is so high in certain hours that inter-connectors tend to get congested. Ref. [7] considers 25 European electricity market areas and finds similar results.

According to our results, policy makers face two opposing trends (convergence/divergence) behind recent events. Fortunately, this does not necessarily mean that they must take some mutually offsetting measures. For one, increasing inter-connection capacity bodes well both for the center and the periphery. Enhanced inter-connections not only contribute to integrating markets, but to the green energy transition as well. Consistent with this view, the European Commission published the fifth list of Projects of Common Interest (PCI) in November 2021. As it turns out, the majority of PCI projects involve just four countries, namely France, Germany, Spain, and Portugal. The largest one in particular concerns the French–Spanish border; Ref. [30] assesses the economics of the French–Spanish inter-connector over the period 2020–2022. Despite these projects, even in 2030 up to eight countries will fall short of the minimum inter-connection requirement (15%), among them France, Germany, Italy, and Spain; see Ref. [31]. Thus, additional investments in inter-connectors are needed. As suggested above, progress must be made in market harmonization as well.

In addition to the above factual issues, our results are also dependent on the particular price model adopted, which has some limitations. Thus, the parameters underlying mean reversion and discrete jumps are time-independent. This precludes any seasonal behavior, which can be restrictive; for instance, jumps may be more frequent at peak hours, or in winter (this is an empirical issue). Maybe continuous changes and discrete jumps revert to their starting levels at different speeds; imposing a single speed can lead to overestimation for the former and underestimation for the latter. In a more general model, they would be time-dependent; volatility would also be stochastic and account for jumps. Our treating national price series independently (as opposed to jointly) is another limitation of the paper. These venues are left for future research.

6. Conclusions

This paper focuses on spot power prices in Western Europe during the COVID-19 pandemic and the outbreak of the Russia–Ukraine war. It aims to assess whether these events have changed the inner dynamics of these prices (and, if so, how).

We introduce a new model for the hourly spot price of electricity. It comprises a deterministic component (with a time trend plus yearly, weekly, and daily seasonalities) and a stochastic component (with both a continuous- and a discrete-time process).

Our sample comprises Belgium, France, Germany, Italy, The Netherlands, Portugal, Spain, and the UK. Drawing on publicly available market prices, within the sample period we identify a 'normal' or 'pre-crisis' period (April 2020–May 2021) and an (energy prices) 'crisis' period (June 2021–May 2023).

A cursory look at hourly prices shows that all of the usual descriptive statistics (price averages, volatilities, skewness, and kurtosis) increase noticeably in the second period (relative to the first one). On the other hand, in the second period the correlation between core countries rises, while it falls between them and the two Iberian countries (which suggests de-coupling). In short, the pandemic and the war seem to have had two opposing impacts on the bumpy road toward an integrated EU internal market for electricity; while core power markets are getting more correlated (thus bringing the EU goal closer to fulfillment), the Iberian Peninsula is farther apart from the mainland.

Regarding estimation of the price model, we start from the deterministic component. In the first period, we find a positive linear time trend which, in the second one, shows up accompanied by a negative quadratic time trend. Not surprisingly, some regularities arise in terms of seasonalities across countries. In the crisis period, all of the seasonality parameters undergo a sizeable increase in their absolute values. Concerning the stochastic component, in the second period the expected number of price jumps per hour decreases everywhere, the average jump turns from negative to positive, and jump volatility rises. On the other hand, the long-term price level switches from positive to negative everywhere, the reversion speed falls, and volatility increases.

Our findings are partly owing to market 'imperfections', which call for further market harmonization. Increasing inter-connection capacity is another venue for improvement. Be that as it may, both private and public agents have prominent roles to play.

Supplementary Materials: The following supporting information can be downloaded at: https://www.mdpi.com/article/10.3390/en17143420/s1, Table S1: Spain deterministic parameters during normal period. Table S2: France deterministic parameters during normal period. Table S3: Portugal deterministic parameters during normal period. Table S4: U.K. deterministic parameters during normal period. Table S5: Italy deterministic parameters during normal period. Table S6: Germany deterministic parameters during normal period. Table S7: Belgium deterministic parameters during normal period. Table S8: Netherlands deterministic parameters during normal period. Table S9: Spain deterministic parameters during the energy crisis period. Table S10: France deterministic parameters during the energy crisis period. Table S11: Portugal deterministic parameters during the energy crisis period. Table S12: U.K. deterministic parameters during the energy crisis period. Table S13: Italy deterministic parameters during the energy crisis period. Table S14: Germany deterministic parameters during the energy crisis period. Table S15: Belgium deterministic parameters during the energy crisis period. Table S16: Netherlands deterministic parameters during the energy crisis period.

Author Contributions: Conceptualization, L.M.A. and J.M.C.; Methodology, L.M.A.; Software, L.M.A.; Validation, L.M.A.; Formal analysis, L.M.A. and J.M.C.; Investigation, L.M.A. and J.M.C.; Resources, L.M.A. and J.M.C.; Data curation, L.M.A.; Writing—original draft, J.M.C.; Writing—review & editing, J.M.C.; Visualization, L.M.A.; Supervision, J.M.C.; Project administration, L.M.A. and J.M.C.; Funding acquisition, J.M.C. All authors have read and agreed to the published version of the manuscript.

Funding: The authors gratefully acknowledge support from the Basque Government (IT1777-22–Environmental and Resources Economics Policy Group) and the Low Carbon Programme.

Data Availability Statement: The authors have drawn on the database of the Spanish power transmission and distribution system operator (Red Eléctrica de España), which is publicly available: https:/www.esios.ree.es/ (accessed on 31 October 2023).

Acknowledgments: The authors thank two anonymous referees for their helpful remarks. The authors remain responsible for any errors.

Conflicts of Interest: Author Luis María Abadie is employed by the Metroeconomica company. Author José Manuel Chamorro declares that the research was conducted in the absence of any commercial or financial relationships that could be construed as a potential conflict of interest.

References

1. EUR-Lex. Single European Act. Available online: https://eur-lex.europa.eu/eli/treaty/sea/sign (accessed on 1 November 2023).
2. EUR-Lex. Directive 96/92/EC of the European Parliament and of the Council of 19 December 1996 Concerning Common Rules for the Internal Market in Electricity. Available online: https://eur-lex.europa.eu/legal-content/EN/TXT/?uri=CELEX:31996L0092 (accessed on 1 November 2023).
3. EUR-Lex. Directive 2003/54/EC of the European Parliament and of the Council of 26 June 2003 Concerning Common Rules for the Internal Market in Electricity and Repealing Directive 96/92/EC—Statements Made with Regard to Decommissioning and Waste Management Activities. Available online: https://eur-lex.europa.eu/legal-content/EN/TXT/?uri=celex:32003L0054 (accessed on 1 November 2023).
4. EUR-Lex. Directive 2009/72/EC of the European Parliament and of the Council of 13 July 2009 Concerning Common Rules for the Internal Market in Electricity and Repealing Directive 2003/54/EC (Text with EEA Relevance). Available online: https://eur-lex.europa.eu/legal-content/en/TXT/?uri=CELEX:32009L0072 (accessed on 1 November 2023).
5. EUR-Lex. Directive (EU) 2019/944 of the European Parliament and of the Council of 5 June 2019 on Common Rules for the Internal Market for Electricity and Amending Directive 2012/27/EU (Recast) (Text with EEA Relevance). Available online: https://eur-lex.europa.eu/legal-content/en/TXT/?uri=CELEX:32019L0944 (accessed on 1 November 2023).
6. De Menezes, L.M.; Houllier, M.A. Reassessing the integration of European electricity markets: A fractional cointegration analysis. *Energy Econ.* **2016**, *53*, 132–150. [CrossRef]
7. Gugler, K.; Haxhimusa, A.; Liebensteiner, M. Integration of European Electricity Markets: Evidence from Spot Prices. *Energy J.* **2018**, *39*, 41–66. [CrossRef]
8. Ioannidis, F.; Kosmidou, K.; Makridou, G.; Andriosopoulos, K. Market design of an energy exchange: The case of Greece. *Energy Pol.* **2019**, *133*, 110887. [CrossRef]
9. Zachmann, G. Electricity wholesale market prices in Europe: Convergence? *Energy Econ.* **2008**, *30*, 1659–1671. [CrossRef]
10. Huisman, R.; Huurman, C.; Mahieu, R. Hourly electricity prices in day-ahead markets. *Energy Econ.* **2007**, *29*, 240–248. [CrossRef]
11. Escribano, A.; Peña, J.I.; Villaplana, P. Modelling electricity prices: International evidence. *Oxf. Bull. Econ. Stat.* **2011**, *73*, 622–650. [CrossRef]

12. Gudkov, N.; Ignatieva, K. Electricity price modelling with stochastic volatility and jumps: An empirical investigation. *Energy Econ.* **2021**, *98*, 105260. [CrossRef]
13. Ioannidis, F.; Kosmidou, K.; Savva, C.; Theodossiou, P. Electricity pricing using a periodic GARCH model with conditional skewness and kurtosis components. *Energy Econ.* **2021**, *95*, 105110. [CrossRef]
14. Deschatre, T.; Féron, O.; Gruet, P. A survey of electricity spot and futures price models for risk management applications. *Energy Econ.* **2021**, *102*, 105504. [CrossRef]
15. Hauser, P.; Schönheit, D.; Scharf, H.; Anke, C.P.; Möst, D. COVID-19's impact on European power sectors: An econometric analysis. *Energies* **2021**, *14*, 1639. [CrossRef]
16. Navon, A.; Machlev, R.; Carmon, D.; Onile, A.E.; Belikov, J.; Levron, Y. Effects of the Covid-19 pandemic on energy systems and electric power grids: A review of the challenges ahead. *Energies* **2021**, *14*, 1056. [CrossRef]
17. Bompard, E.; Mosca, C.; Colella, P.; Antonopoulos, G.; Fulli, G.; Masera, M.; Poncela-Blanco, M.; Vitiello, S. The immediate impacts of COVID-19 on European electricity systems: A first assessment and lessons learned. *Energies* **2021**, *14*, 96. [CrossRef]
18. Kapral, K.; Soetaert, K.; Castro, R. An off-site power purchase agreement (PPA) as a tool to protect against electricity price spikes: Developing a framework for risk assessment and mitigation. *Energies* **2024**, *17*, 2161. [CrossRef]
19. Szeberényi, A.; Bakó, F. Electricity market dynamics and regional interdependence in the face of pandemic restrictions and the Russian-Ukrainian conflict. *Energies* **2023**, *16*, 6515. [CrossRef]
20. World Health Organization. Coronavirus Disease (COVID-19) Pandemic. Available online: https://www.who.int/europe/emergencies/situations/covid-19 (accessed on 1 June 2024).
21. Bosco, B.; Parisio, L.; Pelagatti, M.; Baldi, F. Long-run relations in European electricity markets. *J. Appl. Econom.* **2010**, *25*, 805–832. [CrossRef]
22. Nepal, R.; Jamasb, T. Interconnections and market integration in the Irish Single Electricity Market. *Energy Policy* **2012**, *51*, 425–434. [CrossRef]
23. Christensen, T.S.; Benth, F.E. Modelling the joint behaviour of electricity prices in interconnected markets. *Quant. Financ.* **2020**, *20*, 1441–1456. [CrossRef]
24. Çanakoğlu, E.; Adiyeke, E. Comparison of electricity spot price modelling and risk management applications. *Energies* **2020**, *13*, 4698. [CrossRef]
25. Engel, C.; Rogers, J.H. European product market integration after the Euro. *Econ. Policy* **2004**, *19*, 347–384. [CrossRef]
26. Keles, D.; Dehler-Holland, J.; Densing, M.; Panos, E.; Hack, F. Cross-border effects in interconnected electricity markets—An analysis of the Swiss electricity prices. *Energy Econ.* **2020**, *90*, 104802. [CrossRef]
27. Mas Urquijo, I.; Paraschiv, F. Cross-border Effects between the Spanish and French Electricity Markets: Asymmetric Dynamics and Benefits in the Light of European Market Integration. *Energy J.* **2023**, *44*, 241–276. [CrossRef]
28. REE. Available online: https://www.sistemaelectrico-ree.es/2021/informe-del-sistema-electrico/mercados/mercado-diario (accessed on 1 June 2024).
29. Keppler, J.H.; Phan, F.; Le Pen, Y. The impacts of variable renewable production and market coupling on the convergence of French and German electricity prices. *Energy J.* **2016**, *37*, 343–359. [CrossRef]
30. Abadie, L.M.; Chamorro, J.M. Evaluation of a cross-border electricity interconnection: The case of Spain-France. *Energy* **2021**, *233*, 121177. [CrossRef]
31. Mezősi, A.; Kácsor, E.; Diallo, A. Projects of common interest? Evaluation of European electricity interconnectors. *Util. Policy* **2023**, *84*, 101642. [CrossRef]

energies

MDPI

Article

Spatiotemporal Downscaling Model for Solar Irradiance Forecast Using Nearest-Neighbor Random Forest and Gaussian Process

Shadrack T. Asiedu [1,†], Abhilasha Suvedi [1,†], Zongjie Wang [2], Hossein Moradi Rekabdarkolaee [3] and Timothy M. Hansen[1,*]

1 McComish Department of Electrical Engineering and Computer Science, South Dakota State University, Brookings, SD 57007, USA; shadrack.asiedu@jacks.sdstate.edu (S.T.A.); abhilasha.suvedi@jacks.sdstate.edu (A.S.)
2 Department of Electrical and Computer Engineering, University of Connecticut, Storrs, CT 06268, USA; zongjie.wang@uconn.edu
3 Department of Mathematics and Statistics, South Dakota State University, Brookings, SD 57007, USA; hossein.moradirekabdarkolaee@sdstate.edu
* Correspondence: timothy.hansen@sdstate.edu
† These authors contributed equally to this work.

Abstract: Accurate solar photovoltaic (PV) capacity estimation requires high-resolution, site-specific solar irradiance data to account for localized variability. However, global datasets, such as the National Solar Radiation Database (NSRDB), provide regional averages that fail to capture the fine-scale fluctuations critical for large-scale grid integration. This limitation is particularly relevant in the context of increasing distributed energy resources (DERs) penetration, such as rooftop PV. Additionally, it is critical to the implementation of the U.S. Federal Energy Regulatory Commission (FERC) Order 2222, which facilitates DER participation in U.S. bulk power markets. To address this challenge, this study evaluates Nearest-Neighbor Random Forest (NNRF) and Nearest-Neighbor Gaussian Process (NNGP) models for spatiotemporal downscaling of global solar irradiance data. By leveraging historical irradiance and meteorological data, these models incorporate spatial, temporal, and feature-based correlations to enhance local irradiance predictions. The NNRF model, a machine-learning approach, prioritizes computational efficiency and predictive accuracy, while the NNGP model offers a level of interpretability and prediction uncertainty by numerically quantifying correlations and dependencies in the data. Model validation was conducted using day-ahead predictions. The results showed that the average Goodness of Fit (GoF) of the NNRF model of 90.61% across all eight sites outperformed the GoF of the NNGP of 85.88%. Additionally, the computational speed of NNRF was 2.5 times faster than the NNGP. Finally, the NNGP displayed polynomial scaling while the NNRF scaled linearly with increasing number of nearest neighbors. Additional validation of the model on five sites in Puerto Rico further confirmed the superiority of the NNRF model over the NNGP model. These findings highlight the robustness and computational efficiency of NNRF for large-scale solar irradiance downscaling, making it a strong candidate for improving PV capacity estimation and real-time electricity market integration for DERs.

Keywords: distributed energy resources; FERC 2222; nearest-neighbor Gaussian process; nearest-neighbor random forest; solar irradiance forecasting; spatiotemporal downscaling

1. Introduction

The global push for decarbonization has driven regulatory developments such as the U.S. Federal Energy Regulatory Commission (FERC) Order 2222 [1], which enables the participation of distributed energy resources (DERs), including rooftop solar photovoltaic (PV) systems, in wholesale electricity markets. This regulatory shift is accelerating the integration of renewable energy resources and reshaping both power system operations and market dynamics [2]. However, integrating variable renewable energy resources (VREs) like solar PV presents significant challenges for grid stability, reliability, and dispatch [3]. Unlike conventional generators, VREs exhibit volatile output due to rapid fluctuations in cloud cover, humidity, and other localized weather conditions. As a result, accurate and high-resolution solar irradiance forecasting is essential for managing DERs effectively within the power grid. However, existing global irradiance models typically operate at coarse spatial and temporal resolutions, limiting their utility for site-specific forecasting and real-time grid operations.

Advances in power electronic converter (PEC) technology, such as improved dynamic models [4] and virtual inertia solutions [5], address some stability concerns. However, accurate solar PV forecasting remains a critical challenge for grid operators and market participants. Existing approaches for solar PV forecasting primarily rely on numerical weather prediction (NWP) models [6,7], statistical methods and machine-learning techniques [8–11], deep learning algorithms [12,13], and hybrid approaches [9,14,15]. Some studies incorporate probabilistic methods to quantify uncertainty [16–18], while others consider the spatial influence of neighboring sites on a target location [12,19]. Although these methods have demonstrated success, many rely on an implicit assumption that global solar irradiance datasets provide accurate site-specific representations. This assumption introduces inaccuracies because global datasets, such as those from the National Solar Radiation Database (NSRDB) [20], provide irradiance values averaged over large spatial scales (e.g., 4 km × 4 km), which fail to capture fine-grained local variability at individual DERs.

A key challenge in solar PV forecasting is the need for high-resolution, site-specific solar irradiance data. Current datasets from sources like NSRDB and Open Meteo [21] cover broad spatial regions, with some reaching resolutions as coarse as 11 km × 11 km. This leads to approximation errors that can undermine accurate capacity estimation and power system operations such as dispatch and balancing. While the ideal solution would involve deploying weather stations at every prospective PV site to obtain direct irradiance measurements, this approach is impractical due to cost, maintenance, and logistical constraints. Furthermore, it would be infeasible for forecasting irradiance at prospective solar PV sites where physical measurements are not yet available. Therefore, there is a pressing need for advanced downscaling techniques that can transform coarse global irradiance data into high-resolution, site-specific estimates suitable for real-time market operations and DER integration.

By leveraging historical data on both global and local solar irradiance measurements, models can be developed to learn the mapping of the global values to the local values. Additionally, by incorporating spatial correlations, the dependencies of neighboring sites could be captured to infer forecast values for sites whose local solar irradiance measurements are unknown. As promising as downscaling is for power system operations, there is a dearth of literature on this method. Most solar irradiance downscaling studies focused on generating high-resolution time-scale measurements from coarse-grained solar irradiance forecast values [22–26]. Some of the studies that focused on spatial downscaling include [27], where coarse resolution downward shortwave radiation is disaggregated into a 30 m scale using an atmospheric transmittance-based weighting technique. Although the proposed model achieves reliable downscaling results, especially for mountainous areas, the model heavily depends on a satellite-derived dataset. Similarly, artificial neural networks (ANNs) were used to downscale weather variables from a 1.2 km spatial resolution to a 240 m resolution [28]. The study, however, used a simulated dataset generated from the Weather Research and Forecasting (WRF) model in Large Eddy Simulation mode with a 240 grid, which may bring into question the practical implementation of their model. The study in [29] adopted the nearest-neighbor Gaussian process (NNGP) to downscale solar irradiance from global resolution to a more fine-grained local resolution. However, the model in [29] performed interpolation instead of extrapolation (i.e., forecasting) to obtain day-ahead predictions. By evaluating the NNGP model on a temporal point within the training temporal space, the study failed to validate the capability of the NNGP to accurately forecast downscaled irradiance for time points beyond the training temporal space.

The main contribution of this paper is the development of an accurate, scalable spatiotemporal downscaling model for day-ahead solar irradiance forecast. Our study designs a novel approach to spatiotemporal downscaling of solar irradiance for modeling and forecasting using a Nearest-Neighbor Random Forest (NNRF). The NNRF approach is compared to the performance of NNGP updated from [29] to properly perform forecasting.

In the rest of the paper, the methodology is presented in Section 2, which describes the NNGP and NNRF spatiotemporal models. The simulation setup and data collection are presented in Section 3. Section 4 of this paper discusses the results. Finally, conclusions are made in Section 5, alongside future research directions. Additionally, the data preprocessing steps are shown in Appendix A.

2. Downscaling Theory and Methods

2.1. Spatiotemporal Models

Time series data, used in many forecasts, contains information about *when* each data element was observed. Spatial data, on the other hand, reveals information about *where* each data element was collected [30]. Spatiotemporal forecast models, combining these two dependencies, predict target variables by analyzing the dataset in both space and time dimensions [31]. The combination of temporal and spatial features enables the model to capture both the influence of nearby locations and past observations on the prediction outcome. Such integration of spatial components into temporal forecasts is demonstrated in the literature to improve forecast accuracy [32]. Mathematically, a spatiotemporal model for predicting a variable y can be expressed as:

$$\mathbf{y}_{\mathbf{i},\mathbf{t}} = f(\mathbf{y}_{i,t-1}, \ldots, \mathbf{y}_{i,t-n}, \mathbf{X}_{i,t}, \mathbf{Y}_{\text{neighbors}}, \epsilon_{i,t}), \tag{1}$$

where, at time t and location i, $\boldsymbol{y}_{i,t} \in R^n$ is the target variable, $y_{i,t-1}, \ldots, y_{i,t-n}$ represents n past values (i.e., temporal dependency), $\mathbf{X}_{i,t}$ are exogenous variables (e.g., weather features in the case of solar irradiance forecast), $\mathbf{Y}_{\text{neighbors}}$ are target variable values from neighboring locations (i.e., spatial dependency), and $\epsilon_{i,t} \in R^n$ is the model error term. The spatiotemporal model is then represented as the function f (e.g., statistical, machine learning, deep learning).

2.2. Methods for Downscaling Solar Irradiance

Global solar irradiation data, like those provided by the NSRDB, cover a 4 km by 4 km spatial resolution and mostly in a coarse temporal resolution [33]. The low resolution of the NSRDB poses two significant challenges to the accurate estimation of solar PV output. First, the 30 min time resolution is not suitable for real-time dispatch of solar PV, where output forecasts are needed on a shorter time scale (e.g., 5 min for real-time electricity markets). Second, due to the erratic variability of weather, the averaged solar irradiance over the 4 km by 4 km space does not reflect the individual site-specific measurements, which can contain many different solar PV installations in the 16 km^2 area. Figure 1 shows that the inherent variation between the global solar irradiance and the ground measurements (i.e., the local solar irradiance) can range from very small, negligible deviations to significantly high disparities in a day. Such significant margins could misinform the forecast of solar PV output and lead to inefficient market dispatch or, in the worst case, grid failure and blackouts. These spatiotemporal resolution issues necessitate the development of accurate models to map global irradiance to site-specific solar irradiance for better PV capacity estimation and electricity market integration.

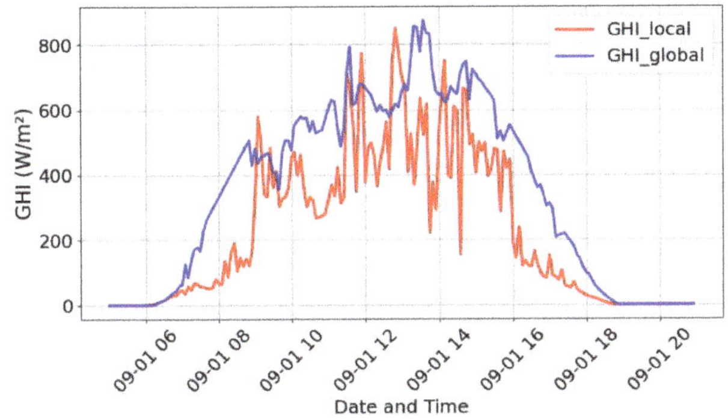

Figure 1. Daily Global and Local Solar Irradiance.

To better estimate the ground measurements, the Gaussian copula is used to downscale global to local global horizontal irradiance (GHI) clear sky indices in [24]. The same method is used in [23] to spatiotemporally downscale solar irradiance from hourly resolution to a 15 s resolution. This method models dependencies between variables with potentially different marginal distributions by transforming them into a common Gaussian space using a copula function. The study in [24] is improved by adopting the T-copula method to downscale the GHI clear sky index in [25]. This improved method offered the advantage of better simulation of binary events caused by the movement of clouds of adjustable frequency. Furthermore, a nearest-neighbor (NN) approach is combined with a Gaussian process (GP) in [29] to downscale solar irradiance from the global level to local resolution. The GP simplifies the Gaussian copula method by assuming the marginal distributions of the variables within the dataset to be uniform or Gaussian. Such an assumption makes the GP model very suitable for a sparse dataset. However, this process still involves the inversion of large covariance matrices, making its application computationally intensive. Based on the concept of the NNGP, our proposed NNRF resolves this challenge by replacing the GP with a random forest (RF) model, eliminating the need to invert a large covariance matrix. Instead, an ensemble of decision trees is used to capture the temporal and spatial relationship between neighboring sites and past observations. We will next describe the underlying theory and application of NNGP and NNRF to spatiotemporal downscaling.

2.3. *Nearest-Neighbor Gaussian Process (NNGP)*

NNGP is an efficient approximation of the traditional GP, designed for large-scale datasets. Traditional GPs involve a covariance matrix of size $n \times n$, where n is the number of training points. This leads to computational complexities of $\mathcal{O}(n^3)$ for matrix inversion and $\mathcal{O}(n^2)$ for storage. NNGP mitigates this limitation using only the k-nearest neighbors for each data point, resulting in a sparse precision matrix. This approximation reduces the computational complexity from $\mathcal{O}(n^3)$ to $\mathcal{O}(k^3 n)$, while preserving local dependencies. As k is a constant parameter, the scaling of NNGP becomes linear with n.

For a random process $f(x)$, a GP is defined as:

$$f(x) \sim \mathcal{GP}(m(x), k(x, x')), \tag{2}$$

where $m(x)$ is the mean function, and $k(x, x')$ is the covariance kernel function that models the similarity between points x and x'. A common kernel is the squared exponential kernel [34,35], given as:

$$k(x, x') = \sigma^2 \exp\left(-\frac{\|x - x'\|^2}{2\ell^2}\right), \tag{3}$$

where σ^2 is the variance and ℓ is the length scale controlling smoothness. NNGP approximates the full GP by considering only the k nearest neighbors (k-NNs) for each data point. The covariance matrix is sparsified as:

$$K_{\mathrm{NN}}(x_i, x_j) = \begin{cases} k(x_i, x_j), & \text{if } x_j \in \mathcal{N}_i, \\ 0, & \text{otherwise,} \end{cases} \tag{4}$$

where $\mathcal{N}_i$ denotes the set of k-NNs of x_i. This informs the computation of the covariance matrices, which incorporate spatial, temporal, and feature-based distances. Thus:

$$k(x, x') = \sigma^2 \exp\left(-\frac{d_s^2}{2\ell_s^2}\right) \exp\left(-\frac{d_t^2}{2\ell_t^2}\right) \exp\left(-\frac{d_g^2}{2\ell_g^2}\right), \tag{5}$$

where

- $d_s = \|x_{\text{spatial}} - x'_{\text{spatial}}\|$ is the spatial distance,
- $d_t = |t - t'|$ is the temporal distance, and
- $d_g = \|x_{\text{global features}} - x'_{\text{global features}}\|$ is the feature-based distance.

Using the GP kernel in (5), we compute the covariance matrix between the test point and its neighbors ($K_{\text{test,train}}$), as well as the covariance matrix among the neighboring points themselves ($K_{\text{train,train}}$). The final downscaled prediction is obtained as:

$$\hat{\mathbf{y}} = \mathbf{X}_{\text{test}}^T \hat{\beta} + \mathbf{K}_{\text{test, train}} \mathbf{K}_{\text{train, train}}^{-1} \left(\mathbf{y}_{\text{train}} - \mathbf{X}_{\text{train}}^T \hat{\beta}\right), \tag{6}$$

where

$$\hat{\beta} = \left(\mathbf{X}^T \mathbf{X}\right)^{-1} \mathbf{X}^T \mathbf{y}. \tag{7}$$

In the above equation, X_{test} represents the input features of the target site, while X_{train} denotes the input features of the neighboring sites. The matrix $K_{\text{test, train}}$ is the covariance matrix between the test site and its neighbors, and $K_{\text{train, train}}$ is the covariance matrix among the neighboring sites. The vector y_{train} contains the target values of the neighboring sites. More generally, X and y denote the input features and target values for all sites, respectively, and $\hat{\beta}$ is the estimated coefficient vector obtained from the linear regression model. Finally, $\hat{y}$ is the predicted downscaled solar irradiance of the target site. The pseudo algorithm for the implementation of this model is described in Algorithm 1.

Algorithm 1 Pseudo Algorithm for Nearest-Neighbor Gaussian Process (NNGP).

Require:
- Training data $\mathcal{D} = \{(\mathbf{x}_i, y_i)\}_{i=1}^{n}$ with spatial coordinates $\mathbf{s}_i$ and global features $\mathbf{f}_i$
- Hyperparameters σ, ℓ_s, ℓ_t, ℓ_g (spatial, temporal, and feature length scales, and signal variance)
- Regularization term ϵ and number of neighbors k

Ensure: Predictions $\hat{y}$ for test points $\mathbf{x}_*$

Step 1: Data Preprocessing

Construct input features $\mathbf{X}$ for training points:

$$\mathbf{X}_i \leftarrow \{t_i, \mathbf{s}_i, \mathbf{f}_i\} \quad \forall i \in \{1, 2, \ldots, n\}$$

Step 2: Kernel Function

Define the combined kernel function:

$$K(\mathbf{x}_i, \mathbf{x}_j) = \sigma^2 \exp\left(-\frac{d_s^2}{2\ell_s^2}\right) \exp\left(-\frac{d_t^2}{2\ell_t^2}\right) \exp\left(-\frac{d_g^2}{2\ell_g^2}\right),$$

where:
- $d_s = \|\mathbf{s}_i - \mathbf{s}_j\|$ is the spatial distance
- $d_t = |t_i - t_j|$ is the temporal distance
- $d_g = \|\mathbf{f}_i - \mathbf{f}_j\|$ is the feature distance

Step 3: Hyperparameter Selection

Select hyperparameters that minimize the validation Mean Absolute Error:

$$\hat{\ell}_s, \hat{\ell}_t, \hat{\ell}_g, \hat{\sigma} = \arg \min_{\ell_s, \ell_t, \ell_g, \sigma} \mathrm{MAE}_{\mathrm{val}}$$

Step 4: Nearest-Neighbor Selection

for each test point $\mathbf{x}_*$ **do**

  Compute spatial distances d_s, temporal distances d_t, and feature distances d_g between $\mathbf{x}_*$ and all training points

  Identify k-NN based on combined distances

end for

Step 5: Covariance Computation

Compute the covariance matrix among neighbors for training points:

$$\mathbf{K}_{\text{train, train}} = K(\mathbf{x}_i, \mathbf{x}_j) + \epsilon \mathbf{I},$$

where:
- $\epsilon \mathbf{I}$ is a penalty term to prevent singularity of the covariance matrix

Compute the covariance vector between the test point and neighbors:

$$\mathbf{K}_{\text{test, train}} = K(\mathbf{x}_*, \mathbf{x}_j) \quad \forall j \in \text{neighbors}$$

Step 6: Prediction

Estimate the coefficients of the linear regression model

$$\hat{\beta} = \left(\mathbf{X}^T \mathbf{X}\right)^{-1} \mathbf{X}^T \mathbf{y}$$

Predict the output for the test point:

$$\hat{\mathbf{y}} = \mathbf{X}_{\text{test}}^T \hat{\beta} + \mathbf{K}_{\text{test, train}} \mathbf{K}_{\text{train, train}}^{-1} \left(\mathbf{y}_{\text{train}} - \mathbf{X}_{\text{train}}^T \hat{\beta}\right)$$

Step 7: Evaluation

Calculate metrics such as Mean Absolute Error (MAE), Root Mean Squared Error (RMSE), and Goodness of Fit (GoF) for predictions

Step 8: Output

Return trained NNGP predictions for each site and evaluation results

2.4. Nearest-Neighbor Random Forest (NNRF)

The RF algorithm is formed by bagging random samples of decision trees to produce an ensemble model that improves the predictions of each tree [36]. RFs are well known for their ability to learn complex non-linear patterns while efficiently handling high-dimensional features and large datasets [37,38]. This makes RF well-suited for scalability as the number of neighbors and data points increases.

An RF model minimizes the Mean Squared Error (MSE) of the ensemble decision trees over the training points such that:

$$\mathrm{MSE} = \frac{1}{N} \sum_{i \in \mathcal{N}_s} \left(y_i - \frac{1}{T} \sum_{t=1}^{T} h_t(\mathbf{f}_i) \right)^2, \tag{8}$$

where T is the total number of trees in the forest and $h_t(\mathbf{f}_i)$ is the prediction from the t-th tree for input $\mathbf{f}_i$. Each decision tree, h_t, is trained by recursively splitting the training data to minimize the *impurity*. A common type of impurity is the entropy, which quantifies the amount of uncertainty or information disorder at each splitting node. The RF model seeks to minimize the average entropy of the decision tree as:

$$\min H(S) = -\sum_{i} P_i(S) log_2 P_i(S), \tag{9}$$

where S represents the set of samples at a particular node in the decision tree, H is the entropy of S, and P_i measures the probability of class i in sample S.

For a given test point $\mathbf{X}_{\text{test}}$, the RF model predicts the normalized local irradiance as:

$$\hat{\mathbf{y}}_{\text{test}} = \frac{1}{T} \sum_{t=1}^{T} h_t(\mathbf{X}_{\text{test}}), \tag{10}$$

where $h_t(\mathbf{X}_{\text{test}})$ is the prediction from the t-th tree for test input $\mathbf{X}_{\text{test}}$.

The NNRF approach enhances traditional RF by incorporating an NN framework, ensuring both spatial and temporal locality in predictions. NNs are selected based on spatial coordinates, allowing the model to capture local dependencies more effectively while maintaining computational efficiency. For a target site s, the k-NNs are determined using the Euclidean distance [39]:

$$d_{\text{spatial}}(s, n) = \|\mathbf{x}_s - \mathbf{x}_n\|_2, \tag{11}$$

where $\mathbf{x}_s$ and $\mathbf{x}_n$ are the spatial coordinates of the target site s and the n-th neighbor, respectively. The k-NNs are stored in the set:

$$\mathcal{N}_s = \{n_1, n_2, \dots, n_k\}.$$

For a target site s with spatial coordinates $\mathbf{x}_s$, the k-NNs are identified as:

$$\mathcal{N}_s = \arg\min_{\mathcal{S}'} \sum_{i \in \mathcal{S}'} \|\mathbf{x}_s - \mathbf{x}_i\|_2, \tag{12}$$

where $|\mathcal{S}'| = k$ and $\mathcal{S}$ is the set of all sites. For each site s, the training data consists of its feature vectors in addition to the global irradiance of the neighboring sites. The target variable for each site is the local irradiance of that particular test site. The pseudo algorithm for the implementation of NNRF is described in Algorithm 2.

Algorithm 2 Pseudo Algorithm for Nearest-Neighbor Random Forest (NNRF).

Require: Training data $\mathcal{D} = \{(\mathbf{x}_i, y_i)\}_{i=1}^{n}$, spatial coordinates $\mathbf{s}_i$, number of neighbors k, features $\mathcal{F}$, target variable y
Ensure: Trained random forest model for predicting y

Step 1: Preprocessing
Combine data from all sites into a single dataset $\mathcal{D}$
Normalize feature values using MinMaxScaler
Step 2: Nearest-Neighbor Identification
for each site s in spatial coordinates $\mathbf{s}_i$ **do**
    Compute spatial distances d_s between site s and all other sites
    Identify k-NNs based on d_s
end for
Step 3: Training Data Preparation
for each site s **do**
    Extract training data $\mathcal{D}_s$ for site s from $\mathcal{D}$
    for each neighbor s_{neighbor} of site s **do**
        Extract neighbor's data $\mathcal{D}_{s_{\text{neighbor}}}$
        Align neighbor's data by datetime with $\mathcal{D}_s$
        Add neighbor's GHI as a new feature: GHI_{neighbor}
    end for
end for
Step 4: Random Forest Training
for each site s **do**
    Define features $\mathcal{F}$, including GHI_{neighbor} for each neighbor
    Split data into $\mathbf{X}_{\text{train}}$ (features) and $mathbfy_{\text{train}}$ (target variable GHI_local)
    Train Random Forest Regressor on $\mathbf{X}_{\text{train}}$ and $\mathbf{y}_{\text{train}}$ with $n_{\text{estimators}}$ trees
end for
Step 5: Prediction
for each test point $\mathbf{x}_*$ at site s **do**
    Align neighbor data by datetime and add GHI_{neighbor} as features
    Predict $\hat{\mathbf{y}}$ using trained random forest model
end for
Step 6: Evaluation
Calculate metrics such as Mean Absolute Error (MAE), Root Mean Squared Error (RMSE), and Goodness of Fit (GoF) for predictions
Step 7: Output
Return trained NNRF predictions for each site and evaluation results

3. Simulation Setup

In this study, we trained the NNGP and NNRF models on datasets from eight different sites and compared their performances in terms of accuracy and computational speed. The datasets were first cleaned of outliers and missing data. Details of the preprocessing steps are provided in the appendix. The temporal features were engineered to capture the periodic patterns in the dataset. The Euclidean distances between the sites were computed to determine the nearest neighboring sites. The NNGP model was first built using the squared exponential kernel, followed by the replacement of the GP with RF to form the NNRF model. Both models were developed in Python version 3.11.7. Major packages used in the code implementation include scikit-learn, mainly for the machine-learning modeling, and geopandas for the spatial analysis of the sites' coordinates. Both models were executed on a local desktop equipped with an Intel® Core™ processor featuring 16 CPUs, each operating at a clock speed of 2.0 GHz and 16 GB of RAM. The rest of the simulation setup describes the dataset used in the study, the implementation details of NNGP and NNRF, the scalability test of the models, and the evaluation metrics.

3.1. Data Collection

Texas has significant solar energy potential and deployed solar resources, with annual average global tilt solar radiation ranging from 4.76 kWh/m^2/day to 6.58 kWh/m^2/day across the state [40]. Its abundant solar resources, combined with the unique characteristics of the isolated ERCOT grid, make it an ideal region for this study. Eight sites across Texas were selected for analysis, with their locations and coordinates listed in Table 1, and their spatial distribution illustrated in Figure 2. In this study, Sites 2 and 7 share the same global 4 km^2 grid space, while each other site has a distinct global region.

Table 1. Selected Sites and their Locations.

Name	Site	Latitude	Longitude
3-Door, Ovalo	1	32.154290	−99.81891
Camp David Dallas	2	32.776600	−96.79706
Hermitage Austin	3	30.26726	−97.74281
Kruckenland San Antonio	4	29.5213	−98.5101
Misty Morn Cedar Park	5	30.49450	−97.82257
Roof Station Houston	6	29.7554	−95.37575
Weather Toy Dallas	7	32.78709	−96.79605
Wiley, Leander	8	30.58262	−97.86712

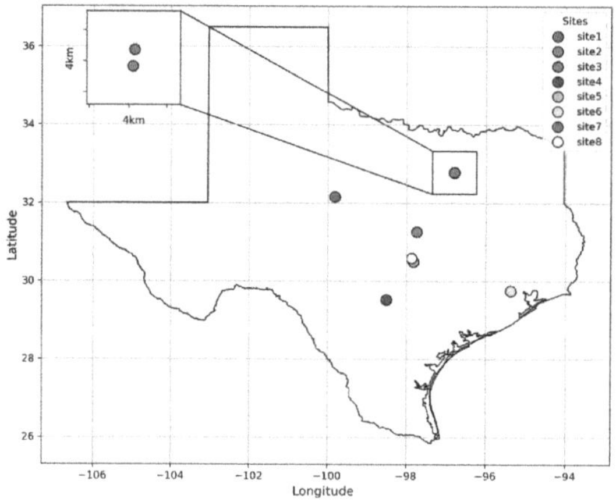

Figure 2. Spatial representation of selected sites in Texas. The inset provides a zoomed-in view of Site 2 and Site 7, located in the same 4 km by 4 km global space.

The spatial distribution of the selected sites, shown in Figure 2, indicates that most of the sites are sparsely distributed across the region. However, certain locations—specifically Sites 3, 4, 5, and 8 form a relatively close cluster, while Sites 2 and 7 share the same global space as shown by the inset plot in Figure 2. Understanding this spatial arrangement is essential for accurately capturing the influence of neighboring sites on local conditions. Table 2 presents the ground distances between the sites in km, further illustrating their relative proximity and separation.

The global dataset for each site was obtained from the NSRDB website [20], while the local dataset was sourced from Ambient Weather [41]. The global dataset consists of 5-min interval data on solar irradiance, temperature, pressure, wind speed, and dew point, spanning the period from 1 January 2022 to 31 December 2022. During the same period and with the same temporal resolution, ground-measured solar irradiance data were collected as the local irradiance dataset. The dataset for each site has 105,120 data points, enough to effectively train a machine-learning model.

Table 2. Distances between sites (in km).

Site	1	2	3	4	5	6	7	8
1	0.00	291.81	219.76	318.32	264.60	500.59	292.17	254.71
2	291.81	0.00	190.04	396.97	271.69	362.07	1.17	264.13
3	219.76	190.04	0.00	207.62	86.26	282.26	191.11	77.05
4	318.32	396.97	207.62	0.00	126.86	304.03	398.07	133.25
5	264.60	271.69	86.26	126.86	0.00	249.26	272.81	10.69
6	500.59	362.07	282.26	304.03	249.26	0.00	363.12	256.55
7	292.17	1.17	191.11	398.07	272.81	363.12	0.00	265.25
8	254.71	264.13	77.05	133.25	10.69	256.55	265.25	0.00

Figures 3 and 4 present the annual profiles of global and local irradiance, respectively, for Site 1 in 2022. The plots show significant differences between the global and local irradiance. This is confirmed by the residual plot in Figure 5, which shows the numerical deviations of the global and local irradiance at each time point for Site 1. The global irradiance, being forecast values by NSRDB, does not accurately represent the ground measurements. In the same manner, the datasets of the other seven sites have similar distribution and deviation between the global and local irradiance. This observation reveals that the use of global solar irradiance for solar PV estimation could result in inaccurate solar PV forecasting, posing risks to power dispatch reliability and even network stability. This affirms the need to downscale the forecast global irradiance to more accurate site-specific values.

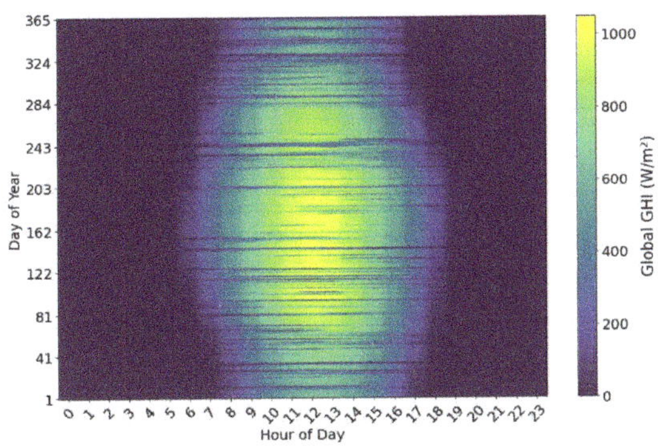

Figure 3. Heatmap of Global Irradiance Profile for Site 1 (2022).

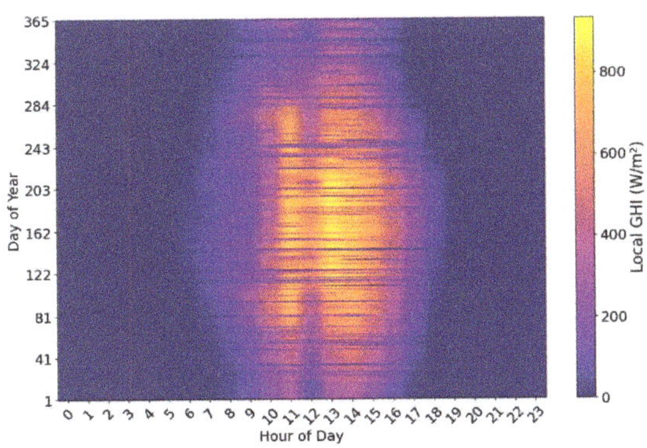

Figure 4. Heatmap of Local Irradiance Profile for Site 1 (2022).

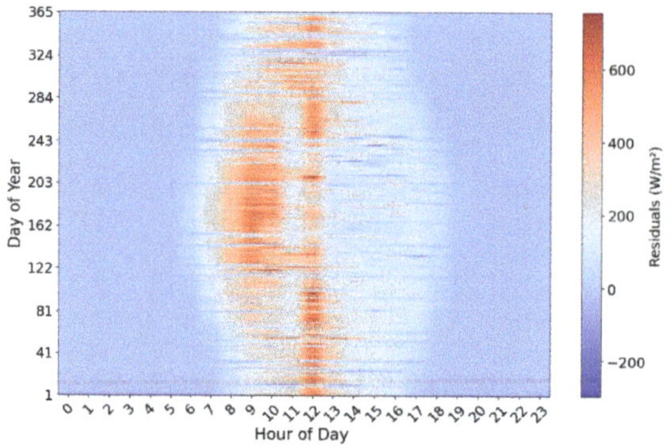

Figure 5. Heatmap of Deviation Between Global and Local Solar Irradiance for Site 1 (2022).

3.2. Feature Engineering

Solar irradiance is influenced by temporal and meteorological factors. Therefore, critical temporal features such as the hour of the day and day of the year were encoded using sine and cosine transformations to capture cyclical trends accurately. This cyclical encoding ensures continuity across periodic boundaries, such as midnight to midnight or December to January transitions. Capturing these cyclical temporal variations is crucial in reducing predictive errors, particularly during transitional periods with rapid changes in solar angles. Meteorological parameters such as temperature, atmospheric pressure, wind speed, and dew point from the global dataset were incorporated to reflect weather-dependent variability in irradiance. Integrating meteorological data provides essential contextual information that improves the model's ability to generalize under diverse weather conditions and thus enhances its robustness. The `Hour_sin` and `Hour_cos` features were computed using equations:

$$\text{Hour\_sin} = \sin\left(\frac{2\pi \cdot \text{Hour}}{24}\right), \tag{13}$$

$$\text{Hour\_cos} = \cos\left(\frac{2\pi \cdot \text{Hour}}{24}\right). \tag{14}$$

Similarly , the `DayOfYear_sin` and `DayOfYear_cos` features were computed as:

$$\text{DayOfYear\_sin} = \sin\left(\frac{2\pi \cdot \text{DayOfYear}}{365}\right), \tag{15}$$

$$\text{DayOfYear\_cos} = \cos\left(\frac{2\pi \cdot \text{DayOfYear}}{365}\right). \tag{16}$$

The input data to the models are the encoded hour and day values, temperature, pressure, wind speed, dew point, and the global solar irradiance of the target and neighboring sites. These features were scaled using scikit-learn's *MinMax Scaler*. The model was then trained to predict local solar irradiance as the output.

3.3. NNGP Implementation

The NNGP model was implemented to predict site-specific GHI values using a combination of spatial, temporal, and feature-based distances. For each site, its k-NNs were identified based on Euclidean distance in the spatial domain, with geographic coordinates serving as inputs. By focusing solely on these neighbors, the NNGP framework effectively localized the GP. This significantly reduced the dimensionality of the covariance matrices. Temporal dependencies were incorporated by considering the time difference between the current observation and historical data from neighboring sites. Temporal distances were calculated in minutes to ensure fine-grained temporal resolution suitable for solar irradiance modeling. Furthermore, feature-based distances were included to account for meteorological conditions such as temperature, pressure, wind speed, and dew point. These feature-based distances were computed in the feature-space to capture similarities and differences in environmental conditions across sites.

The squared exponential kernel in (5), used for the computation of the covariances incorporated spatial, temporal, and feature-based distances, with separate length-scale parameters (ℓ_s, ℓ_t, ℓ_g, σ) for each distance type. Larger values of these parameters implied that distance neighbors and observations had a significant influence on the prediction outcome. These parameters, including the variance (σ), were tuned to balance their influence on the predictions. The final values of the hyperparameters used for the results in this paper are presented in Table 3.

Table 3. Values of Hyperparameters for the NNGP model.

Hyperparameter	Value
`number of neighbors` (k)	3
σ	2.0
ℓ_s	0.1
ℓ_t	288 (1 day)
ℓ_g	0.5
penalty term	0.2

To enhance the predictive capacity of the model, cyclical encoding of time was applied to represent the periodic nature of hours and days of the year. The covariance matrix among training points ($K_{\text{train, train}}$) was regularized for numerical stability during inversion. Regularization was achieved by adding a small offset penalty term to the training covariance matrix to prevent singularities. Predictions were made using the GP regression formula provided in (6). This was achieved by combining the computed covariance matrices to scale the residuals between the neighbors' local GHI and their predicted values from the linear regression model computed in (7). The resulting scaled residuals, known as the Gaussian correction term, were added to the linear regression estimated values of the target site to obtain the final downscaled GHI.

3.4. NNRF Implementation

Similar to the NNGP model, the NNRF model was developed to predict site-specific GHI values by leveraging spatial, temporal, and meteorological information. For each site, the model utilized its k-NNs, determined using Euclidean distances computed from the geographic coordinates of the sites. A k value of 3 was used in this model. This nearest-neighbor framework enabled the model to incorporate local spatial relationships, ensuring that predictions were informed by the most relevant neighboring data. Temporal dependencies were integrated by incorporating the encoded Hour and Day as part of the input features. This was carried out to ensure that training data reflected consistent diurnal and seasonal patterns. Additionally, meteorological features such as temperature, pressure, wind speed, and dew point were included to capture the environmental factors influencing solar irradiance variability.

The global GHI values of neighboring sites were also included as additional features to provide spatial context for each prediction. These neighbor features were aligned temporally to ensure consistency between the target site and its neighbors. The RF model, comprising an ensemble of decision trees, was trained using these features to predict the local GHI values. The model was initialized with 150 estimators to balance accuracy and computational efficiency. The ensemble approach enabled the RF model to handle complex non-linear relationships between input features and the target variable.

3.5. Scalability Test

Additionally, we investigated how well both models scale in terms of accuracy and computational speed with varying numbers of k-NNs. We achieved this by adding seven more sites to the existing ones to make fifteen, and retrained the models on a new dataset. To ensure adequate evaluation of the model's scaling performance, the new training dataset was collected from 1 January to 30 December 2023, different from the former training dataset. Similar to earlier simulations, the dataset for 31 December 2023 was excluded to validate the accuracy of the scaled models. The k-NNs were varied from 2 to 14 for each of the sites. As a result of the increased data size, the scalability test was performed on one compute node of the South Dakota State University's Innovator's Cluster equipped with 2 Intel Xeon Gold 6342 CPUs with 48 cores. Each processor operated at a clock speed of 2.80 GHz and was supported by 256 GB of RAM.

3.6. Evaluation of Models

After the initial training of the two models on the dataset from 1 January 2022 to 30 December 2022, the models were validated by making day-ahead predictions of the downscaled local GHI for all eight sites for 31 December 2022. The evaluation of the model's performance was achieved using Mean Absolute Error (MAE) and the Goodness of Fit (GoF). The MAE was computed as:

$$\text{MAE} = \frac{1}{N}\sum_{i=1}^{n}|y_i - \hat{y}_i|, \tag{17}$$

where y_i is the actual or true value of the i-th observation, $\hat{y}_i$ is the predicted value corresponding to the i-th observation, and N denotes the sample size in the testing set. To measure how well the estimated models capture the patterns of the dataset, the GoF is computed using the normalized Root Mean Squared Error (NRMSE), such that:

$$\text{GoF}(Percent) = (1 - \text{NRMSE}) * 100, \tag{18}$$

where NRMSE is given by:

$$\text{NRMSE} = \frac{\sqrt{\frac{1}{n}\sum_{i=1}^{n}(y_i - \hat{y}_i)^2}}{y_{\max} - y_{\min}}. \tag{19}$$

The use of the NRMSE to compute the GoF offers the advantage of normalizing the error to make it dimensionless, allowing easier comparison of the model's performance across the dataset with different units. Additionally, while MAE, NRMSE, and GoF measure the effectiveness of the model, we assessed the model's efficiency by measuring the computational time, as this is required for real-time power system operations.

3.7. Case Study on Puerto Rico

To evaluate the generalizability of the proposed data-driven models, it is essential to test their performance not only on unseen data but also across distinct climate zones. To this end, we conduct an additional validation using a new dataset from Puerto Rico, which represents a markedly different climatic environment compared to Texas. Puerto Rico was chosen due to its highly variable and erratic weather patterns, providing a test of the models' robustness. Successful performance in this setting demonstrates the models' ability to adapt to diverse environmental conditions. As with the Texas simulation, comparable datasets including global and local solar irradiance, as well as meteorological variables, were collected for this analysis. The data spans from 1 February 2023 to 31 December 2023, across five geographically distributed sites in Puerto Rico. Figure 6 presents the map of the selected locations.

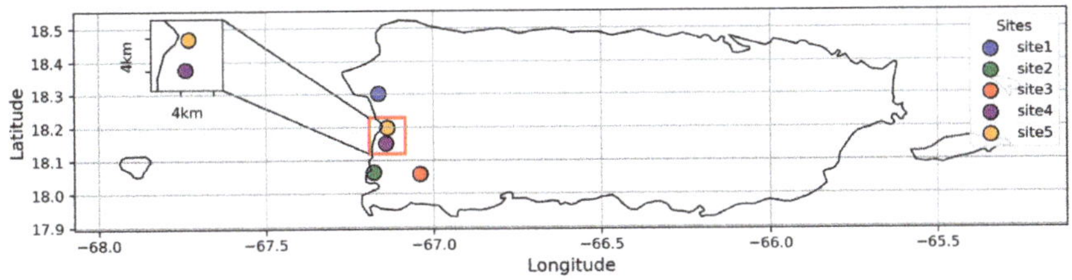

Figure 6. Spatial representation of selected sites in Puerto Rico. The inset provides a zoomed-in view of Site 4 and Site 5 located in the same 4 km by 4 km global space.

Both the NNGP and the NNRF were trained on the dataset from Puerto Rico from 1 February 2023 to 31 December 2023. Similar to the earlier simulations, predictions were made for 31 December 2023. The hyperparameters of both models were maintained, but the number of decision trees for the RF was tuned from 100 trees to 150 trees.

4. Results and Discussion

4.1. Training Performance of NNRF and NNGP

Both the NNGP and NNRF perform very well during the training phase; however, the NNRF shows slightly superior accuracy over the NNGP. The performance of NNRF is consistently higher than NNGP across all eight sites, as could be seen from Table 4. The NNRF records an average GoF of 97.63% across all sites, compared to 95.87% of the NNGP. This superior performance of the NNRF model can be attributed to RF's ability to learn non-linear patterns in the dataset, in contrast to the NNGP's inherent linearity. Furthermore, averaging the predictions of several decision trees makes the RF model robust to noise and outliers. This makes it comparatively more suitable for non-smooth data such as solar irradiance. Visual displays of the training performance of both models for Site 8 are shown in Figure 7a,b.

A careful observation of the two plots reveals how the NNRF model accurately captures the variability of the local irradiance better than the NNGP model. However, even though the RF model gives some partial interpretability in terms of the number of decision tree estimators and tree depth, the GP offers a better interpretability in terms of the correlation and influence of neighboring sites on predictions. The interpretability of NNRF can be further explored using explainable AI (xA), such as SHapley Additive exPlanations (SHAP) values [42]. The GP's hyperparameters, such as the temporal, spatial, and feature length scales, and the variance, give visibility of how far in time-space or feature-space the model looks for correlations. Similarly, the covariance matrix's structure reveals how points influence each other's predictions.

This improved interpretability of the NNGP, however, comes with increased computational cost. The NNGP took 1004.19 s to train on the eight sites, compared to 407.63 s of the NNRF model. By avoiding the computation and inversion of covariance matrices, the NNRF model speeds up computational speed by 2.5 times more than the NNGP model. While these results are promising, validation is required to confirm how well they generalize and scale with an increasing number of neighboring sites, which is discussed later in the results.

Table 4. Comparison of NNRF and NNGP Training Performance.

Site	NNRF			NNGP		
	MAE	NRMSE	GoF (%)	MAE	NRMSE	GoF (%)
1	13.6480	0.0278	97.22	19.8886	0.0479	95.21
2	8.1693	0.0178	98.22	12.7651	0.0303	96.97
3	13.5003	0.0256	97.44	20.5084	0.0446	95.54
4	11.9800	0.0271	97.29	19.0394	0.0475	95.25
5	12.0539	0.0245	97.55	20.7114	0.0441	95.59
6	13.6236	0.0250	97.50	20.4696	0.0432	95.68
7	5.9689	0.0183	98.17	9.2296	0.0316	96.84
8	12.4045	0.0235	97.65	19.6315	0.0410	95.90

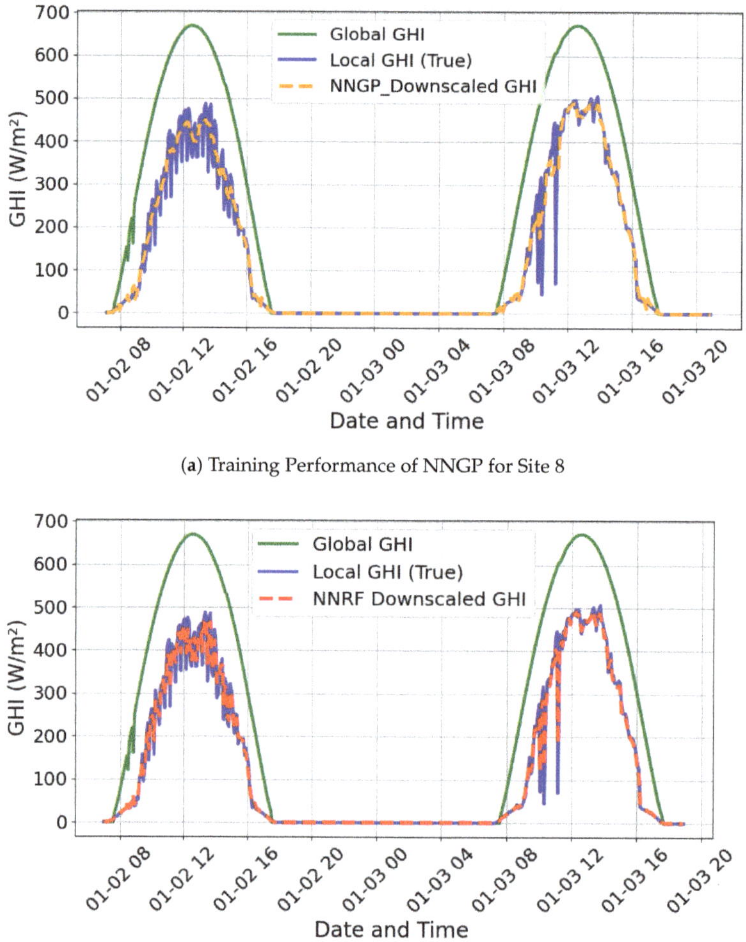

(**a**) Training Performance of NNGP for Site 8

(**b**) Training Performance of NNRF for Site 8

Figure 7. Training performance comparison for Site 8.

4.2. Validation Performance of NNRF and NNGP

The final evaluation of the NNRF was made by predicting the downscaled local solar irradiance for the next day to be used in day-ahead power system operations. The dataset for 31 December 2022, used for this validation, was excluded from the training. The features of the global data and the global solar irradiance of the selected neighboring sites were used as input variables for the predictions. The day-ahead forecast was made for each site in a 5 min time step, and compared to the actual local solar irradiance of each site. The plots of the predicted downscaled solar irradiance for Sites 1, 2, 7, and 8 are presented in Figure 8. The orange line represents the NNGP's predictions, and the red line depicts the predictions of the NNRF. Table 5 displays the model's evaluation results for each site.

Table 5. Validation Results of NNRF and NNGP for Day-Ahead Forecast.

Site	NNRF			NNGP		
	MAE	NRMSE	GoF (%)	MAE	NRMSE	GoF (%)
1	32.4447	0.1162	88.38	36.4148	0.1197	88.03
2	22.1226	0.0986	90.14	50.8843	0.2216	77.84
3	35.8097	0.1115	88.85	43.9536	0.1385	86.15
4	21.4388	0.0728	92.72	25.2055	0.0835	91.65
5	26.0342	0.1047	89.53	29.8221	0.1187	88.13
6	18.8950	0.0748	92.52	28.3873	0.1185	88.15
7	14.4113	0.1209	87.91	29.4329	0.2304	76.96
8	16.3520	0.0594	94.06	25.8150	0.0946	90.54

Once again, the NNRF performed better across all eight sites with an average GoF of 90.61%, compared to 86.31% of the NNGP model. This is confirmed by the plots in Figure 8, where the NNRF's predictions closely follow the trajectory of the local GHI better than the NNGP model. The boxplots in Figure 9 show the residuals of the actual and predicted values. While each model's residual has an approximate mean of 0, the variance of the NNGP is much larger than the NNRF (which appears in the other error metrics). Additionally, there are larger outliers predicted by NNGP, such as the −400 point in Site 1. All these observations confirm the superior performance of the NNRF model over the NNGP model.

Aside from these observations, the NNGP predictions show an interesting trend where it performed better for sites with less rapid variation in the local irradiance profile. Likewise, its performance is relatively lower for sites with rapid local irradiance variation, which deviates more from the shape of the global solar irradiance. For instance, as can be seen in Figure 8, Site 1 and Site 7, which recorded the lowest GoF, also display higher deviations from the profile of the global irradiance. This can be seen from Table 6, where they show relatively lower correlation values of 0.7635 and 0.5797, respectively. Similarly, with the exception of Site 3, it is observed that the more the shape of the local and global irradiance is highly correlated, the higher the NNRF prediction's accuracy. This implies that, generally, the alignment of the shape profiles of the global and local irradiance has a higher impact on the accuracy of prediction than the value of their variance. The MAE and the NRMSE also follow a similar pattern, where Site 7, whose profile's shape deviates most from that of the global irradiance, recorded the highest prediction error. In sharp contrast, Site 8 records the minimum prediction error due to its high correlation value of 0.9707 between the global and local GHI.

This high correlation makes the global solar irradiance a good predictor of the local solar irradiance. However, when the local irradiance differs significantly from the global profile due to localized weather conditions such as cloud cover, terrain shading, or microclimate effects, the global irradiance features are unable to fully explain these local variations. The departure of Site 3 from this trend could be the result of a lower influence of the other features, such as wind, temperature, pressure, and dew point, on the predictions.

Table 6. Correlations Between Global and Local Solar Irradiance.

Site	1	2	3	4	5	6	7	8
Correlation	0.7635	0.9346	0.9543	0.9226	0.6487	0.9686	0.5797	0.9707

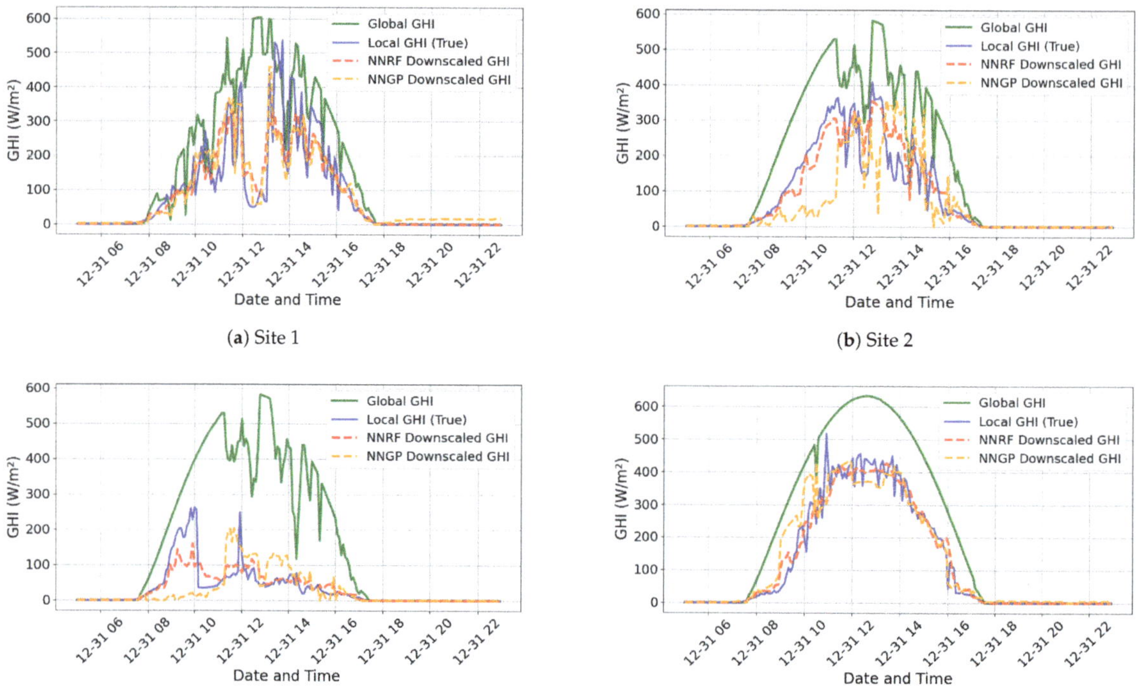

(**a**) Site 1 (**b**) Site 2

(**c**) Site 7 (**d**) Site 8

Figure 8. Day-Ahead GHI Prediction at Different Sites.

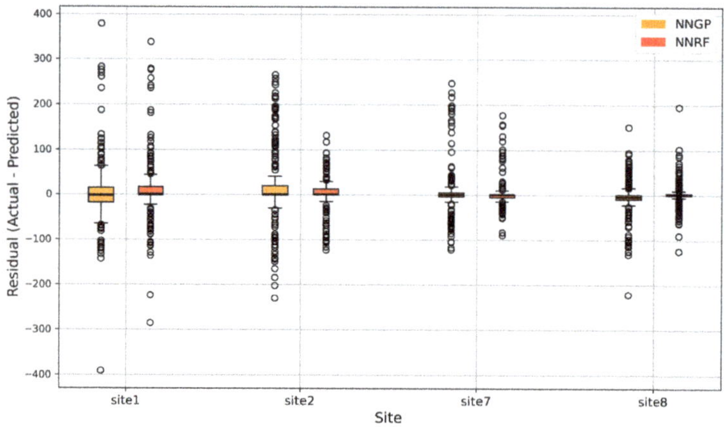

Figure 9. Boxplots of NNGP and NNRF Prediction Residuals for Sites 1, 2, 7, and 8.

While the models' performances are satisfactory, the variability in solar irradiance demands quantification of prediction uncertainty to improve grid planning. The NNRF model, though robust and more accurate, produces point estimates only. This makes it unable to give any insight into the prediction tolerance for effective planning or operation under uncertainties. The NNGP model, however, displays high strength in this regard. By using the values of the Gaussian corrections as variance, upper and lower bound prediction values could be generated by addition and subtraction with the mean values. These upper and lower bounds give some insight into the accompanying uncertainties of the predicted values. The results of the NNGP's predictions for Site 3 and Site 4 are shown in Figure 10. This makes the NNGP model very suitable for probabilistic forecasts. Based on these results, the decision to choose NNRF or NNGP does not depend solely on factors such as accuracy, computational speed, and interpretability, but also, whether point forecasts or probabilistic forecasts are desirable.

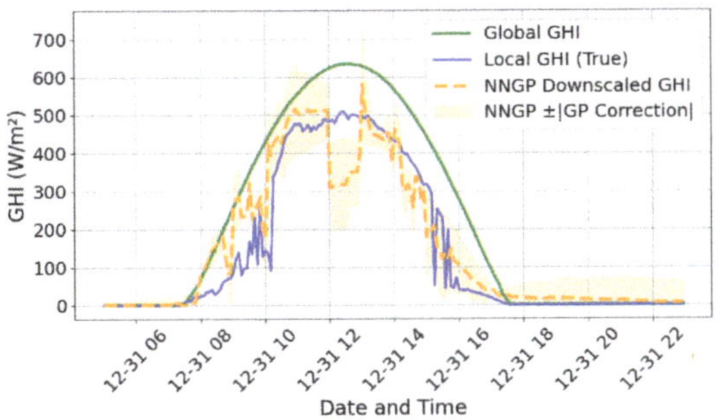

(a) Site 3

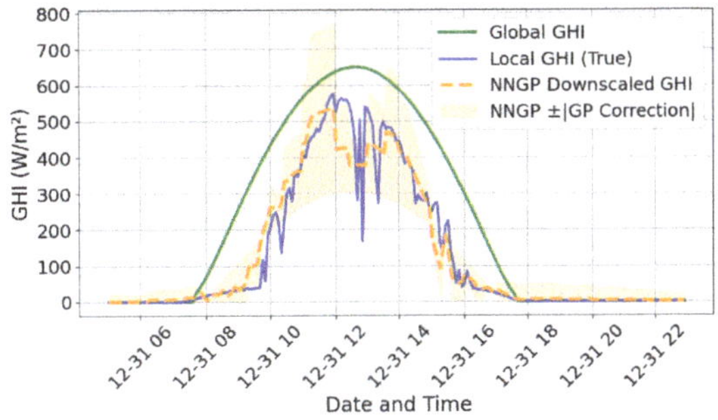

(b) Site 4

Figure 10. NNGP's Day-Ahead GHI Predictions with Gaussian Uncertainty.

4.3. Scalability of the NNGP and NNRF Models

The scalability results as presented in Figure 11 reveal how the NNRF consistently performs better than the NNGP model for different numbers of k-NNs. These accuracy values are the GoF of the day-ahead predictions of the NNGP and NNRF for 31 December 2023. It can be seen from Figure 11a that the NNRF requires as few as 3 k-NNs to achieve very good accuracy. Beyond 3 k-NNs, the NNRF model does not seem to record any significant improvement in prediction accuracy. Conversely, the NNGP accuracy scales linearly from 2 to 6 k-NNs, beyond which the accuracy shows very minimal improvement. If we select 3 k-NNs and 6 k-NNs as the local optimum values for the NNRF and NNGP, respectively, the corresponding accuracy is 90.13% for NNRF and 88.74% for NNGP. The equivalent computational time is 582.92 s for the NNRF and 1578.60 s for the NNGP. This clearly shows that besides requiring a smaller number of k-NNs to achieve a locally optimum accuracy, the NNRF also takes comparably less computational time (in this case, almost three times as fast as the NNGP model). Additionally, based on the results presented in Figure 11b, the NNRF scales linearly, whiles the NNGP displays a polynomial scaling with increasing number of k-NNs. This implies the NNRF responds to a step increase in the number of k-NNs with a scalar multiple of the computational time, while the NNGP responds with an exponential increase in the computational time. This further proves the effectiveness and efficiency of the NNRF model for large-scale solar irradiance downscaling tasks, making it more suitable for real-time grid integration of DERs.

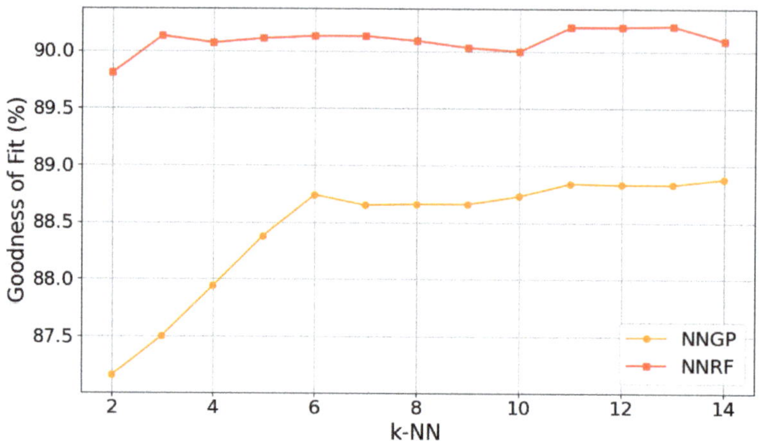

(**a**) GoF with Varying Number of k-NNs

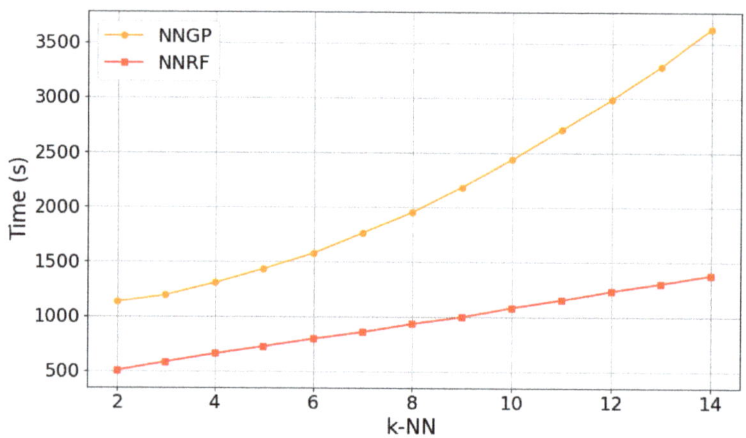

(**b**) Training Time with Varying Number of k-NNs

Figure 11. Scalability Performance of NNGP and NNRF.

4.4. Results of Puerto Rico Case Study

The NNRF model continued to outperform the NNGP model in both predictive accuracy and computational efficiency. Validation results yielded a Goodness of Fit (GoF) of 85.83% for NNRF compared to 84.79% for NNGP. Although the NNRF performance showed a slight decrease relative to its results in the Texas simulation, this reduction may be attributed to the increased volatility and rapid fluctuations in local solar irradiance observed in Puerto Rico. Nevertheless, the NNRF maintained a significant advantage in computational speed, completing the simulation in 276.1 s—nearly half the time required by the NNGP model, which took 530.8 s. These results further underscore the effectiveness of the proposed NNRF model in achieving comparable or superior accuracy while substantially reducing computational overhead. Figure 12 presents a visual comparison of the prediction performance of both models.

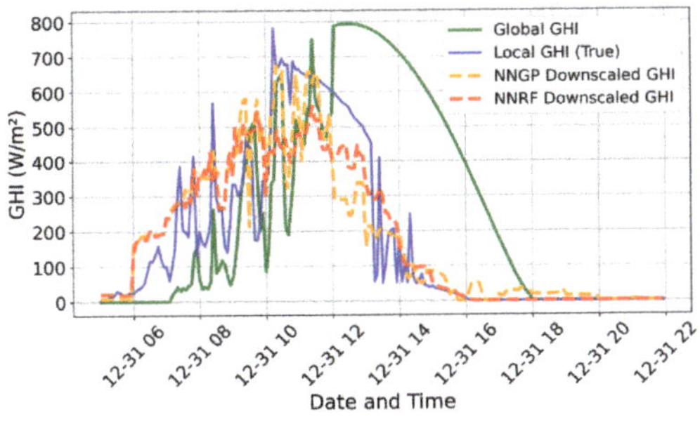

(a) Site 1

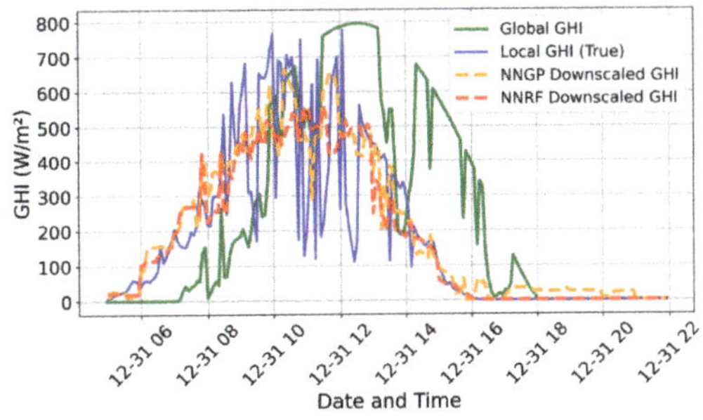

(b) Site 2

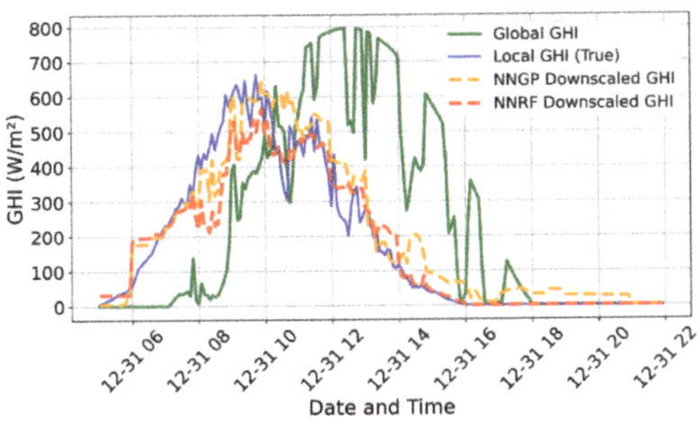

(c) Site 3

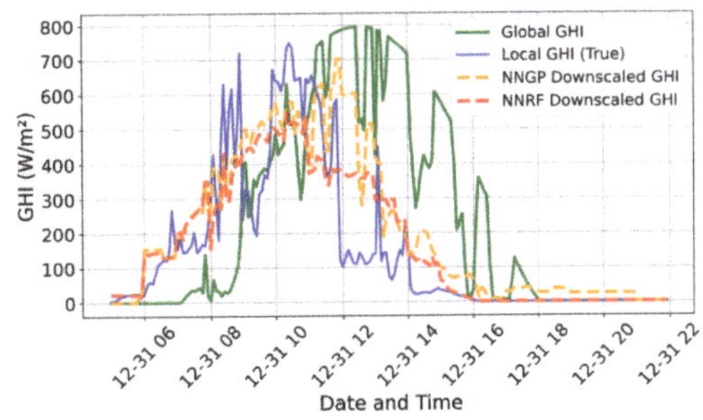

(d) Site 4

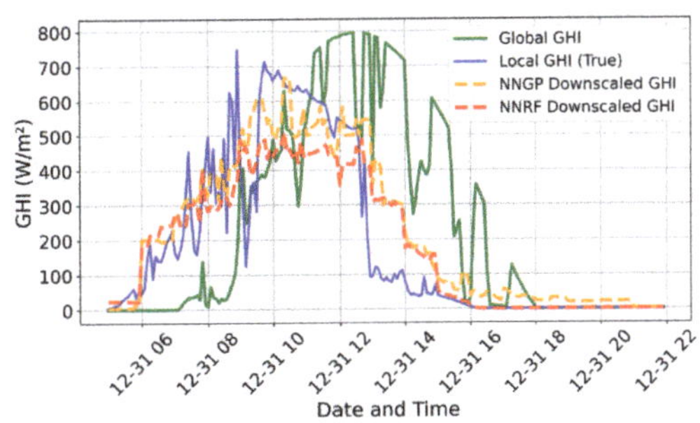

(e) Site 5

Figure 12. Day-Ahead GHI prediction for 5 sites in Puerto Rico.

The results displayed in Figure 12 above are the point estimates of NNRF and NNGP. However, as stated earlier, the NNGP has the advantage of giving a probabilistic forecast that shows the prediction uncertainties to be expected. These are displayed in Figure 13 for Site 1 and Sites 5 in Puerto Rico.

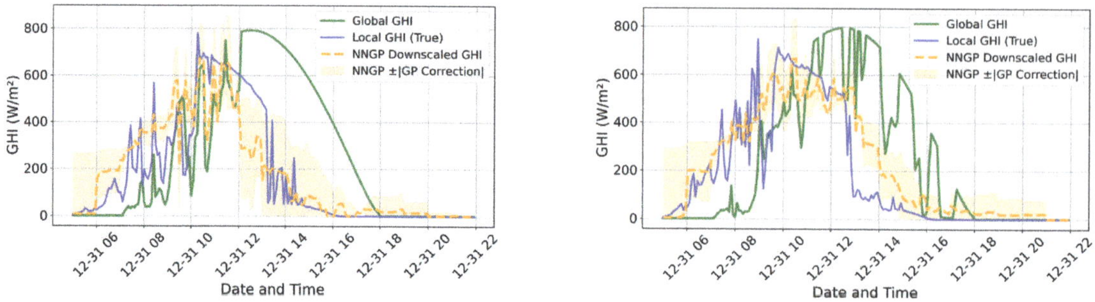

(**a**) Site 1

(**b**) Site 5

Figure 13. NNGP's Day-Ahead GHI Predictions with Gaussian Uncertainty for Sites in Puerto Rico.

Finally, we present the residual plot in Figure 14 to once again demonstrate how the NNRF gives lower prediction errors compared to the NNGP model.

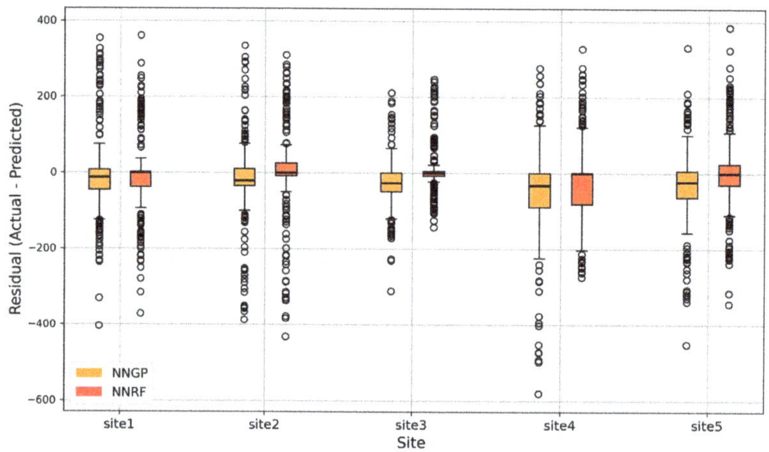

Figure 14. Boxplots of NNGP and NNRF Prediction Residuals for Sites 1, 2, 3, 4, and 5 in Puerto Rico.

5. Conclusions

This study evaluates the performance of the NNRF and NNGP models in downscaling solar irradiance to localized measurements. The NNRF model performs better than the NNGP model in terms of accuracy and computational time during both training and validation. The NNRF recorded an average validation accuracy of 90.61% which outperformed the 85.88% recorded by the NNGP model. The NNRF model also improved computational speed by 2.5 times over the NNGP model. A case study on five sites in Puerto Rico further confirmed the superior performance of the NNRF model over the NNGP model in terms of accuracy and computational speed. Although the NNGP lags behind in terms of accuracy and computational speed, it shows strength in terms of interpretability and prediction uncertainty quantification. Its hyperparameters, such as the variance (σ), temporal (ℓ_t), spatial (ℓ_s), and feature-based (ℓ_g) scale parameters, give insight into how far in time and space past observations and neighboring sites influence the prediction outcome. This makes the NNGP more suitable for probabilistic estimates, which demands more transparency of the modeling process.

Furthermore, scalability tests that measured both models' performance and computational speed with varying numbers of k-NNs from 2 to 14 showed a linear scaling for

NNRF and a polynomial scaling for NNGP in terms of computational time. Similarly, the NNRF achieved a local optimal accuracy with 3 k-NNs, while the NNGP took 6 k-NNs to obtain a local optimal accuracy. Irrespective of the increased number of k-NNs, the NNGP still could not outperform the NNRF, which had a scaling accuracy of 90.13% compared to 88.74% of the NNGP. These findings prove the superiority of the NNRF for large-scale solar irradiance downscaling tasks. Further studies could investigate improving the NNRF's accuracy by combining the nearest-neighbor model with either artificial neural networks, the extreme gradient boosting (XGboost) method, or other advanced machine-learning models.

These findings could be very useful in the implementation of the FERC order 2222, by enabling real-time, accurate estimation of solar PV capacity for day-ahead dispatch. Additionally, due to the improved accuracy and computational speed of the NNRF model, its application can be extended to the real-time electricity market with accurate PV estimates on a 5 min rolling basis. Finally, the use of downscaled solar irradiance forecasts reduces the level of solar PV generation uncertainties with its associated reserve requirements and instability issues in power dispatch planning.

Author Contributions: Conceptualization: S.T.A., H.M.R. and T.M.H.; methodology: S.T.A., A.S., H.M.R. and T.M.H.; software: S.T.A. and A.S.; validation: S.T.A., A.S. and T.M.H.; formal analysis: S.T.A. and A.S.; data curation: S.T.A. and A.S.; writing—original draft preparation: S.T.A., A.S. and T.M.H.; writing—review and editing: Z.W., H.M.R. and T.M.H.; visualization: S.T.A. and A.S.; supervision: Z.W., H.M.R. and T.M.H.; funding acquisition: T.M.H. All authors have read and agreed to the published version of the manuscript.

Funding: This work is supported by the SDSU Richard McComish, Ph.D. Fellowship in Energy Infrastructure, the U.S. Department of Energy Office of Science, Office of Electricity Microgrid R&D Program, and Office of Energy Efficiency and Renewable Energy Solar Energy Technology Office under the EPSCoR grant number DE-SC0020281, and the U.S. National Science Foundation grant no. 2316400.

Institutional Review Board Statement: Not applicable.

Informed Consent Statement: Not applicable.

Data Availability Statement: The dataset utilized in this study, along with the corresponding code, results, and visualizations, are publicly accessible on GitHub. Interested readers can access these resources at the following link: https://github.com/ShadrackTAsiedu/Downscaling-NNGP-NNRF, accessed on 3 May 2025.

Conflicts of Interest: The authors declare no conflicts of interest.

Abbreviations

The following abbreviations are used in this manuscript:

ANN	Artificial Neural Network
CPU	Central Processing Unit
DER	Distributed Energy Resource
ERCOT	Electric Reliability Council of Texas
FERC	Federal Energy Regulatory Commission
GB	Giga Bytes
GHI	Global Horizontal Irradiance
GHz	Gigahertz
GoF	Goodness of Fit
GP	Gaussian Process

MAE	Mean Absolute Error
NN	Nearest-Neighbor
NNGP	Nearest-Neighbor Gaussian Process
NNRF	Nearest-Neighbor Random Forest
NRMSE	Normalized Root Mean Squared Error
NSRDB	National Solar Radiation Database
NWP	Numerical Weather Prediction
PEC	Power Electronic Converter
PV	Photovoltaic
RAM	Random Access Memory
RF	Random Forest
RMSE	Root Mean Squared Error
VREs	Variable Renewable Energy Resources

Appendix A. Data Preprocessing

The preprocessing steps include handling raw data inconsistencies, ensuring uniform temporal resolution, addressing missing data, and engineering features critical for downstream modeling. Below, we detail each stage of the data preprocessing pipeline.

Appendix A.1. Raw Data Cleaning

The raw irradiance data from eight sites were provided as individual CSV files. Each dataset contained a timestamp and solar irradiance measurements, with inconsistencies in date formats and row order. To address these challenges, each dataset was loaded, and the Date column was converted to a standardized *datetime* format for uniformity. The rows were reversed to ensure chronological order, starting from the earliest date, and the indices were reset to sequential integers starting from 1. The column containing solar irradiance data, originally named inconsistently across files, was renamed to GHI_local for uniformity. The cleaned datasets were stored in a dictionary, with site names as keys for efficient access.

Appendix A.2. Temporal Alignment and Missing Data Handling

Global solar irradiance data are often recorded at irregular intervals, leading to temporal misalignment and missing values. For accurate spatiotemporal modeling, a complete time series with a 5-min resolution was generated for the entire year (2022), yielding a total of 105,120 expected timestamps. Each site's dataset was reindexed to align with the expected date range, and missing timestamps were identified by comparing the reindexed dataset to the complete time series. Missing irradiance values in the local irradiance data were imputed using a trend-based approach. For each missing value, data from the same time of day in the previous or subsequent valid days were averaged over a 7-day window. If sufficient historical data were unavailable, forward imputation was used as a fallback. Additionally, the cleaned datasets were capped at a maximum irradiance value of 1200 W/m^2 to avoid outliers resulting from sensor anomalies.

Appendix A.3. Data Validation and Visualization

To verify the integrity of the cleaned data, temporal trends in solar irradiance were visualized for each site. The local irradiance datasets were plotted over time for the entire year, revealing daily and seasonal variability across the sites. Each site's dataset was inspected for remaining inconsistencies, such as abrupt changes in irradiance values, which could indicate potential anomalies. The annual local solar irradiance for Site 1, before and after cleaning and filling missing data, is shown in Figures A1 and A2.

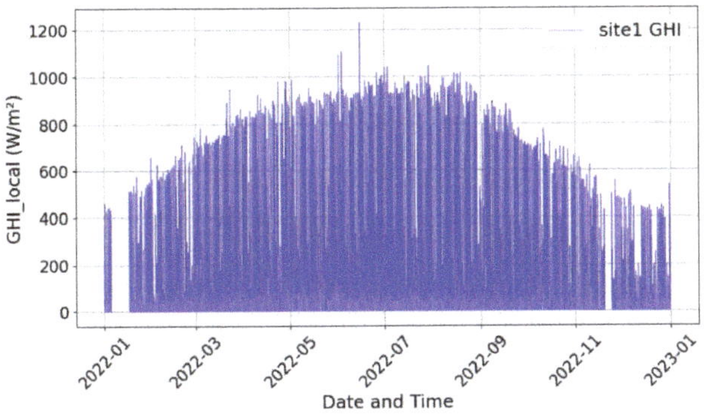

Figure A1. Raw Data on Local Solar Irradiance for Site 1 (2022).

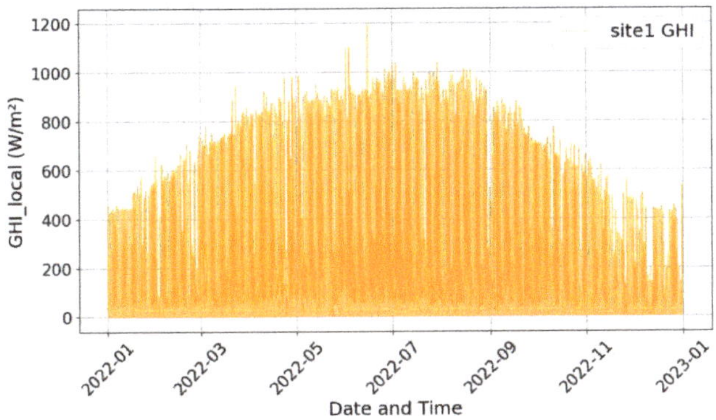

Figure A2. Cleaned Data on Local Solar Irradiance for Site 1 (2022).

References

1. National Renewable Energy Laboratory (NREL). *Technical Report NREL/TP-5C00-80166*; Technical Report; U.S. Department of Energy Office of Energy Efficiency & Renewable Energy: Golden, CO, USA, 2021.
2. Eldridge, B.C.; Somani, A. *Impact of FERC Order 2222 on DER Participation Rules in US Electricity Markets*; Technical Report; Pacific Northwest National Laboratory (PNNL): Richland, WA, USA, 2022.
3. Garg, S.; Tyagi, S. A Comprehensive Review on Opportunities and Challenges of Grid Integration of Renewable Energy Resources. In Proceedings of the 2024 Second International Conference on Smart Technologies for Power and Renewable Energy (SPECon), Ernakulam, India, 2–4 April 2024; pp. 1–6. [CrossRef]
4. Subedi, S.; Poudel, B.; Aslami, P.; Fourney, R.; Rekabdarkolaee, H.M.; Tonkoski, R.; Hansen, T.M. Automated Data–Driven Model Extraction and Validation of Inverter Dynamics with Grid Support Function. *E-Prime—Adv. Electr. Eng. Electron. Energy* **2023**, *6*, 100365. [CrossRef]
5. Tamrakar, U.; Shrestha, D.; Maharjan, M.; Bhattarai, B.P.; Hansen, T.M.; Tonkoski, R. Virtual Inertia: Current Trends and Future Directions. *Appl. Sci.* **2017**, *7*, 654. [CrossRef]
6. Tiwari, S.; Sabzehgar, R.; Rasouli, M. Short Term Solar Irradiance Forecast Using Numerical Weather Prediction (NWP) with Gradient Boost Regression. In Proceedings of the 2018 9th IEEE International Symposium on Power Electronics for Distributed Generation Systems (PEDG), Charlotte, NC, USA, 25–28 June 2018; pp. 1–8. [CrossRef]
7. Dou, W.; Wang, K.; Shan, S.; Zhang, K.; Wei, H. Day-Ahead Correction of Numerical Weather Prediction Solar Irradiance Forecasts Based on Similar Day Analysis. In Proceedings of the 2023 35th Chinese Control and Decision Conference (CCDC), Yichang, China, 20–22 May 2023; pp. 2554–2559. [CrossRef]
8. De Giorgi, M.G.; Congedo, P.M.; Malvoni, M. Photovoltaic power forecasting using statistical methods: Impact of weather data. *IET Sci. Meas. Technol.* **2014**, *8*, 90–97. [CrossRef]
9. Asiedu, S.T.; Nyarko, F.K.; Boahen, S.; Effah, F.B.; Asaaga, B.A. Machine learning forecasting of solar PV production using single and hybrid models over different time horizons. *Heliyon* **2024**, *10*, e28898. [CrossRef] [PubMed]

10. Gayathry, V.; Kaliyaperumal, D.; Salkuti, S.R. Seasonal solar irradiance forecasting using artificial intelligence techniques with uncertainty analysis. *Sci. Rep.* **2024**, *14*, 17945. [CrossRef] [PubMed]
11. Garg, S.; Agrawal, A.; Goyal, S.; Verma, K. Day Ahead Solar Irradiance Forecasting using Different Statistical Techniques. In Proceedings of the 2020 IEEE International Conference on Power Electronics, Drives and Energy Systems (PEDES), Jaipur, India, 16–19 December 2020; pp. 1–4. [CrossRef]
12. Ziyabari, S.; Du, L.; Biswas, S. A Spatio-temporal Hybrid Deep Learning Architecture for Short-term Solar Irradiance Forecasting. In Proceedings of the 2020 47th IEEE Photovoltaic Specialists Conference (PVSC), Calgary, AB, Canada, 15 June–21 August 2020; pp. 0833–0838. [CrossRef]
13. Hari, N.G.; G, J. Solar Irradiance Prediction using Deep Learning-Based Approaches. In Proceedings of the 2022 IEEE Asia-Pacific Conference on Computer Science and Data Engineering (CSDE), Gold Coast, Australia, 18–20 December 2022; pp. 1–5. [CrossRef]
14. Kartini, U.T.; Hariyati; Aribowo, W.; Wardani, A.L. Development Hybrid Model Deep Learning Neural Network (DL-NN) For Probabilistic Forecasting Solar Irradiance on Solar Cells To Improve Economics Value Added. In Proceedings of the 2022 Fifth International Conference on Vocational Education and Electrical Engineering (ICVEE), Surabaya, Indonesia, 10–11 September 2022; pp. 151–156. [CrossRef]
15. Zafar, R.; Chung, I.Y. Day-Ahead Solar Irradiance Forecasting using a Hybrid Weather-Based Attention BiLSTM Approach for Power System Operation Scheduling. In Proceedings of the 2023 IEEE PES Innovative Smart Grid Technologies Europe (ISGT EUROPE), Grenoble, France, 23–26 October 2023; pp. 1–5. [CrossRef]
16. Panamtash, H.; Zhou, Q. Coherent Probabilistic Solar Power Forecasting. In Proceedings of the 2018 IEEE International Conference on Probabilistic Methods Applied to Power Systems (PMAPS), Boise, ID, USA, 24–28 June 2018; pp. 1–6. [CrossRef]
17. Kakimoto, M.; Endoh, Y.; Shin, H.; Ikeda, R.; Kusaka, H. Probabilistic Solar Irradiance Forecasting by Conditioning Joint Probability Method and Its Application to Electric Power Trading. *IEEE Trans. Sustain. Energy* **2019**, *10*, 983–993. [CrossRef]
18. Opoku, K.; Lucemo, S.; Zhou Sun, Q.; Dimitrovski, A. A Bayesian Approach to Probabilistic Solar Irradiance Forecasting. In Proceedings of the 2022 North American Power Symposium (NAPS), Salt Lake City, UT, USA, 9–11 October 2022; pp. 1–6. [CrossRef]
19. Ziyabari, S.; Du, L.; Biswas, S.K. Multibranch Attentive Gated ResNet for Short-Term Spatio-Temporal Solar Irradiance Forecasting. *IEEE Trans. Ind. Appl.* **2022**, *58*, 28–38. [CrossRef]
20. National Renewable Energy Laboratory (NREL). National Solar Radiation Database (NSRDB), 2024. Available online: https://nsrdb.nrel.gov/ (accessed on 2 December 2024).
21. Zippenfenig; Patrick. Open-Meteo.com Weather API, 2023. Available online: https://zenodo.org/records/14582479 (accessed on 22 January 2025).
22. Yang, J.; Sengupta, M.; Bailey, M.; Nychka, D.; Habte, A.; Bandyopadhyay, S.; Xie, Y. Bias Correction and Statistical Downscaling of Future Solar Irradiance Projections Using the NSRDB. In Proceedings of the 2024 IEEE 52nd Photovoltaic Specialist Conference (PVSC), Seattle, WA, USA, 9–14 June 2024; pp. 793–795. [CrossRef]
23. Widén, J.; Munkhammar, J. Spatio-Temporal Downscaling of Hourly Solar Irradiance Data Using Gaussian Copulas. In Proceedings of the 2019 IEEE 46th Photovoltaic Specialists Conference (PVSC), Chicago, IL, USA, 16–21 June 2019; pp. 3172–3178.
24. Huang, J.; Perez, M.; Perez, R.; Yang, D.; Keelin, P.; Hoff, T. Nonparametric Temporal Downscaling of GHI Clear-sky Indices using Gaussian Copula. In Proceedings of the 2022 IEEE 49th Photovoltaics Specialists Conference (PVSC), Philadelphia, PA, USA, 5–10 June 2022; pp. 0654–0657.
25. Huang, J.; Perez, M. Temporal Downscaling of GHI Clear-Sky Indices Using T-Copula. In Proceedings of the 2023 IEEE 50th Photovoltaic Specialists Conference (PVSC), San Juan, PR, USA, 11–16 June 2023; pp. 1–3. [CrossRef]
26. Munkhammar, J.; Widén, J. Downscaling global, beam and diffuse horizontal irradiance based on hour resolution global horizontal irradiance using Markov mixture distribution modeling. In Proceedings of the 21st Wind & Solar Integration Workshop (WIW 2022), The Hague, The Netherlands, 12–14 October 2022; Volume 2022, pp. 662–667. [CrossRef]
27. Wang, W.; Yin, G.; Zhao, W.; Wen, F.; Yu, D. Spatial Downscaling of MSG Downward Shortwave Radiation Product Under Clear-Sky Condition. *IEEE Trans. Geosci. Remote Sens.* **2020**, *58*, 3264–3272. [CrossRef]
28. Di Paola, F.; Cimini, D.; De Natale, M.P.; Gallucci, D.; Geraldi, E.; Gentile, S.; Genzano, N.; Larosa, S.; Nilo, S.T.; Ricciardelli, E.; et al. Weather Forecast Downscaling for Applications in Smart Agriculture and Precision Farming using Artificial Neural Networks. In Proceedings of the IGARSS 2023—2023 IEEE International Geoscience and Remote Sensing Symposium, Pasadena, CA, USA, 16–21 July 2023; pp. 2791–2794. [CrossRef]
29. Wright, T.; Aslami, P.; Dentlinger, M.P.; Hansen, T.M.; Kimn, J.H.; Andrade, F.; Rekabdarkolaee, H.M. Nearest-Neighbor Gaussian Process to Downscale Solar Forecasting at the Grid-Edge for Increased Situational Awareness. In Proceedings of the 2023 North American Power Symposium (NAPS), Asheville, NC, USA, 15–17 October 2023; pp. 1–6. [CrossRef]
30. Berkani, S.; Guermah, B.; Zakroum, M.; Ghogho, M. Spatio-Temporal Forecasting: A Survey of Data-Driven Models Using Exogenous Data. *IEEE Access* **2023**, *11*, 75191–75214. [CrossRef]

31. Xu, L.; Chen, N.; Chen, Z.; Zhang, C.; Yu, H. Spatiotemporal forecasting in earth system science: Methods, uncertainties, predictability and future directions. *Earth-Sci. Rev.* **2021**, *222*, 103828. [CrossRef]
32. Tascikaraoglu, A.; Sanandaji, B.M.; Chicco, G.; Cocina, V.; Spertino, F.; Erdinc, O.; Paterakis, N.G.; Catalão, J.P. Compressive Spatio-Temporal Forecasting of Meteorological Quantities and Photovoltaic Power. *IEEE Trans. Sustain. Energy* **2016**, *7*, 1295–1305. [CrossRef]
33. Buster, G.; Rossol, M.; Maclaurin, G.; Xie, Y.; Sengupta, M. A physical downscaling algorithm for the generation of high-resolution spatiotemporal solar irradiance data. *Sol. Energy* **2021**, *216*, 508–517. [CrossRef]
34. Ulapane, N.; Thiyagarajan, K.; Kodagoda, S. Hyper-Parameter Initialization for Squared Exponential Kernel-based Gaussian Process Regression. In Proceedings of the 2020 15th IEEE Conference on Industrial Electronics and Applications (ICIEA), Kristiansand, Norway, 9–13 November 2020; pp. 1154–1159. [CrossRef]
35. Rasmussen, C.E.; Williams, C.K.I. *Gaussian Processes for Machine Learning*; MIT Press: Cambridge, MA, USA, 2006.
36. Cutler, A.; Cutler, D.; Stevens, J. *Random Forests*; Springer: New York, NY, USA, 2011; Volume 45, pp. 157–176. [CrossRef]
37. Ahmed, H.; Nandi, A.K. Decision Trees and Random Forests. In *Condition Monitoring with Vibration Signals: Compressive Sampling and Learning Algorithms for Rotating Machines*; John Wiley & Sons: Hoboken, NJ, USA, 2019; pp. 199–224. [CrossRef]
38. Ramasamy, S.; Kantharaju, H.C.; Bindu Madhavi, N.; Haripriya, M.P. 8 Meta-learning through ensemble approach: Bagging, boosting, and random forest strategies. In *Toward Artificial General Intelligence: Deep Learning, Neural Networks, Generative AI*; De Gruyter: Berlin, Germany, 2024; pp. 167–188.
39. Putro Dirgantoro, G.; Soeleman, M.A.; Supriyanto, C. Smoothing Weight Distance to Solve Euclidean Distance Measurement Problems in K-Nearest Neighbor Algorithm. In Proceedings of the 2021 IEEE 5th International Conference on Information Technology, Information Systems and Electrical Engineering (ICITISEE), Purwokerto, Indonesia, 24–25 November 2021; pp. 294–298. [CrossRef]
40. Kjellsson, J.; Webber, M. The Energy-Water Nexus: Spatially-Resolved Analysis of the Potential for Desalinating Brackish Groundwater by Use of Solar Energy. *Resources* **2015**, *4*, 476–489. [CrossRef]
41. Ambient Weather. Ambient Weather Data and Tools. 2024. Available online: https://ambientweather.com/ (accessed on 2 December 2024).
42. Das, A.; Rad, P. Opportunities and challenges in explainable artificial intelligence (xAI): A survey. *arXiv* **2020**, arXiv:2006.11371.

Disclaimer/Publisher's Note: The statements, opinions and data contained in all publications are solely those of the individual author(s) and contributor(s) and not of MDPI and/or the editor(s). MDPI and/or the editor(s) disclaim responsibility for any injury to people or property resulting from any ideas, methods, instructions or products referred to in the content.

energies

MDPI

Article

Long-Term Multi-Resolution Probabilistic Load Forecasting Using Temporal Hierarchies

Shafie Bahman and Hamidreza Zareipour *

Department of Electrical and Software Engineering, University of Calgary, Calgary, AB T2N 1N4, Canada; shafie.bahman@ucalgary.ca

* Correspondence: hzareipo@ucalgary.ca

Abstract: Accurate long-term electricity load forecasting is critical for energy planning, infrastructure development, and risk management, especially under increasing uncertainty from climate and economic shifts. This study proposes a multi-resolution probabilistic load forecasting framework that leverages temporal hierarchies to generate coherent forecasts at hourly, daily, monthly, and yearly levels. The model integrates climate and economic indicators and employs tailored forecasting techniques at each resolution, including XGBoost and ARIMAX. Initially incoherent forecasts across time scales are reconciled using advanced methods such as Ordinary Least Squares (OLS), Weighted Least Squares with Series Variance Scaling (WLS_V), and Structural Scaling (WLS_S) to ensure consistency. Using historical data from Alberta, Canada, the proposed approach improves the accuracy of deterministic forecasts and enhances the reliability of probabilistic forecasts, particularly when using the OLS reconciliation method. These results highlight the value of temporal hierarchy structures in producing high-resolution long-horizon load forecasts, providing actionable insights for utilities and policymakers involved in long-term energy planning and system optimization.

Keywords: load forecasting; probabilistic forecasts; hierarchical forecasting; temporal aggregation; forecast reconciliation

1. Introduction

Long-term load forecasting (LTLF) is critical for ensuring the efficient operation and expansion of grid transmission and distribution infrastructure, which is essential in response to future power demand [1]. It plays a key role in decisions related to capacity expansion, demand response planning, and infrastructure investment while also supporting the integration of renewable energy sources and shaping energy policies [2]. By offering insights into potential power reliability issues, LTLF aids in comprehensive infrastructure planning efforts [3]. Defined as forecasts with horizons longer than a year [4–7], long-term load forecasts are vital for electric utility planning, where forecast inaccuracies can lead to significant financial risks. For instance, a 1% error in mean absolute percentage error (MAPE) could result in financial losses of hundreds of thousands of dollars per gigawatt (GW) of peak demand for utility companies [8]. These risks highlight the need for accurate and robust long-term forecasting models that account for the increasing uncertainties in modern power systems.

As countries and regions strive to mitigate climate change, achieving carbon neutrality represents a critical objective [9]. As uncertainties from global warming and extreme temperature changes grow, forecasting electricity demand under carbon peaking and carbon neutral strategies is becoming harder [10]. In addition, increasing penetration of distributed energy resources (DERs) such as renewable energy resources (RES), battery storage systems, electric vehicles (EVs), and flexible loads leads to increased uncertainties in load forecasting [11]. Taking into account the new uncertainties introduced in load forecasting when making probabilistic forecasts has become increasingly important in LTLF to better manage these risks. These evolving dynamics necessitate more flexible and uncertainty-aware forecasting frameworks.

Probabilistic forecasting has emerged as a key tool in managing uncertainty in energy markets, where evolving challenges such as aging infrastructure and the integration of renewable sources are creating new complexities [12]. In this dynamic environment, probabilistic load forecasting (PLF) plays an increasingly critical role in the planning and management of energy systems [13]. By quantifying the range of possible outcomes, probabilistic forecasts provide a range of possible outcomes, leading to more informed decision-making in energy management compared to point forecasts [14,15]. The hierarchical nature of the load demand data combined with the need for scalability in forecasts further emphasizes the importance of adopting probabilistic approaches in forecasting. Therefore, probabilistic methods are becoming a standard approach in long-term forecasting applications.

In many forecasting applications, including load forecasting, time series data naturally follow a hierarchical structure [16]. In forecasting, a hierarchy refers to the organization of data or forecasts into different levels of aggregation [17]. These hierarchies can be categorized into spatial, cross-sectional, or temporal. Spatial hierarchies refer to data organized by geographic divisions, such as load demand measured at the the country, system, area, and Point of Delivery (POD) levels [17]. Cross-sectional hierarchies emerge from organizational divisions [18]. Temporal hierarchies are particularly relevant to load forecasting, as different functions and planning activities require forecasts at varying time resolutions and planning horizons [19,20]. Among these, temporal hierarchies are especially well-suited to address the challenges of long-term multi-resolution forecasting.

Temporal hierarchies are constructed through non-overlapping temporal aggregation, which allows for the integration of high-level managerial forecasts that incorporate complex and unstructured information with lower-level statistical forecasts [21]. This approach is independent of specific forecasting models, enabling organizations to exploit information across all levels of a given hierarchy [21]. Utilizing temporal hierarchies can improve forecasting accuracy over conventional methods, especially in situations characterized by increased modeling uncertainty. This enhanced accuracy is achieved by combining forecasts at different aggregation levels and then reconciling them, ultimately supporting aligned decision-making across various planning horizons [21,22]. This makes hierarchical modeling a powerful tool for both operational and strategic energy planning.

Hierarchical forecasting structures are founded on the principle of coherency, which dictates that lower-level forecasts should sum to the higher-level forecasts [19]. This requirement is met through a set of reconciliation methods that adjust the original forecasts to align with the aggregation constraints of the hierarchical structure. Forecast reconciliation is crucial, as it ensures that forecasts at all levels of the hierarchy are consistent by adjusting them to maintain accuracy and coherence across the entire hierarchy [17]. By effectively implementing reconciliation methods, organizations can enhance the reliability of their forecasts, making them more useful for strategic decision-making in energy management. Ensuring coherency across time resolutions is particularly important for long-term planning scenarios that require high-fidelity forecasts across multiple time scales.

To the best of our knowledge, this is the first study to apply temporal hierarchy reconciliation methods to long-term and multi-resolution deterministic and probabilistic load forecasting. In this paper, we propose a long-term deterministic and probabilistic load forecasting model with a temporal hierarchy structure, deploying various deterministic reconciliation methods to produce ex post and ex ante coherent forecasts. In addition, we combine a probabilistic reconciliation method based on Gaussian Distribution linearity with deterministic reconciliation methods to produce coherent long-term probabilistic load forecasts. The effects of each reconciliation method on long-term load forecasting are examined. The proposed long-term load forecasting model achieves three key objectives: (1) producing one-year-ahead deterministic and probabilistic forecasts at four different resolutions (hourly, daily, monthly, yearly) while using different types of information for each resolution; (2) applying both deterministic and probabilistic reconciliation methods within a temporal hierarchy to generate coherent forecasts across all resolutions and extracting peak load forecasts at different levels; and (3) identifying the most effective reconciliation method for improving the accuracy and reliability of long-term load forecasts, especially for hourly load forecasting.

Thus, this work makes three key contributions to the literature. First, we propose using a temporal hierarchy structure for a multi-resolution long-term load forecasting framework that generates coherent forecasts across hourly to yearly resolutions, unlike prior studies that have applied temporal hierarchies mainly in short-term forecasting contexts [18,20,23]. Second, the proposed model is capable of producing high-resolution year-ahead deterministic and probabilistic forecasts both ex post and ex ante, addressing a gap between typically high-resolution short-term and typically low-resolution long-term forecasting approaches [18,20,23–26]. Third, we evaluate not only the effects of reconciliation on aggregated forecasts, as is typical in prior work [18,20,23–28], but also its impact on peak load forecasting, which is an important factor for planning and reliability.

The rest of this paper is structured as follows: in Section 2, a review of the state-of-the-art literature is provided; Section 3 presents the model and describes the training procedure; details of the experimental setup and empirical evaluation results are presented in Section 4; finally, Section 5 concludes the paper.

2. Literature Review

2.1. Long-Term Load Forecasting (LTLF)

Long-term load forecasting (LTLF) plays a vital role in capacity expansion, investment planning, and infrastructure reliability. Accurate LTLF is necessary to identify potential power reliability issues and optimize infrastructure planning efforts [2,3]. Most traditional LTLF methods rely on statistical techniques such as regression and time series models, often operating at coarse resolutions such as monthly or yearly [6,8,25]. However, long-term forecasting is subject to increasing uncertainties from economic shocks, climate change, and policy interventions [29]. To address these challenges, recent studies have explored advanced and hybrid modeling approaches aimed at enhancing forecast accuracy and resilience.

Various techniques have been explored in the literature. For example, ref. [4] presented a comprehensive survey of forecasting models in smart grids, highlighting their respective strengths and limitations. Deep learning methods have shown high potential due to their ability to model complex, nonlinear relationships [30]. In [31], the authors used a double-layer CatBoost model for daily load forecasting, while [32,33] applied ANN and LSTM networks for year-ahead forecasts. In another approach, ref. [34] introduced a linear clustering method to forecast long-term system load, benchmarking it against traditional ARIMA and exponential smoothing.

To further enhance accuracy, researchers have used data reshaping [35], decomposition and dimensionality reduction [36], and hybrid learning methods [37–39]. These hybrid approaches combine machine learning, ensemble techniques, and domain-specific preprocessing to improve model performance in monthly and yearly peak load forecasting.

2.2. Probabilistic Load Forecasting (PLF)

Probabilistic load forecasting (PLF) addresses the growing uncertainties in power systems by generating forecast distributions instead of point estimates. This probabilistic approach is essential for risk-aware decision-making under climate change, demand volatility, and renewable integration [13–15]. While PLF has been widely used in short-term load forecasting, its application to long-term contexts is still evolving.

Forecasts that incorporate uncertainty are especially useful in the presence of DERs, EVs, and flexible loads, which introduce nonlinear effects and stochasticity into power consumption patterns [11]. Tools such as GEFCom2017 have showcased the complexity of generating probabilistic month-ahead forecasts across spatial hierarchies [40]. However, most of the literature focuses on single-resolution probabilistic forecasts, typically for short-term planning horizons.

In the long-term context, PLF can account for climate scenarios and socioeconomic pathways. Studies such as [9,41] have demonstrated probabilistic projections over 40-year horizons, modeling future temperature and carbon-neutral policy impacts. These works highlight how PLF can be extended to integrate long-term uncertainties; however, they also expose a lack of hierarchical coherence in most probabilistic frameworks. This creates a need for unified multi-resolution PLF systems capable of capturing forecast distributions across temporal scales.

2.3. Hierarchical Forecasting and Reconciliation Methods

Hierarchical forecasting organizes data across multiple levels, for instance spatial, cross-sectional, or temporal, and enforces coherency across them. In the context of cross-sectional and spatial hierarchies, many studies have applied reconciliation techniques for improved accuracy. For instance, ref. [24] proposed a hybrid bottom-up/top-down method for spatial hierarchies, while [26] introduced robust estimators for monthly forecasts. In the probabilistic domain, Gaussian assumption reconciliation techniques have been developed by [27] and extended in the open-source hierarchical forecast library [42]. Furthermore, ref. [28] proposed score-optimized reconciliation methods to improve distribution accuracy. Despite these advances, almost all reconciliation methods are applied to short-term horizon forecasting with spatial or cross-sectional data.

2.4. Temporal Hierarchies in Forecasting

Temporal hierarchies allow for coherent forecasting across different time resolutions, such as hourly, daily, monthly, and yearly. In [21], the authors introduced variance-based reconciliation methods (WLS_H, WLS_V, WLS_S) tailored to temporal structures. Later, refs. [18,20] applied temporal hierarchies to short-term applications, including day-ahead and hour-ahead solar or wind forecasting. In [23], the authors addressed the probabilistic extension by developing methods for hierarchical aggregation of full distributions, particularly under temporal or spatial structures. Although these studies validate the usefulness of temporal hierarchies, they are limited to high-frequency short-horizon forecasts.

Although temporal hierarchies have been successfully applied in short-term forecasting settings, such as hour-ahead or day-ahead predictions, these applications typically involve closely spaced resolutions (e.g., 15 min to hourly) with similar time series behavior and shared feature sets. In contrast, long-term forecasting involves fundamentally different patterns and data drivers across resolutions; the hourly load is sensitive to weather and calendar effects, whereas the yearly load reflects economic and demographic trends. This resolution-specific variability requires distinct models and feature engineering at each level, making coherent integration through reconciliation both more challenging and more valuable in the long-term setting.

2.5. Research Gap

Despite progress in hierarchical and probabilistic forecasting, several important gaps remain. First, temporal hierarchies have not been applied to long-term forecasting across multiple resolutions. Second, the integration of probabilistic reconciliation methods into multi-resolution long-horizon frameworks is largely unexplored. Third, the effects of reconciliation on peak load forecasts are critical for planning under extreme conditions, but have received limited attention. The existing literature typically focuses on aggregated loads or assumes a fixed forecast horizon. To the best of our knowledge, no study has yet proposed a unified framework for generating coherent long-term forecasts across resolutions (hourly to yearly) with both deterministic and probabilistic reconciliation while also evaluating peak forecast performance.

3. Materials and Methods

Unlike prior studies focusing predominantly on short-term horizons or spatial hierarchies, our approach leverages temporal hierarchies to produce coherent and probabilistic forecasts across multiple time resolutions for long-term forecasting. This integration of various time scales is crucial for addressing uncertainties due to climate change, policy shifts, and economic trends, which are more pronounced in long-term horizons.

This section outlines our approach to generating independent probabilistic forecasts across hourly, daily, monthly, and yearly resolutions. We first describe the methodology for producing original forecasts, then detail the reconciliation methods for deterministic and probabilistic forecasts within a temporal hierarchy structure.

3.1. Original Forecasts

Forecasts can be categorized as either ex ante or ex post. Ex ante forecasts rely solely on the information available at the time of the forecast, while ex post forecasts incorporate data that become available later, including observed values of predictors [43]. For example, if a weather-dependent load forecast for the year 2025 were made at the beginning of 2025 using only historical data up to 2024, this would be classified as an ex ante forecast. In contrast, if the load for 2024 were forecasted using the actual weather data from 2024, this would be considered an ex post forecast. Ex post forecasting is a common approach reported in most load forecasting studies [37] when the goal is comparing model performance by removing other sources of error. In this paper, we start with ex post forecasting, then move on to the more practical ex ante forecasting.

The same models and feature sets are used for both ex post and ex ante forecasts. The only difference is that ex post forecasting uses observed feature values, while ex ante forecasting uses scenario-generated inputs. For ex post forecasts, actual feature values are used to produce deterministic out-of-sample electricity load forecasts for each level within the temporal hierarchy. In contrast, ex ante forecasting relies on scenarios generated from historical weather and economic variables in order to create probabilistic long-term load forecasts. The mean value of the probabilistic forecasts is taken as the deterministic forecast. Initially, the original forecasts are not coherent across the temporal hierarchy; thus, reconciliation methods are applied to generate coherent forecasts.

Figure 1 illustrates the block diagram of the proposed method. First, feature engineering is performed across all resolutions. Following Algorithm A1, new features are generated from the available data. Because an excessive number amount of features results in overfitting on the part of the models, a feature selection step in included in Algorithm A2. Feature selection is conducted independently for each resolution, acknowledging that different temporal scales (e.g., hourly, yearly) are influenced by distinct underlying drivers. For example, hourly forecasts may rely on weather and calendar effects, while yearly forecasts emphasize economic and demographic indicators. For each resolution, we use XGBoost's gain-based feature importance to rank the predictors and apply a recursive elimination process similar to recursive feature elimination (RFE), removing the least important features one at a time. Model performance is evaluated after each iteration using rolling-origin backtesting with five years of training and one year of validation, which is repeated 13 times by monthly shifting of the origin. This approach mitigates overfitting and allows out-of-sample performance to guide feature selection. In the yearly resolution, the number of observations is small; thus, we begin with a limited number of aggregated features and apply a conservative pruning strategy to ensure model simplicity and reduce the risk of overfitting.

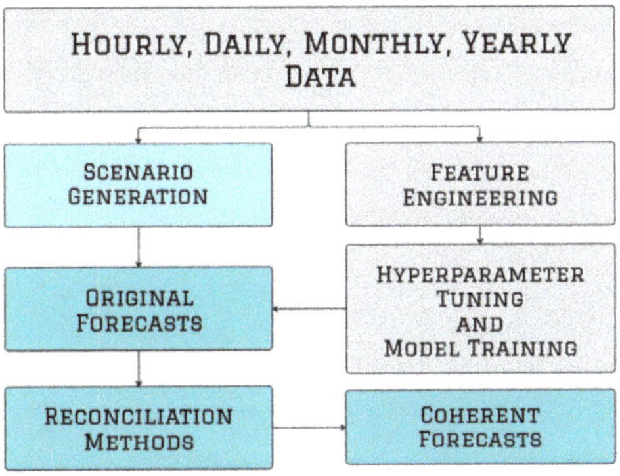

Figure 1. Illustration of the multi-resolution forecasting pipeline. Separate models and feature sets are used at each resolution, and reconciliation is applied to ensure the resulting forecasts are temporally coherent.

Each resolution is modeled independently using the features and algorithms best suited to its temporal scale while reflecting the distinct patterns and drivers at each level. Rather than assuming consistency across resolutions, we enforce it through a reconciliation step that adjusts the independently generated forecasts into a coherent system, as illustrated in Figure 1. XGBoost and ARIMAX are selected as the base forecasting models for the temporal hierarchy due to their demonstrated effectiveness in long-term load forecasting applications. XGBoost is a tree-based ensemble learning algorithm that has been widely used for its ability to model nonlinear relationships and interactions in high-resolution time series data. Electricity load data are influenced by weather, calendar effects, and behavioral factors [44–46]. XGBoost's scalability and interpretability make it well suited for applications requiring fine-grained forecasts, including prediction of hourly, daily, and monthly loads. In contrast, annual load forecasts typically contain fewer observations and exhibit smoother long-term trends driven by macroeconomic indicators. For this reason, we employ ARIMAX at the annual resolution. ARIMAX is a classical time series model that incorporates exogenous variables. It remains a competitive choice for long-horizon forecasting due to its robustness, simplicity, and strong performance in capturing long-term trends [34,47]. Together, these two models provide a stable and interpretable foundation for the hierarchical reconciliation framework proposed in this work. Algorithm A3 shows the procedure for tuning and training the models for all resolutions.

Temperature and economic scenarios generated with Algorithm A4 are used to generate the ex ante probabilistic forecasts, with deterministic forecasts derived as the mean using Algorithm A5. Economic scenarios are based on historical trends and growth scenarios, while temperature scenarios are based on historical values and shifting. After the original forecasts are generated, reconciliation methods are applied following Algorithm A6 to ensure coherence across the various temporal resolutions.

3.2. Hierarchical Modeling and Reconciliation

When data have a hierarchical structure, such as the structure displayed in Figure 2, hierarchical modeling and reconciliation methods can be used to improve forecasts. The figure illustrates a temporal hierarchy; in this case, the highest level is a time series with a frequency of one week, while the second and third levels have frequencies of days and hours, respectively.

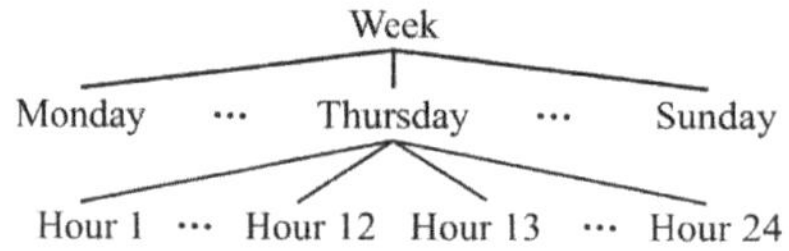

Figure 2. A simple temporal hierarchy in a power system.

Considering a time series $\{y_t^{[1]}; t = 1,2,3,\ldots,T\}$ with the highest resolution, in this case hourly, the number of series at the bottom level is $m = 168$. The k-aggregates can be constructed from y_t, where k is a factor of m. The total number of series in the hierarchy is $n = 1 + 7 + 168 = 176$. The various aggregated series can be written as [21]

$$y_j^{[k]} = \sum_{t=t^*+jk-k}^{t^*+jk-1} y_t, \tag{1}$$

where j ranges from 1 to $\lfloor T/k \rfloor$, k belongs to $\{k_1, k_2 \ldots, k_p = m\}$, and $t^* = T - \lfloor T/k \rfloor m + 1$. Using Equation (1), each series in the hierarchy in Figure 2 can be expressed as $y^{[k]}_{M_k(i-1)+\delta_k}$, where i ranges from 1 to $\lfloor T/m \rfloor$, $M_k = m/k$ and $\delta_k = 1,2,\ldots,M_k$. Figure 3 shows the new expression for the hierarchy using the common index i. We obtain the vector of all observations in the hierarchy as $\boldsymbol{y}_i = [y^{[k]}_{M_k(i-1)+\delta_k}]$, where $\delta_k \in \{1,2,\ldots,M_k\}$ and the vector of all observations in the lowest level of hierarchy is $\boldsymbol{\beta}_i = [y^{[1]}_{m(i-1)+\delta_1}]$, where $\delta_1 \in \{1,2,\ldots,m\}$. Now, we have

$$\boldsymbol{y}_i = \mathbf{S}\boldsymbol{\beta}_i, \tag{2}$$

where $\mathbf{S}$ is an $n \times m$ matrix called the summing matrix [43].

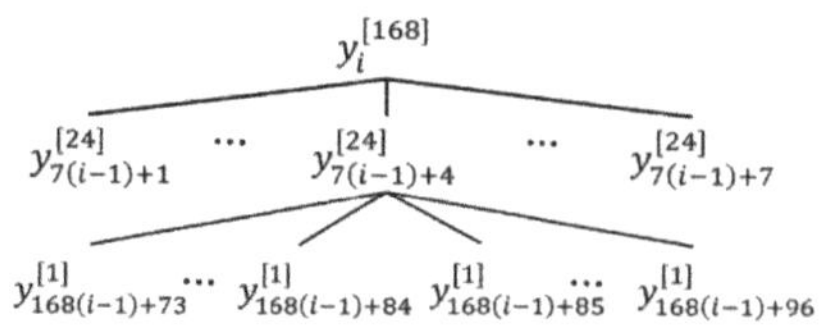

Figure 3. The simple temporal hierarchy from Figure 2 using the common index i for all levels of aggregation.

The reconciliation problem can be considered as a combination of direct forecasts of each time series, with additional indirect forecasts derived from the linear constraints resulting from the relation between the time series; in other words, each aggregation level in the hierarchy is forecasted separately and the results are reconciled to obtain coherent forecasts. Here, $\hat{\boldsymbol{y}}_h$ is the vector of all forecasts for any h-step horizon at the highest level of the hierarchy and $\tilde{\boldsymbol{y}}_h$ is the vector of reconciled forecasts:

$$\tilde{y}_h = \mathbf{SG}\hat{y}_h \tag{3}$$

where $\mathbf{G}$ represents the mapping matrix applied in different reconciliation methods, as described by [17]. The bottom-up method starts with forecasts at the lowest level of the hierarchy and aggregates them to produce forecasts for higher levels; conversely, the top-down method begins with a forecast at the highest level, which is then disaggregated down to lower levels. In the middle-out approach, forecasts are generated at an intermediate level of the hierarchy, then disaggregated to lower levels and aggregated to higher levels as needed. Lastly, the optimal reconciliation method independently forecasts each hierarchical level and then adjusts these forecasts to ensure coherency across levels. This adjustment is optimized to minimize the overall forecast error.

3.3. Deterministic Forecast Reconciliation

The objective of a reconciliation method is to determine the optimal matrix $\mathbf{G}$. In the case of optimal reconciliation, the best matrix $\mathbf{G}$ is the one that minimizes the error variances of the coherent forecasts. With h^* as the maximum required forecast horizon at the lowest level of the hierarchy (highest resolution) and h as the forecast horizon required at the highest level of the hierarchy (lowest resolution), the variance–covariance matrix of the h-steps-ahead coherent forecast errors is provided as follows [43]:

$$V_h = Var[y_{T+h} - \tilde{y}_h] = \mathbf{SGW}_h\mathbf{G}'\mathbf{S}' \tag{4}$$

where $\mathbf{W}_h = Var[(y_{T+h} - \hat{y}_h)]$ is the variance–covariance matrix of the original forecast errors. Now, we have

$$\mathbf{G} = (\mathbf{S}'\mathbf{W}_h^{-1}\mathbf{S})^{-1}\mathbf{S}'\mathbf{W}_h^{-1}. \tag{5}$$

To find the best matrix $\mathbf{G}$, we need the variance–covariance matrix $\mathbf{W}_h$ in Equation (5). If $\mathbf{W}_h$ were known, the problem could be easily solved using a generalized least squares (GLS) estimator. However, knowing the exact $\mathbf{W}_h$ relies on knowing the actual forecast error values, which is not possible. Thus, we instead estimate $\mathbf{W}_h$. A straightforward approach involves using an ordinary least squares (OLS) estimator and assuming $\mathbf{W}_h = \sigma^2\mathbf{I}$.

An alternative approach is to use weighted least squares (WLS), which uses in-sample forecast errors to estimate $\mathbf{W}_h$. Defining the in-sample one-step-ahead original forecast errors as $e_i = (e_i^{[m]'}, \ldots, e^{[k_2]'}, e^{[k_1]'})'$ for $i = 1, \ldots, \lfloor T/m \rfloor$, we have

$$\mathbf{W}_h = k_h\mathbf{\Lambda}, \tag{6}$$

where $k_h \geq 0$ and

$$\mathbf{\Lambda} = \frac{1}{\lfloor T/m \rfloor}\sum_{i=1}^{\lfloor T/m \rfloor} e_i e_i. \tag{7}$$

Two specific alternatives to the WLS method for temporal hierarchies were introduced by [21]. The series variance scaling (WLS_V) method assumes that the forecast error variances are consistent within each level of a temporal hierarchy, as they correspond to the same time series. This assumption reduces the number of variances that need to be estimated to the number of aggregation levels. Here, $\mathbf{\Lambda}_V$ is a diagonal matrix that contains estimates of the in-sample one-step-ahead error variances across each level. In addition to the assumptions from (WLS_V), the structural scaling (WLS_S) method assumes that the bottom-level original forecast error variance is σ^2 and that $\mathbf{W}_h$ is set to $\sigma^2 \times \mathbf{\Lambda}_S$, where $\mathbf{\Lambda}_S$ is a diagonal matrix with each element containing the number of forecast errors contributing to that aggregation level:

$$\mathbf{\Lambda}_S = diag(\mathbf{S1}) \tag{8}$$

where $\mathbf{1}$ is a unit column vector with a dimension of m. In this way, only one variable estimate is needed.

Two other alternatives for $\mathbf{W}_h$ were presented by [43], called Mint(sample) and Mint(shrink). In the MinT(sample) method, $\mathbf{W}_h = k_h \mathbf{W}_1$. This approach is not recommended when the number of bottom-level series m is large relative to the series length T. In the MinT(shrink) method, $\mathbf{W}_h = k\mathbf{W}^*_{1,D}$, where $\mathbf{W}^*_{1,D} = \lambda_D \mathbf{W}_D + (1-\lambda_D)\mathbf{W}_{1,D}$. Here, $\mathbf{W}_D$ is a diagonal matrix comprising the diagonal entries of $\mathbf{W}_1$, while λ_D is the shrinkage intensity parameter. when the shrinkage intensity parameter is high, this method is similar to the WLS_V method.

In this paper, the MinT(shrink) method is deployed along with OLS, WLS_S, WLS_V, and a simple bottom-up reconciliation approach as benchmarks in a temporal hierarchy structure for the proposed long-term load forecasting model. Our goal is to produce a one-year-ahead load forecast with four aggregation levels—hourly, daily, monthly, and yearly—leveraging information from each resolution.

We exclude the MinT(sample) method from our analysis, as in our case $h = 1$, making MinT(sample) equivalent to the WLS approach. Additionally, given that $m = 8760$ and T = 87,600, MinT(sample) is unsuitable. We implement these reconciliation methods using the Python Hierarchical Forecast library developed by [42].

The reconciliation methods used in this study rely on key assumptions about the statistical properties of forecast errors. In particular, WLS-based methods assume that reconciliation weights can be estimated using a known or estimable variance–covariance matrix of base forecast errors. We follow standard practice by estimating this matrix using in-sample residuals under the assumption of approximate normality and temporal independence [48]. To evaluate the impact of this assumption, we also explore an alternative approach that estimates the variance structure from out-of-sample forecast errors. This method, labeled $WLS_{V\text{-}OSE}$, provides a more realistic treatment of forecast uncertainty in long-horizon settings.

This analysis aims to evaluate the effectiveness of reconciliation methods in producing coherent long-term deterministic load forecasts in both ex post and ex ante approaches. In addition to deterministic forecasts, our approach produces probabilistic forecasts, which also require reconciliation to ensure consistency across temporal levels. Although real-world load forecast errors may exhibit skewness or seasonality, the Gaussian assumption enables computationally efficient reconciliation and has shown strong empirical performance in prior studies.

3.4. Probabilistic Forecast Reconciliation

According to [27], for an h-steps-ahead probabilistic forecast presented as $\mathcal{N}(\hat{y}_T + h, \mathbf{W}_h)$, where $\hat{y}_T + h$ is the mean h-steps-ahead forecast and

$$\mathbf{W}_h = E[y_T + h - \hat{y}_T + h][y_T + h - \hat{y}_T + h]^T, \tag{9}$$

there is a matrix **G** that produces

$$\mathcal{N}(\mathbf{SG}\hat{y}_T + h,\ \mathbf{SGW}_h\mathbf{G}^T\mathbf{S}^T), \tag{10}$$

which represents the h-steps-ahead reconciled forecast. Assuming that the predictive densities of the original forecasts follow a Gaussian distribution, the same matrix **G** used for reconciling point forecasts can also be employed for probabilistic forecasts. Therefore, the reconciliation methods used for deterministic forecasts are extended to probabilistic forecast reconciliation in this study.

Our probabilistic reconciliation is based on the Gaussian assumption, as proposed by [27], where the forecast distribution is assumed to follow a multivariate normal distribution. This permits closed-form reconciliation through linear algebraic transformation.

4. Experimental Setup and Empirical Evaluation Results

This section outlines the experimental setup and empirical evaluation of our proposed long-term load forecasting model. We utilize electricity demand data from the cities of Calgary and Edmonton as well as from the province of Alberta, Canada to assess the model's performance. The evaluation focuses on various forecasting scenarios, encompassing both deterministic and probabilistic approaches. We detail the data sources and characteristics, the generation of weather and economic scenarios, and the evaluation metrics employed to measure the model's reliability. The results from deterministic and probabilistic forecasts are discussed, highlighting the effectiveness of reconciliation approaches in producing coherent forecasts in a temporal hierarchy.

4.1. Data

We extracted the historical hourly load data for the cities of Calgary and Edmonton and for the province of Alberta from [49,50] for the period of 2012–2024. For our analysis, we used six years of data for training and applied a monthly rolling validation strategy over the following year. This process was repeated to span three years of validation, enabling effective model tuning. The hourly load from the next available year (2020) was used for out-of-sample testing. This process was repeated for the next three years to produce one-year ahead out-of-sample load forecasts for four years, and the results are reported for the combination of all four years.

The three datasets used in this study—Calgary, Edmonton, and the province of Alberta—each represent distinct load characteristics. Calgary and Edmonton primarily reflect residential and commercial consumption patterns typical of urban centers, while the Alberta-wide dataset is dominated by industrial demand. This diversity in load composition allows for the evaluation of forecasting and reconciliation methods across a wide range of demand behaviors. Although the data are geographically limited to one jurisdiction, the variation in load types makes the case study more broadly informative.

The available information is presented at various resolutions. Historical economic variables are typically available at yearly, quarterly, and monthly frequencies. We incorporated economic data from [51] alongside long-term climate trends for monthly and yearly load forecasting. Weather-related variables [52] are available at higher resolutions, and are utilized in forecasting hourly and daily loads.

Table 1 outlines the information groups used for load forecasting at each resolution. Each group consists of one or more variables of a similar nature; for instance, solar radiation and visibility are categorized together. To enhance the model's predictive capability, we included squared and cubed forms of weather variables as well as their interactions with dummy variables such as hour of the day and day of the week.

To ensure data quality and robust evaluation, we performed preprocessing on all input series. Missing values were filled using forward-filling and seasonal mean backfilling. Outliers were detected for the training period only using interquartile range (IQR)-based filtering and replaced with local median values. Numerical features were standardized within each resolution level to control for scale effects. Importantly, the one-year-ahead forecasting process was repeated across four independent test years for each region (Calgary, Edmonton, and Alberta), providing a robust basis for generalization and mitigating the risk of overfitting to a single forecast year. The consistency of performance improvements across these multi-year and multi-region experiments reinforces the robustness of the framework under varying conditions and input distributions.

Table 1. Information used for forecasting loads in different resolutions.

Feature Type	Feature Group	Hourly Load	Daily Agg	Monthly Agg	Yearly Agg
Time Related	Calendar dummy variables	✓	✓	✓	✓
	Day light	✓	✓	✗	✗
Weather related	Temperature	✓	a,b,c	a,b,c	a
	Humidity	✓	a	a	✗
	Pressure	✓	a	a	✗
	Solar radiation	✓	a	a	✗
	Wind	✓	a	a	✗
	Precipitation	✓	a	a	✗
Economy related	GDP	✗	✗	✗	✓
	Population	✗	✗	✗	✓
	Productivity	✗	✗	✗	✓
	Income	✗	✗	✓	✗
	Inflation	✗	✗	✓	✗
	Unemployment	✗	✗	✓	✗
	Housing	✗	✗	✓	✗
	Drills	✗	✗	✓	✗
	Oil & Gas production	✗	✗	✓	✗

Note: a: Averaged value of a variable in specific periods. b: Maximum value of a variable in specific periods. c: Minimum value of a variable in specific periods.

4.2. Scenario Generation

Probabilistic forecasts can be derived from various scenarios, especially when producing ex ante forecasts where input data such as weather variables for the forecasting period are unavailable. In this context, scenario generation methods are employed to facilitate the production of probabilistic forecasts. Weather-related variables are particularly crucial in load forecasting, and using simulated weather scenarios in point load forecasting models has gained widespread acceptance in the industry thanks to its simplicity and ease of interpretation.

As outlined by [12], weather scenario simulation methods range from straightforward to more complex approaches. The simplest method directly uses the original historical weather data, while a moderate approach involves reorganizing the data through techniques such as rotation or bootstrapping. The most advanced methods entail simulating historical weather patterns using mathematical models, such as applying Fourier transformation to the original weather series to generate additional scenarios. The first two methods are practical and widely used in industry [12]. However, given the limited availability of historical hourly data, the first approach results in a restricted number of scenarios. Therefore, we opt here for the modest approach to temperature scenario generation.

Specifically, we used rotation of historical weather data to generate 200 weather scenarios for hourly and daily forecasts. The same method was applied to produce 50 monthly and yearly weather scenarios alongside four economic growth scenarios (normal growth, double growth, no growth, and negative growth), leading to a total of 200 scenarios. The resulting forecasts from these scenarios for each resolution were then utilized to construct the probability density function for the probabilistic forecasts, with the mean value of these probability density functions serving as the deterministic forecasts.

4.3. Evaluation Metrics

Evaluation of deterministic forecasts is conducted using the root mean square error (RMSE) metric. The RMSE is particularly sensitive to outliers, such as sudden peak loads in our case, making it an appropriate choice for assessing our results. For the evaluation of probabilistic forecasts, we utilize the Pinball loss function and Winkler score, both of which are widely recognized in the field of probabilistic forecast evaluation.

The Pinball loss function serves as an error measure that accounts for both sharpness and resolution in quantile forecasts. Let $\hat{y}_{t,q}$ denote the load forecast at the q_{th} quantile and let y_t represent the actual load value observed; then, the Pinball loss function can be expressed as follows [13].

$$P(\hat{y}_{t,q}, y_t, q) = \begin{cases} (1-q)(\hat{y}_{t,q} - y_t) & y_t < \hat{y}_{t,q} \\ q(y_t - \hat{y}_{t,q}) & y_t \geq \hat{y}_{t,q} \end{cases} \tag{11}$$

The Winkler score allows for a joint assessment of unconditional coverage and interval width. For a central $(1-\alpha) \times 100\%$ prediction interval, $[l_{\alpha,t}, u_{\alpha,t}]$ it is defined as follows [43].

$$W = \begin{cases} (u_{\alpha,t} - l_{\alpha,t}) + \frac{2}{\alpha}(l_{\alpha,t} - y_t) & y_t < l_{\alpha,t} \\ (u_{\alpha,t} - l_{\alpha,t}) & l_{\alpha,t} \leq y_t \leq u_{\alpha,t} \\ (u_{\alpha,t} - l_{\alpha,t}) + \frac{2}{\alpha}(y_t - u_{\alpha,t}) & y_t > u_{\alpha,t} \end{cases} \tag{12}$$

We calculate both the Pinball and Winkler scores for all four resolutions as well as for the daily, monthly, and yearly peak loads.

4.4. Deterministic Ex Post Forecasts

Three years of out-of-sample feature values used for forecasting load are utilized to generate forecasts across various resolutions. The hourly load forecast is subsequently used to extract daily, monthly, and yearly peak loads. A range of reconciliation methods, including bottom-up, MinT(shrink), ordinary least squares (OLS), weighted least squares $(WLS)_S$, and WLS_V, are employed to generate coherent forecasts. These methods leverage in-sample forecast errors; additionally, we apply out-of-sample errors with the WLS_V method to reconcile the forecasts. The performance of each method is evaluated by comparing the RMSE of the coherent forecasts against the RMSE of the original forecasts. The Clark–West (CW) test is used to check the statistical significance of the improvements in the forecasts after reconciliation compared to the original forecasts. Ex post and ex ante forecast results are reported separately and should not be compared directly, as they rely on different information availability at forecast time.

Tables 2–4 present the relative percentage improvements in RMSE scores for ex post deterministic forecasts across different reconciliation methods compared to the original forecasts. A positive percentage indicates a decrease in the RMSE value, suggesting that the original forecast is superior to the reconciled forecast, while a negative percentage reflects an increase in RMSE. Notably, the MinT(shrink) and WLS_V methods yield remarkably similar results, with less than a 0.001% difference across all datasets. This can be a result of a high shrinkage intensity parameter. Thus, only the results for WLS_V are reported in the tables.

Table 2. Percentage improvements in RMSE for ex post deterministic forecast for the City of Calgary.

Load Type	Bottom-Up	OLS	WLS_S	WLS_V	$WLS_{V\text{-}OSE}$
Hourly Load	0.00	0.13 *	**0.71** *	0.56 *	−0.37 *
Daily Aggregated	0.84 *	1.02 *	**1.78** *	1.64 *	0.65 *
Monthly Aggregated	2.27	1.92	4.02 *	**5.02** *	1.28
Yearly Aggregated	−65.06	**3.37**	−28.09	−23.37	−62.71
Daily Peak	0.00	**0.91** *	0.69 *	0.26 *	−0.34 *
Monthly Peak	0.00	−0.09 *	−1.99 *	−0.92 *	**0.55**
Yearly Peak	0.00	1.40	−1.39	−3.33	**3.87**

Note: Each value shows the RMSE percentage improvement compared to the original forecast. * denotes statistical significance at the 0.05 level (Clark–West test). Bold values indicate the best performance (highest improvement) for each load type. The out-of-sample error is used for $WLS_{V\text{-}OSE}$.

Table 3. Percentage improvements in RMSE for ex post deterministic forecast for the City of Edmonton.

Load Type	Bottom-Up	OLS	WLS_S	WLS_V	$WLS_{V\text{-}OSE}$
Hourly Load	0.00	−1.19 *	**0.83** *	−0.14 *	−0.15 *
Daily Aggregated	0.92	−0.57	**1.96** *	0.83 *	0.88
Monthly Aggregated	−4.87 *	−9.63	**−1.52** *	−5.21 *	−4.66 *
Yearly Aggregated	21.34	6.02 *	**30.47** *	22.32	23.63
Daily Peak	0.00	−0.16	**0.91** *	−0.13 *	−0.33 *
Monthly Peak	0.00	**3.50**	1.23 *	−0.26 *	−0.17 *
Yearly Peak	0.00	**3.88**	−0.81	−1.38	−0.06

Note: Each value shows the RMSE percentage improvement compared to the original forecast. * denotes statistical significance at the 0.05 level (Clark–West test). Bold values indicate the best performance (highest improvement) for each load type. The out-of-sample error is used for $WLS_{V\text{-}OSE}$.

Table 4. Percentage improvements in RMSE for ex post deterministic forecast for the Province of Alberta.

Load Type	Bottom-Up	OLS	WLS_S	WLS_V	$WLS_{V\text{-}OSE}$
Hourly Load	0.00	−10.50 *	**12.80** *	1.32 *	0.80 *
Daily Aggregated	−10.90 *	−24.17 *	**5.58** *	−9.16 *	−8.42 *
Monthly Aggregated	−28.58 *	−47.80	**6.55** *	−25.36 *	−21.77 *
Yearly Aggregated	23.31 *	5.70 *	**59.84** *	26.43 *	30.31 *
Daily Peak	0.00	−6.77 *	**14.90** *	1.81 *	−1.32 *
Monthly Peak	0.00	−1.17 *	**19.49** *	2.85 *	−12.69 *
Yearly Peak	0.00	**55.42** *	37.13 *	3.95	−76.82 *

Note: Each value shows the RMSE percentage improvement compared to the original forecast. * denotes statistical significance at the 0.05 level (Clark–West test). Bold values indicate the best performance (highest improvement) for each load type. The out-of-sample error is used for $WLS_{V\text{-}OSE}$.

While improvements in forecast accuracy through reconciliation are certainly beneficial, they are not the primary objective. The main goal of reconciliation methods is to ensure coherence across forecasts. Therefore, the focus is on comparing the results of various reconciliation methods with the simple bottom-up approach.

Table 2 presents the results for ex post deterministic forecasts in Calgary. The results show that all methods except $WLS_{V\text{-}OSE}$ achieve statistically significant improvements in RMSE at the hourly, daily aggregated, and daily peak levels, as indicated by the dagger symbols. The best performance at each frequency is generally achieved by either WLS_S or WLS_V, particularly for hourly and daily resolutions, with the OLS and bottom-up methods also showing significant but smaller improvements. For monthly and yearly aggregates and peaks, improvements are less often statistically significant. While both WLS_V and $WLS_{V\text{-}OSE}$ show similar trends, there is no clear evidence of consistent superiority of either method in this setting.

Overall, these results support the benefit of reconciliation for improving the coherence and accuracy of ex post deterministic forecasts in Calgary, particularly OLS and WLS_S, with statistical evidence showing that most improvements are meaningful rather than due to random variability. Notably, these results highlight that the extent of improvement and significance can vary substantially by aggregation level and load type.

The ex post deterministic forecasting results for the City of Edmonton, shown in Table 3, reveal meaningful improvements in RMSE across several reconciliation methods and aggregation levels. The WLS_S method yields the highest and statistically significant RMSE improvements for most load types, particularly for hourly, daily aggregated, and yearly aggregated loads. In the case of monthly aggregated loads, WLS_S still provides the least negative (best) result among methods, with significance also observed for bottom-up, WLS_V, and $WLS_{V\text{-}OSE}$. For hourly, daily, and monthly peak loads, WLS_S again shows the largest improvements, most of which are statistically significant. The OLS method occasionally shows significance, notably in yearly aggregated and monthly peak loads, but with lower improvements compared to WLS-based approaches.

When comparing WLS_V and $WLS_{V\text{-}OSE}$ (the latter using out-of-sample error for reconciliation), their performance is generally similar across most load types. The difference in RMSE improvement between WLS_V and $WLS_{V\text{-}OSE}$ remains small for most metrics, and both methods show comparable significance patterns. These results suggest that in the context of Edmonton's long-term forecasting, employing out-of-sample errors for WLS-based reconciliation does not lead to substantial changes in performance.

The ex post deterministic results for the province of Alberta, shown in Table 4, reveal that reconciliation methods can provide substantial improvements in RMSE for many load types and aggregation levels. This is particularly the case for WLS_S, with the majority of improvements being statistically significant ($p < 0.05$). WLS_S delivers the highest improvements overall, achieving double-digit gains for hourly, yearly aggregated, daily peak, monthly peak, and yearly peak forecasts. Similar to the other two cases, employing out-of-sample errors in the reconciliation process does not substantially alter the forecast improvements compared to using in-sample errors.

The results indicate that employing a temporal hierarchy structure can enhance the accuracy of ex post deterministic forecasts while ensuring coherence across forecasts. It is obvious that the performance of reconciliation methods varies across datasets. The difference in performance can be related to the composition of load in each dataset, although this needs to be investigated further in the future. Considering all of these results, the WLS_S method consistently improves forecast performance for hourly and daily loads as well as daily peak loads in all three datasets.

Although methods such as WLS_V are typically designed for situations where using out-of-sample errors is challenging or infeasible, our findings indicate that incorporating out-of-sample errors only rarely improves forecast performance, and in many cases can actually reduce forecast accuracy.

It is important to note when using reconciliation methods that there can be a deterioration in forecast performance compared to the original forecasts, particularly at higher levels of the temporal hierarchy. Nonetheless, the primary objective is to produce coherent forecasts across all hierarchy levels. The simplest way to achieve this is through the bottom-up approach. Our results demonstrate that optimal reconciliation methods such as WLS_S yield better forecasts compared to the bottom-up approach.

4.5. Deterministic Ex Ante Forecasts

While ex post forecasts are excellent for comparing model performance by isolating the external error effects, ex ante forecasts are more practical and useful in industrial contexts. While ex post forecasts are affected by errors in the forecasting model, ex ante forecasting must additionally face the problem of input errors. Ex ante deterministic forecasts are derived by calculating the mean values of the probability distributions from the scenario-based probabilistic forecasts. The combined results for four years of one-year-ahead ex ante load forecasts are presented in Tables 5–7.

Table 5. Percentage improvements in RMSE for ex ante deterministic forecast for the City of Calgary.

Load Type	Bottom-Up	OLS	WLS_S	WLS_V	$WLS_{V\text{-}OSE}$
Hourly Load	0.00	2.83 *	**4.69** *	1.44 *	1.08 *
Daily Aggregated	−6.76	−2.84 *	**−0.24** *	−4.77	−5.12
Monthly Aggregated	−5.02 *	1.16	**9.06** *	−0.86 *	−1.39 *
Yearly Aggregated	−4.67 *	−0.30	**40.75**	12.01	0.53
Daily Peak	0.00	1.44 *	**3.61** *	1.04 *	1.37 *
Monthly Peak	0.00	2.29	3.47 *	1.43 *	**6.77** *
Yearly Peak	0.00 *	**1.01** *	−0.57 *	−0.19 *	0.53 *

Note: Each value shows the RMSE percentage improvement compared to the original forecast. * denotes statistical significance at the 0.05 level (Clark–West test). Bold values indicate the best performance (highest improvement) for each load type. The out-of-sample error is used for $WLS_{V\text{-}OSE}$.

Table 6. Percentage improvements in RMSE for ex ante deterministic forecast for the City of Edmonton

Load Type	Bottom-Up	OLS	WLS_S	WLS_V	$WLS_{V\text{-}OSE}$
Hourly Load	0.00	4.45 *	**4.97** *	1.43 *	0.56 *
Daily Aggregated	−4.95	0.85 *	**1.53** *	−3.01	−4.19
Monthly Aggregated	−9.91	0.30	**4.77** *	−5.17	−8.35
Yearly Aggregated	−30.20	1.36	**6.57**	−14.35	−31.43
Daily Peak	0.00	−0.77 *	**2.77** *	0.88 *	0.77 *
Monthly Peak	0.00	−5.93	−0.21	0.28	**3.50** *
Yearly Peak	**0.00**	−11.04	−6.47	−0.89	−0.13

Note: Each value shows the RMSE percentage improvement compared to the original forecast. * denotes statistical significance at the 0.05 level (Clark–West test). Bold values indicate the best performance (highest improvement) for each load type. The out-of-sample error is used for $WLS_{V\text{-}OSE}$.

Table 7. Percentage improvements in RMSE for ex ante deterministic forecast for the Province of Alberta.

Load Type	Bottom-Up	OLS	WLS_S	WLS_V	$WLS_{V\text{-}OSE}$
Hourly Load	0.00	**27.48** *	22.70 *	0.09 *	−0.75
Daily Aggregated	13.45 *	**39.53** *	34.92 *	13.54 *	12.81 *
Monthly Aggregated	−50.42	**6.12** *	−5.01	−50.36	−51.78
Yearly Aggregated	−157.86	**7.25** *	−57.79	−157.88	−161.01
Daily Peak	0.00	**40.77** *	23.19 *	0.26 *	−0.34
Monthly Peak	0.00	**44.57** *	20.86 *	0.63 *	0.79
Yearly Peak	0.00	**54.97** *	18.04 *	0.84	−0.09

Note: Each value shows the RMSE percentage improvement compared to the original forecast. * denotes statistical significance at the 0.05 level (Clark–West test). Bold values indicate the best performance (highest improvement) for each load type. The out-of-sample error is used for $WLS_{V\text{-}OSE}$.

Table 5 presents RMSE improvements for ex ante deterministic forecasts in Calgary. The results indicate that the reconciliation methods deliver significant improvements in RMSE at several aggregation levels, particularly WLS_S and OLS. Most *p*-values are below 0.05, confirming statistical significance. WLS_S provides the largest and most consistent gains, especially for hourly (4.69%) and monthly (9.06%) aggregated loads. In particular, improvements are also significant for daily and monthly peaks in most reconciliation approaches. In contrast, negative or small improvements are observed for daily and monthly aggregates under some methods, with these changes generally being statistically significant as well. The difference between WLS_V and $WLS_{V\text{-}OSE}$ is minimal, with both delivering similar results. In general, these findings highlight the strong benefits of reconciliation, especially WLS_S, in improving the precision of long-term ex ante forecasts for most resolutions and load types.

Table 6 shows RMSE improvements for ex ante deterministic forecasts in Edmonton. Notable improvements in RMSE are apparent, especially for hourly and monthly aggregated loads. WLS_S consistently provides the largest percentage improvement for most load types (e.g., 4.97% for hourly and 4.77% for monthly), with OLS also yielding statistically significant gains across hourly and daily aggregated categories. The improvements for hourly, daily, and daily peak loads are generally supported by p-values below 0.05, indicating robust statistical significance. Some negative improvements are observed for daily and monthly aggregates under certain methods; however, they are not statistically significant. The results for WLS_V and $WLS_{V\text{-}OSE}$ are similar, and neither approach provides substantial additional benefit over WLS_S. Overall, the findings for Edmonton highlight the effectiveness of reconciliation, particularly WLS_S and OLS, for improving ex ante forecast accuracy across most resolutions.

For ex ante deterministic forecasts in Alberta (Table 7), the reconciliation methods deliver large and statistically significant improvements in RMSE at most resolutions, particularly OLS and WLS_S. OLS achieves the highest gains in nearly every category, notably for hourly (27.48%), daily (39.53%), and yearly peak (54.97%) loads. All corresponding p-values are below 0.05, confirming the significance of these improvements. WLS_S also yields substantial and significant improvements, especially for hourly, daily aggregated, and peak loads in every level. For aggregated monthly and yearly series, most methods actually show negative improvements, reflecting a loss of accuracy, though OLS still provides small but significant gains. The WLS_V and $WLS_{V\text{-}OSE}$ methods generally do not offer additional improvement over OLS or WLS_S, and sometimes result in negative or negligible changes. These results highlight the robust advantage of OLS and WLS_S reconciliation for high- and mid-resolution load forecasting in Alberta, with performance on monthly and yearly aggregation being more method-dependent.

While reconciliation typically provides overall coherence and often enhances forecast accuracy across levels, there are cases where performance at a specific resolution may degrade. For example, when the original forecast at a particular level (e.g., yearly) is already highly accurate, the reconciliation step may shift values to ensure consistency with forecasts at other levels, resulting in a tradeoff. This behavior is expected and aligns with findings in the prior hierarchical forecasting literature. Furthermore, large percentage increases in RMSE may correspond to relatively small absolute errors in certain cases, particularly when the baseline RMSE is low. These effects are not anomalies but are rather a consequence of the reconciliation objective prioritizing structural coherence over localized optimization.

The results indicate that there is no clear winner among the reconciliation methods for all levels of load forecasting. The only method that consistently improves hourly load and daily peak load forecasts across all datasets is the WLS_S method. In addition, it is important to note that using out-of-sample forecast errors instead of in-sample errors similar to ex post forecast reconciliation does not guarantee better forecasts.

4.6. Probabilistic Ex-Ante Forecasts

The Pinball and Winkler scores for all four resolution levels of the hierarchy as well as for peak load forecasts are calculated below. Tables 8–10 show the percentage relative improvements in the Pinball scores of different probabilistic reconciliation methods compared to the original forecasts. Because the Clark–West test is not designed for probabilistic forecasts, we use the Diebold–Mariano (DM) test to assess statistical significance. Unlike deterministic forecast reconciliation, where no single method consistently outperforms the others across datasets, the OLS method enhances the original hourly load forecast as well as daily, monthly, and yearly peak load forecasts in all three datasets.

Table 8. Percentage improvements in Pinball score for ex ante probabilistic forecast for the City of Calgary.

Load Type	Bottom-Up	OLS	WLS_S	WLS_V	$WLS_{V\text{-}OSE}$
Hourly Load	−4.86 *	**12.19** *	6.37 *	−2.33 *	−3.20 *
Daily Aggregated	−44.35	**3.23** *	−13.08 *	−39.03	−40.54
Monthly Aggregated	−55.77	**6.86**	−11.23	−46.04	−48.76
Yearly Aggregated	−48.83	9.05	**27.79**	−11.12	−39.34
Daily Peak	−5.19	**7.31** *	2.70 *	−3.44 *	−3.41
Monthly Peak	−4.92	5.02	1.31 *	−2.63 *	**14.46** *
Yearly Peak	−5.83 *	1.19 *	−5.45 *	−6.19 *	**15.79** *

Note: Each value shows the Pinball score percentage improvement compared to the original forecast. * denotes statistical significance at the 0.05 level (Diebold–Mariano test). Bold values indicate the best performance (highest improvement) for each load type. The out-of-sample error is used for $WLS_{V\text{-}OSE}$.

Table 9. Percentage improvements in Pinball score for ex ante probabilistic forecast for the City of Edmonton.

Load Type	Bottom-Up	OLS	WLS_S	WLS_V	$WLS_{V\text{-}OSE}$
Hourly Load	−3.63 *	**17.33** *	8.89 *	−0.16 *	−2.81 *
Daily Aggregated	−47.67 *	**9.64** *	−12.60 *	−39.98	−45.34
Monthly Aggregated	−64.01 *	**8.27**	−17.92	−50.81	−61.30
Yearly Aggregated	−120.75	**9.43**	−28.58	−70.56	−121.73
Daily Peak	−3.19 *	**8.62** *	5.31 *	−0.90 *	−2.18
Monthly Peak	−3.15 *	−5.60 *	−2.72	−2.83	**10.02** *
Yearly Peak	−0.63	−7.57	−7.83	−2.02	**11.95**

Note: Each value shows the Pinball score percentage improvement compared to the original forecast. * denotes statistical significance at the 0.05 level (Diebold–Mariano test). Bold values indicate the best performance (highest improvement) for each load type. The out-of-sample error is used for $WLS_{V\text{-}OSE}$.

Table 10. Percentage improvements in Pinball score for ex ante probabilistic forecast for the Province of Alberta.

Load Type	Bottom-Up	OLS	WLS_S	WLS_V	$WLS_{V\text{-}OSE}$
Hourly Load	−2.60 *	**41.06** *	33.43 *	−2.44 *	−3.07 *
Daily Aggregated	−25.66 *	**46.62** *	34.38 *	−25.21 *	−25.04 *
Monthly Aggregated	−131.95 *	**17.64**	−24.54	−131.68 *	−133.63 *
Yearly Aggregated	−377.10	**11.92**	−141.07	−376.28	−382.34
Daily Peak	−4.40	**52.50** *	33.41 *	−4.10 *	−3.67 *
Monthly Peak	−4.96	**59.16** *	29.32 *	−4.08 *	−5.95
Yearly Peak	−6.75	**68.51** *	23.08 *	−5.97	17.49

Note: Each value shows the Pinball score percentage improvement compared to the original forecast. * denotes statistical significance at the 0.05 level (Diebold–Mariano test). Bold values indicate the best performance (highest improvement) for each load type. The out-of-sample error is used for $WLS_{V\text{-}OSE}$.

Ex ante probabilistic forecasts in Calgary as measured by percentage improvements in Pinball score are presented in Table 8. OLS consistently delivers the most substantial and statistically significant gains, especially for hourly, daily, and peak load forecasts.

For example, OLS improves the hourly Pinball score by 12.19%, daily aggregate by 0.23%, and daily peak by 7.31%, with all improvements being statistically meaningful at the 0.05 level. WLS_S also yields significant but smaller improvements for hourly, daily and monthly peak loads. In contrast, the bottom-up and WLS_V-based methods frequently result in negative improvements, with significantly worse Pinball scores for many horizons, particularly at aggregated and peak levels. Notably, OLS and $WLS_{V\text{-}OSE}$ show meaningful positive gains for peak forecasts. Overall, OLS emerges as the most reliable reconciliation approach for improving the probabilistic accuracy of ex ante forecasts in Calgary, while other methods offer less consistent or even detrimental results across most resolutions.

Table 9 reports percentage improvements in Pinball score for ex ante probabilistic forecasts in Edmonton. Similar to the results for Calgary, OLS stands out by providing the largest and most statistically significant improvements in Pinball score across almost all horizons. OLS achieves an improvement of 17.33% for hourly load and 9.64% for daily aggregated load. WLS_S also demonstrates significant improvement for hourly and daily peak loads, with 8.89% and 5.31% improvement, respectively. However, for most aggregated and peak horizons, both the bottom-up and WLS_V-based methods yield negative or non-significant changes in Pinball score, often performing worse than the unreconciled forecast. Notably, the only consistent positive improvement for monthly and yearly peaks is observed for $WLS_{V\text{-}OSE}$ (e.g., 10.02% for monthly peak). These results confirm that OLS is generally the most robust reconciliation method for enhancing the probabilistic accuracy of ex ante load forecasts for higher resolutions in Edmonton, while the other methods show inconsistent or often detrimental effects, especially for longer aggregation intervals.

Table 10 shows Pinball score improvements for ex ante probabilistic forecasts in Alberta. Both the OLS and WLS_S reconciliation methods achieve substantial and statistically significant improvements in Pinball score at nearly all resolutions, with OLS consistently providing the largest gains of over 40% improvement for hourly, daily, and peak intervals. The p-values for these methods are almost universally below 0.05, confirming that these improvements are meaningful. In contrast, WLS_V and $WLS_{V\text{-}OSE}$ produce little to no improvement or even negative improvements across all resolutions and aggregation levels, and their performance does not significantly differ from each other. Bottom-up reconciliation generally results in negative improvements, especially at higher aggregation levels (e.g., monthly and yearly). These findings reinforce the effectiveness of OLS-based reconciliation in improving the accuracy of multi-resolution probabilistic load forecasts for large geographic regions.

Compared to the bottom-up method, the four reconciliation methods provide better results in almost all resolutions for the three datasets. While the OLS and WLS_S methods both improve upon the original hourly and daily peak forecasts, the OLS method is the only method that improves the forecasts for all aggregation levels. In addition, this method improves the peak load forecasts in different levels for two of the datasets.

The results for the Winkler score follow the same patterns as the results for the Pinball score. The OLS method demonstrates superior performance in improving the Winkler score in most cases, with WLS_S providing the second-best results. The Winkler score improvement results are reported in Tables 11–13. The OLS method shows excellent results in the probabilistic forecasts. This is because of the increasing effect on forecast intervals of probabilistic reconciliation with the OLS method, which results in higher reliability. The scoring methods used for probabilistic forecast evaluation focus on reliability and sharpness, which result in higher scores for the OLS method.

Table 11 summarizes the percentage improvements and p-values for the Winkler score in ex ante probabilistic forecasts for Calgary. the OLS and WLS methods consistently deliver better Winkler score improvements compared to the bottom-up and WLS_V variants across most load types. Statistically significant improvements (p-value < 0.05) are observed for nearly all models and load types, except for some aggregated and peak categories where improvements are smaller or not significant. The largest improvements tend to occur at the hourly and monthly levels, with WLS_V yielding the highest gain for monthly aggregation and OLS performing best at the higher resolutions. Overall, the results highlight the advantage of reconciliation-based approaches for enhancing probabilistic forecast sharpness and reliability.

Table 11. Percentage improvements in Winkler score for ex ante probabilistic forecast for the city of Calgary.

Load Type	Bottom-Up	OLS	WLS_S	WLS_V	$WLS_{V\text{-}OSE}$
Hourly Load	−5.52 *	**8.24** *	3.53 *	−3.42 *	−4.23 *
Daily Aggregated	−27.16 *	**2.72**	−8.35 *	−23.63 *	−24.86 *
Monthly Aggregated	−34.81 *	3.88 *	−5.58 *	**28.36** *	−29.79 *
Yearly Aggregated	−51.30	−8.31	**17.98**	−19.86	−42.82
Daily Peak	−4.37 *	**5.17** *	1.87 *	−3.00 *	−3.45 *
Monthly Peak	−3.18 *	3.04 *	2.13	−1.33 *	**9.36** *
Yearly Peak	−5.02 *	−1.70	−5.38 *	−5.36 *	**4.44** *

Note: Each value shows the Winkler score percentage improvement compared to the original forecast. * denotes statistical significance at the 0.05 level (Diebold–Mariano test). Bold values indicate the best performance (highest improvement) for each load type. The out-of-sample error is used for $WLS_{V\text{-}OSE}$.

Table 12. Percentage improvements in Winkler score for ex ante probabilistic forecast for the City of Edmonton.

Load Type	Bottom-Up	OLS	WLS_S	WLS_V	$WLS_{V\text{-}OSE}$
Hourly Load	−7.06 *	**13.08** *	4.34 *	−3.97	−6.37
Daily Aggregated	−35.78 *	**5.01** *	−12.38 *	−30.34 *	−34.35 *
Monthly Aggregated	−40.96 *	**6.11** *	−10.72 *	−32.67 *	−39.06 *
Yearly Aggregated	−120.43	**−7.88**	−49.21	−80.37	−121.75
Daily Peak	−6.58 *	**5.16** *	1.49 *	−4.47 *	−5.81 *
Monthly Peak	−2.33 *	−7.99	−2.91	−1.99 *	**5.61** *
Yearly Peak	−10.98	−22.64	−19.61	−12.47	**−8.96**

Note: Each value shows the Winkler score percentage improvement compared to the original forecast. * denotes statistical significance at the 0.05 level (Diebold–Mariano test). Bold values indicate the best performance (highest improvement) for each load type. The out-of-sample error is used for $WLS_{V\text{-}OSE}$.

Table 13. Percentage improvements in Winkler score for ex ante probabilistic forecast for the Province of Alberta.

Load Type	Bottom-Up	OLS	WLS_S	WLS_V	$WLS_{V\text{-}OSE}$
Hourly Load	−0.15 *	**41.81** *	33.08 *	−0.10 *	−0.92 *
Daily Aggregated	−3.90 *	**49.11** *	38.36 *	−3.74 *	−4.20 *
Monthly Aggregated	−103.37 *	**10.57** *	−16.79 *	−103.34 *	−105.37 *
Yearly Aggregated	−342.56 *	**0.11**	−138.56	−342.36 *	−347.75 *
Daily Peak	−1.45 *	**55.41** *	33.67 *	−1.24 *	−1.65 *
Monthly Peak	−2.68 *	**58.83** *	28.55 *	−1.94	1.47
Yearly Peak	−8.92	**65.70** *	17.54 *	−8.14	−1.82

Note: Each value shows the Winkler score percentage improvement compared to the original forecast. * denotes statistical significance at the 0.05 level (Diebold–Mariano test). Bold values indicate the best performance (highest improvement) for each load type. The out-of-sample error is used for $WLS_{V\text{-}OSE}$.

Table 12 presents the percentage improvements and p-values for the Winkler score across ex ante probabilistic forecasts for Edmonton. The OLS reconciliation method produced the most consistent and significant improvements, especially at high-frequency and daily aggregation levels, with the exception of yearly and monthly peak where improvements are less pronounced or not statistically significant. WLS_S offers some improvement for hourly, daily peak, and monthly aggregated levels, with significance for most but limited effectiveness at higher temporal resolutions. Both WLS_V and $WLS_{V\text{-}OSE}$ rarely results in meaningful improvements and often shows negative or negligible changes, with only the monthly peak showing a statistically significant gain for $WLS_{V\text{-}OSE}$. Bottom-up results are generally negative and significant, confirming the advantage of optimal reconciliation methods. Overall, OLS reconciliation stands out as the most robust method for improving interval forecast accuracy in Edmonton, with other reconciliation methods offering inconsistent or marginal benefits depending on the temporal resolution.

Table 13 reports the percentage improvements in Winkler score for ex ante probabilistic forecasts in Alberta, with statistical significance (p-values) from the Diebold–Mariano test indicated below each value. The OLS and WLS_S reconciliation methods consistently deliver the largest and most significant improvements in interval forecast accuracy, particularly at high-frequency (hourly, daily) and peak levels. OLS shows the strongest and most stable performance across all aggregation levels, with improvements exceeding 40% for hourly and daily forecasts and 65.7% for the yearly peak. WLS_S also performs well, with notable improvements for daily, monthly, and yearly peaks.

In contrast, the bottom-up and variance-based (WLS_V and $WLS_{V\text{-}OSE}$) methods generally result in either negligible or negative improvements. Overall, the results indicate that OLS is the most reliable approach for enhancing the sharpness and calibration of interval forecasts in large-scale multi-resolution load forecasting settings, while variance-based approaches offer limited benefit in this provincial context.

4.7. Discussion

Based on our comparative evaluation of reconciliation methods across multiple temporal resolutions and forecast types, several city-specific recommendations emerge. For Calgary, WLS_S is highly effective for ex post deterministic forecasts at fine resolutions (hourly, daily), supporting its use in hourly and daily operational planning. For ex ante deterministic forecasts, WLS_S also performs well at monthly and yearly levels, while OLS is more robust for yearly peak forecasts, making it suitable for seasonal demand and capacity planning. In probabilistic settings, OLS consistently improves forecast reliability across all resolutions, while $WLS_{V\text{-}OSE}$ shows some advantage for peak forecasts, indicating its potential for risk-aware extreme event forecasting.

In Edmonton, WLS_S also performs well in ex post forecasts at both hourly and yearly levels. For ex ante forecasts, WLS_S is preferred for finer resolutions, while OLS is more effective for peak loads. Probabilistically, OLS is again the most stable choice, particularly at daily and monthly levels, while $WLS_{V\text{-}OSE}$ offers modest gains for monthly and yearly peak forecasting, albeit inconsistently.

At the provincial level (Alberta), OLS is the most effective and reliable reconciliation method across nearly all load types and resolutions for both deterministic and probabilistic ex ante forecasts, especially for peak and aggregated long-term planning. In contrast, the bottom-up and variance-based methods (WLS_V and $WLS_{V\text{-}OSE}$) consistently degrade performance at higher levels of aggregation, and should generally be avoided in Provincial-level applications. These findings support method selection tailored to geography, resolution, and forecast type, enabling more informed and accurate long-term energy planning.

Considering the effectiveness of the optimal reconciliation methods for ex ante long-term probabilistic hourly load forecasts, Figure 4 presents the deterministic and probabilistic averaged daily load profiles extracted from the original forecasts as well as the reconciled forecasts using the OLS, WLS_S, and WLS_V methods for the city of Calgary for the last year of the forecasting horizon.

Generally, the original forecasting model under-forecasts the hourly load, and reconciliation results in increased hourly forecasts. The WLS_S and OLS methods provide deterministic daily load profiles which are very close to the actual averaged daily load profile. The larger area inside the prediction intervals for the OLS method results in higher reliability while lowering the sharpness. However, the positive effect of reliability is bigger than the negative effect of sharpness, which results in better probabilistic scores.

The empirical results demonstrate that reconciliation methods can significantly influence both the accuracy and coherence of long-term load forecasts. In particular, the OLS method consistently improves forecast accuracy in probabilistic settings across all datasets and time resolutions, outperforming WLS_S and WLS_V in most cases. This aligns with findings from [20,27], where OLS reconciliation methods were noted for their stability and simplicity, especially under Gaussian assumptions. However, unlike short-term forecasting contexts in prior studies, the long-horizon and multi-resolution nature of this framework introduces more pronounced variability, especially in ex ante forecasts.

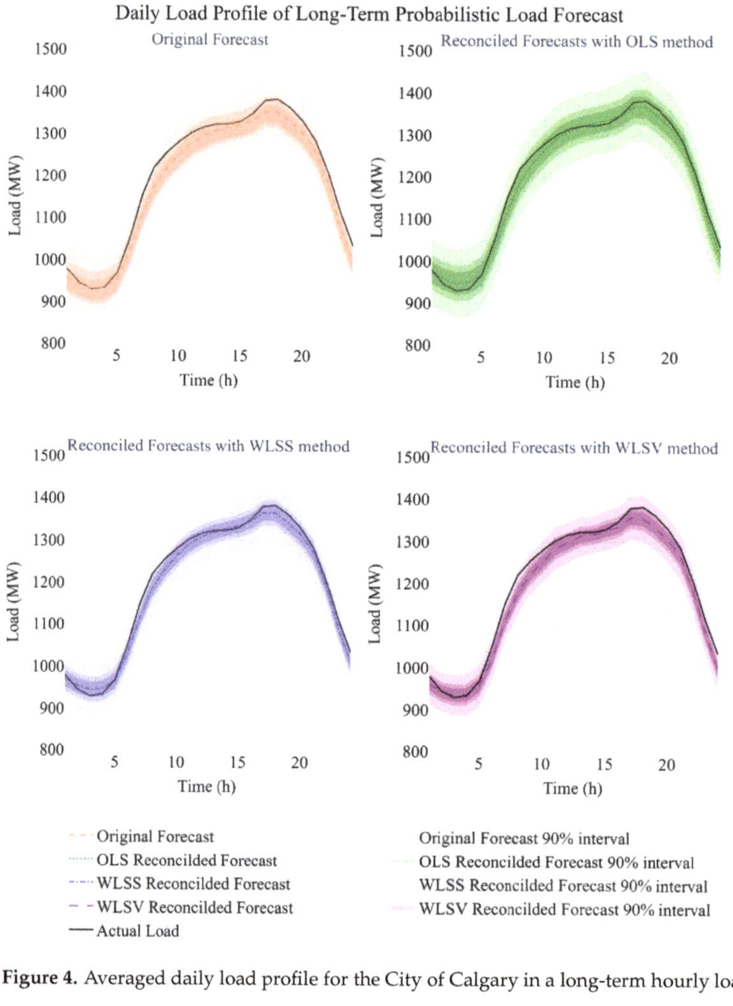

Figure 4. Averaged daily load profile for the City of Calgary in a long-term hourly load forecast.

WLS methods rely on assumptions about the underlying error distributions. Our experiments spanning multiple test years and three different regions demonstrate that the reconciliation framework performs consistently well even under moderate violations of these assumptions. The exploration of both in-sample and out-of-sample error-based reconciliation (WLS_V vs. $WLS_{V\text{-}OSE}$) also provides insight into the tradeoff between theoretical rigor and practical robustness. Further work could explore more flexible nonparametric approaches to estimating the error covariance structure, or could employ resampling techniques to assess the sensitivity of reconciliation outcomes.

Interestingly, WLS_S performed competitively in deterministic forecasting, particularly for hourly and daily forecasts. This suggests that structural scaling, which assumes error consistency within levels, may be better suited to high-resolution load patterns. WLS_V, which is designed for situations where bottom-level variance dominates, showed mixed performance, sometimes worsening aggregated forecasts, especially under out-of-sample error scenarios.

Moreover, differences in reconciliation effectiveness across cities may relate to variations in load composition and demand volatility. Calgary, for example, exhibited stronger improvement under OLS at the daily peak level, potentially due to more predictable consumption patterns compared to the Province-wide Alberta dataset. Overall, these findings underscore that the choice of reconciliation method is not one-size-fits-all. Forecasting context, including resolution, geography, forecast horizon, and data quality, should drive method selection. This flexibility is particularly important in energy systems, where different regions operate under distinct market rules, regulatory environments, and weather-driven variability. These variations indicate that the spatial and temporal granularity of demand data significantly affect reconciliation outcomes, which is another factor warranting further study.

While the proposed method offers significant accuracy gains and coherent forecasts, it assumes that the forecast error distributions are stable and approximately Gaussian, an assumption that may not hold for all regions or forecasting resolutions. Future work should consider non-Gaussian reconciliation strategies and explore hybrid or adaptive methods that respond dynamically to forecast uncertainty structures.

5. Conclusions

In this study, we present a novel approach to long-term electricity load forecasting by integrating temporal hierarchies with advanced reconciliation methods. Our methodology utilizes historical weather data and economic variables to generate probabilistic forecasts at multiple resolutions (hourly, daily, monthly, and yearly) enhancing the reliability and coherence of the predictions.

The results demonstrate the effectiveness of reconciliation methods in improving forecast accuracy. Notably, the OLS method emerges as a standout performer for ex ante probabilistic forecasts, consistently enhancing the Pinball and Winkler scores across all datasets. While the WLS_S shows promise in deterministic forecasts, it did not perform better than the OLS method in probabilistic scenarios.

There was a gap between the performances of different methods in ex post and ex ante forecasting. The introduction of a new source of error by generating input scenarios affected the behavior of different reconciliation methods. The results also indicate that there may be a load composition effect in reconciliation quality. The difference in customer behavior, and consequently in load patterns, introduces a different type of error into the model. Because the error is very important in reconciliation, the effects of customer behavior on load needs to be considered in future works.

The proposed probabilistic multi-resolution long-term forecasting framework provides utilities and system planners with more granular and coherent demand forecasts that can support a range of operational and strategic decisions. For instance, hourly and daily forecasts at the one-year horizon can inform demand response planning, peak load resource scheduling, and seasonal reliability assessments. Probabilistic forecasts made by quantifying forecast uncertainty can enable risk-informed decision-making such as planning based on high-percentile demand scenarios (e.g., P90) rather than single-point estimates. The ability to reconcile forecasts across resolutions further supports consistency in planning activities at both operational and policy-making levels. Our findings underscore the importance of utilizing coherent forecasting approaches that account for the hierarchical nature of electricity demand. Adopting such methods can better inform decision-makers and stakeholders in the energy sector, ultimately leading to more efficient energy management and planning. Future work might explore the integration of this forecasting framework into optimization models for resource allocation, capacity expansion, and regulatory planning under uncertainty.

In conclusion, our work contributes to the growing body of literature on long-term probabilistic electricity load forecasting by demonstrating the utility of temporal hierarchies and reconciliation methods, providing a robust framework for more accurate and coherent load forecasts in the face of an evolving energy landscape.

In future work, we would like to expand the proposed methodology to explore a spatiotemporal hierarchy structure in probabilistic long-term ex ante load forecasting, which would add scalability to the existing model. Forecasting loads for large systems requires a preprocessing step involving fast and global weather station selection and scenario generation.

Author Contributions: Conceptualization, S.B. and H.Z.; methodology, S.B.; validation, S.B.; formal analysis, S.B.; investigation, S.B.; data curation, S.B.; writing—original draft preparation, S.B.; writing—review and editing, S.B. and H.Z.; visualization, S.B.; supervision, H.Z.; project administration, H.Z. All authors have read and agreed to the published version of the manuscript

Funding: This research was funded by NSERC Alliance Grants.

Data Availability Statement: The load data used in this study are available from [49,50]. The economic data were derived from the following resources available in the public domain: [51]. Restrictions apply to the availability of the weather data. Data were obtained from [52] and are available with the permission of [52].

Conflicts of Interest: The authors declare no conflicts of interest.

Abbreviations

The following abbreviations are used in this manuscript:

LTLF	Long-term load forecasting
DER	Distributed energy resources
RES	Renewable energy resources
EV	Electric vehicle
POD	Point of delivery
GNN	Graph neural network
OLS	Ordinary least squares
WLS	Weighted least squares
WLS_S	Structural scaling weighted least squares
WLS_V	Series variance scaling weighted least squares
RMSE	Root mean squared error
MAPE	Mean absolute percentage error

Appendix A

This appendix provides pseudocode for each key step in the proposed forecasting framework. Algorithm A1 outlines the procedure for generating resolution-specific features using weather, calendar, and economic data. Algorithm A2 describes the recursive feature elimination process, which is based on the XGBoost importance scores and the backtesting performance. Algorithm A3 covers hyperparameter tuning and model training, with ARIMAX used for yearly forecasts and XGBoost for higher resolution forecasts. Algorithm A4 details the generation of input scenarios for ex ante forecasting, including both weather and economic assumptions. Algorithm A5 presents the procedure for producing original ex post and ex ante forecasts based on trained models and generated scenarios. Finally, Algorithm A6 shows how reconciliation methods are applied to generate coherent forecasts across different temporal resolutions.

Algorithm A1 Feature generation for different resolutions

Require: Load Dataset *L*, Calendar Dataset *C*, Weather Dataset *W*, Economic Dataset *E*, list of resolutions *resolutions*, resampling function *resample*, combining datasets function *combine*, list of calendar variables used for feature generation *lC*[*r*], list of temperature variables used for feature generation *lW*[*r*], list of lags for each resolution *lags*[*r*], lag function *lag*, list of moving averages for each resolution *MAs*[*r*], moving average function *MA*, adding a new feature to the feature dataset function *add*

Ensure: features *F*[*r*] and load *L*[*r*] for all resolutions

```
Step 1: Resampling and combining the datasets for each resolution
for r in resolutions do
  L[r] ← resample (L, r, [sum])
  W[r] ← resample (W, r, [mean, min, max])
  if r is Hourly or Daily then
    F[r] ← combine (C[r], W[r])
  end if
  if r is Monthly or Yearly then
    F[r] ← combine (C[r], W[r], E[r])
  end if
end for
Step 2: Generating new features for each resolution
for r in resolutions do
  for w in lW[r] do
    F[r] ← add(F[r,w]^2, F[r,w]^3)
    new_lW[r] ← add(F[r,w], F[r,w]^2, F[r,w]^3)
  end for
  for w in new_lW[r] do
    for c in lC[r] do
      F[r] ← add(F[r,c] × F[r,w])
    end for
    for l in lags[r] do
      F[r] ← add(lag(F[r,w], l))
    end for
    for m in MAs[r] do
      F[r] ← add(MA(F[r,w], m))
    end for
  end for
end for
```

Algorithm A2 Feature selection for different resolutions

Require: load for each resolution *L*[*r*], features for each resolution *F*[*r*], train end date *train_end*, list of resolutions *resolutions*, XGBoost models *model*[*r*], XGBoost feature importance function *feature_ranking*, sort function *sort*, RMSE function *rmse*, remove a feature function *remove*, backtesting function *backtest*

Ensure: selected features for each resolution *SF*[*r*]

```
Step 1: Splitting the datasets
for r in resolutions do
  L_train[r] ← L[r][0: train_end]
  F_train[r] ← F[r][0: train_end]
end for
Step 2: Feature selection
score = 0
for r in resolution do
  features_importance ← feature_rank (F_train[r], L_train[r])
  sorted_features ← sort (features_importance)
  for (X_train, Y_train, X_valid, Y_valid) in Back-test(F_train[r], L_train[r]) do
    model[r].fit(X_train, Y_train)
    Ŷ_valid ← model[r].predict(X_valid)
    score = score + rmse (Y_valid, Ŷ_valid)
  end for
  SF[r] ← F[r]
  for f in sorted_features do
    F_valid[r] ← remove(F_valid[r], f)
    F[r] ← remove(F[r], f)
    error = 0
    for (X_train, Y_train, X_valid, Y_valid) in Back-test(F_train[r], L_train[r]) do
      model[r].fit(X_train, Y_train)
      Ŷ_valid ← model[r].predict(X_valid)
      error = error + rmse (Y_valid, Ŷ_valid)
    end for
    if error ≤ score then
    score ← error
    SF[r]← F[r]
    end if
  end for
end for
```

Algorithm A3 Hyperparameter tuning and model training

Require: list of resolutions *resolutions*, load for each resolution *L*[*r*], features for each resolution *SF*[*r*], training end date *train_end*, XGBoost hyper-parameter tuning function with Optuna *optuna*, backtesting function *backtest*, hyper-parameter set values *hyper*[*r*], XGBoost regressor *xgboost*, ARIMAX model *arimax*

Ensure: tuned and trained models *trained_model*[*r*]

```
Step 1: Splitting the datasets
for r in resolutions do
  L_train[r] ← L[r][0: train_end]
  SF_train[r] ← SF[r][0: train_end]
end for
Step 2: Hyper-parameter tuning
for r in resolutions do
  if r is Yearly then
    model = arimax
    for each parameter_set in hyper[r] do
```

Algorithm A3 *Cont.*

```
      model_fit ← model.fit(SF_train[r], L_train[r])
      bic_values [parameter_set] ← model_fit.bic
    end for
    find index of the minimum value in bic_values
    params[r] ← parameter_set corresponding to the minimum bic
    trained_model[r] ← arimax(params[r]).fit(SF_train[r], L_train[r])
  end if
  else then
    params[r] ← optuna(xgboost, hyper[r], back-test(SF_train[r], L_train[r]))
    trained_model[r] ← xgboost(params[r]).fit(SF_train[r], L_train[r])
  end else
end for
```

Algorithm A4 Scenario generation

Require: list of resolutions *resolutions*, features for each resolution *SF*[*r*], validation end date *valid_end*, test end date *test_end*, list of calendar variables for each resolution *lc*[*r*], 4 groups of weather variables for each resolution *lw*[*r*, *i*], list of economic variables for monthly and yearly resolutions *le*[*r*], number of scenarios *ns*, list of 5 years used for weather scenario generation *years*, combining datasets function *combine*, random integer function *random*

Ensure: generated scenarios *scenarios*

```
Step 1: Splitting the datasets
for r in resolutions do
  SF_train[r] ← SF[r][0: valid_end]
  SF_test[r] ← SF[r][valid_end: test_end]
end for
Step 2: Scenario generation
for r in resolution do
  for i in range (4) do
    lsw[r, i] ← []
    for y in years do
      for shift in range (-3, 4) do
        lsw[r, i].append(SF_train[r][lw[r, i][SF_train.index = y].shift(i × days)
      end for
    end for
  end for
  if r is Hourly or Daily then
    for s in range ns do
      SW[r, s] ← combine(lsw[r, i][random(35)] for i in range (4))
      scenarios[r, s] ← combine(SF_test[r][lc[r]], SW[r, s])
    end for
  end if
  else then
    SE[r, 0] ← normal_growth(SF_train[r][le[r]])
    SE[r, 1] ← no_growth(SF_train[r][le[r]])
    SE[r, 2] ← negative_growth(SF_train[r][le[r]])
    SE[r, 3] ← double_growth(SF_train[r][le[r]])
    for s in range ns/4 do
      SW[r, s] ← combine(lsw[r, i][random(35)] for i in range (4))
      for e in range (4) do
        scenarios[r, 4s+1] ← combine(SF_test[r][lc[r]], SW[r, s], SE[r, e])
      end for
    end for
  end else
end for
```

Algorithm A5 Generating original forecasts

Require: list of resolutions *resolutions*, load for each resolution *L*[*r*], features for each resolution *SF*[*r*], validation end date *valid_end*, test end date *test_end*, trained models *trained_model*[*r*], number of scenarios *ns*, generated scenarios *scenarios*[*r*], Probability Density Function *PDF*

Ensure: original ex-post forecasts *ex-post_forecasts*[*r*], original ex-ante deterministic forecasts *ex-ante_point-forecasts*[*r*], original ex-ante probabilistic forecasts *ex-ante_prob-forecasts*[*r*]

```
Step 1: Actual test features for ex-post forecasts
for r in
  SF_train[r] ← SF[r][0: valid_end]
  SF_test[r] ← SF[r][valid_end: test_end]
end for
Step 2: Ex-post deterministic forecasts
for r in resolutions do
  ex-post_forecasts[r] ← trained_model[r].predict(SF_test[r])
end for
Step 3: Ex-ante probabilistic forecasts
for r in resolutions do
  for s in range ns do
    scenario-forecasts[r][s] ← trained_model[r].predict(scenarios[r, s])
  end for
  ex-ante_prob-forecasts[r] ← PDF(scenario-forecasts[r])
end for
Step 4: Ex-ante deterministic forecasts
for r in resolutions do
  ex-ante_point-forecasts[r] ← ex-ante_prob-forecasts[r].mean()
end for
```

Algorithm A6 Reconciliation

Require: load for each resolution *L*[*r*], selected features for each resolution *SF*[*r*], train end date *train_end*, validation end date *valid_end*, list of resolutions *resolutions*, tuned and trained models *trained_model*[*r*], original ex-post forecasts *ex-post_forecasts*[*r*], original ex-ante probabilistic forecasts *ex-ante_forecasts*[*r*], function for building the hierarchy tree and matrix **S** *aggregation*, hierarchy levels *levels*, list of reconciliation methods *recon_list*, reconciliation function *reconciliation*, list of intervals for probabilistic forecasts *int_list*

Ensure: reconciled ex-post forecast with in-sample error *reconciled_P_I*, reconciled ex-post forecast with out-of-sample error *reconciled_P_O*, reconciled ex-ante forecast with in-sample error *reconciled_A_I*, reconciled ex-ante forecast with out-of-sample error *reconciled_A_O*

```
Step 1: Splitting the dataset
for r in resolutions do
  L_train[r] ← L[r][0: train_end]
  L_valid[r] ← L[r][train_end: valid_end]
  SF_train[r] ← SF[r][0: train_end]
  SF_valid[r] ← SF[r][train_end: valid_end]
end for
Step 2: In-sample and out-of-sample validation
for r in resolutions do
  L̂_train[r] ← trained_model[r].predict(SF_train[r])
  L̂_valid[r] ← trained_model[r].predict(SF_valid[r])
  in-sample_error [r] ← L̂_train[r] - L_train[r]
  out-of-sample_error [r] ← L̂_valid[r] - L_valid[r]
end for
```

Algorithm A6 *Cont.*

Step 3: Building and fitting the hierarchy tree
hierarchy_tree, **S** ← *aggregation*(*L_valid*, *levels*)
Step 4: Reconciliation with in-sample errors
for *rl* in *recon_list* **do**
  reconciled_P_I ← *reconciliation*(*rl*, *hierarchy_tree*, **S**, *in-sample_error*, *ex-post_forecasts*[*r*])
  reconciled_A_I ← *reconciliation*(*rl*, *hierarchy_tree*, **S**, *in-sample_error*, *ex-ante_forecasts*[*r*], *int_list*)
end for
Step 5: Reconciliation with out-of-sample errors
for *rl* in *recon_list* **do**
  reconciled_P_O ← *reconciliation*(*rl*, *hierarchy_tree*, **S**, *out-of-sample_error*, *ex-post_forecasts*[*r*])
  reconciled_A_O ← *reconciliation*(*rl*, *hierarchy_tree*, **S**, *ou-of-sample_error*, *ex-ante_forecasts*[*r*], *int_list*)
end for

References

1. Cho, J.; Yoon, Y.; Son, Y.; Kim, H.; Ryu, H.; Jang, G. A Study on Load Forecasting of Distribution Line Based on Ensemble Learning for Mid- to Long-Term Distribution Planning. *Energies* **2022**, *15*, 2987. [CrossRef]
2. Lindberg, K.B.; Seljom, P.; Madsen, H.; Fischer, D.; Korpås, M. Long-term electricity load forecasting: Current and future trends. *Util. Policy* **2019**, *58*, 102–119. [CrossRef]
3. Burillo, D.; Chester, M.V.; Pincetl, S.; Fournier, E.D.; Reyna, J. Forecasting peak electricity demand for Los Angeles considering higher air temperatures due to climate change. *Appl. Energy* **2019**, *236*, 1–9. [CrossRef]
4. Habbak, H.; Mahmoud, M.; Metwally, K.; Fouda, M.M.; Ibrahem, M.I. Load Forecasting Techniques and Their Applications in Smart Grids. *Energies* **2023**, *16*, 1480.
5. Burg, L.; Gürses-Tran, G.; Madlener, R.; Monti, A. Comparative analysis of load forecasting models for varying time horizons and load aggregation levels. *Energies* **2021**, *14*, 7128. [CrossRef]
6. Alfares, H.K.; Nazeeruddin, M. Electric load forecasting: Literature survey and classification of methods. *Int. J. Syst. Sci.* **2002**, *33*, 23–34. [CrossRef]
7. Chow, J.H.; Wu, F.F.; Momoh, J.A. *Applied Mathematics for Restructured Electric Power Systems Optimization, Control, and Computational Intelligence*, 1st ed.; Power Electronics and Power Systems; Springer: New York, NY, USA, 2005.
8. Hong, T.; Wang, P.; White, L. Weather station selection for electric load forecasting. *Int. J. Forecast.* **2015**, *31*, 286–295. [CrossRef]
9. Li, J.; Luo, Y.; Wei, S. Long-term electricity consumption forecasting method based on system dynamics under the carbon-neutral target. *Energy* **2022**, *244*, 122572. [CrossRef]
10. Zhou, J.; He, Y.; Lyu, Y.; Wang, K.; Che, Y.; Wang, X. Long-term electricity forecasting for the industrial sector in western China under the carbon peaking and carbon neutral targets. *Energy Sustain. Dev.* **2023**, *73*, 174–187. [CrossRef]
11. Venkataramanan, V.; Kaza, S.; Annaswamy, A.M. DER Forecast using Privacy Preserving Federated Learning. *IEEE Internet Things J.* **2022**, *10*, 2046–2055. [CrossRef]
12. Xie, J.; Hong, T. Temperature Scenario Generation for Probabilistic Load Forecasting. *IEEE Trans. Smart Grid* **2018**, *9*, 1680–1687. [CrossRef]
13. Hong, T.; Fan, S. Probabilistic electric load forecasting: A tutorial review. *Int. J. Forecast.* **2016**, *32*, 914–938. [CrossRef]
14. Hong, T. Energy forecasting: Past, present and future. *Foresight Int. J. Appl. Forecast.* **2014**, 43–48.
15. Gneiting, T.; Katzfuss, M. Probabilistic forecasting. *Annu. Rev. Stat. Its Appl.* **2014**, *1*, 125–151. [CrossRef]
16. Petropoulos, F.; Apiletti, D.; Assimakopoulos, V.; Babai, M.Z.; Barrow, D.K.; Ben Taieb, S.; Bergmeir, C.; Bessa, R.J.; Bijak, J.; Boylan, J.E.; et al. Forecasting: Theory and practice. *Int. J. Forecast.* **2022**, *38*, 705–871. [CrossRef]
17. Hollyman, R.; Petropoulos, F.; Tipping, M.E. Understanding forecast reconciliation. *Eur. J. Oper. Res.* **2021**, *294*, 149–160. [CrossRef]
18. Rangapuram, S.S.; Kapoor, S.; Nirwan, R.S.; Mercado, P.; Januschowski, T.; Wang, Y.; Bohlke-Schneider, M. Coherent Probabilistic Forecasting of Temporal Hierarchies. In Proceedings of the 26th International Conference on Artificial Intelligence and Statistics, Valencia, Spain, 25–27 April 2023; Ruiz, F., Dy, J., van de Meent, J.W., Eds.; Proceedings of Machine Learning Research; PMLR; Volume 206, pp. 9362–9376.
19. Pritularga, K.F.; Svetunkov, I.; Kourentzes, N. Stochastic coherency in forecast reconciliation. *Int. J. Prod. Econ.* **2021**, *240*, 108221. [CrossRef]

20. Jeon, J.; Panagiotelis, A.; Petropoulos, F. Probabilistic forecast reconciliation with applications to wind power and electric load. *Eur. J. Oper. Res.* **2019**, *279*, 364–379. [CrossRef]
21. Athanasopoulos, G.; Hyndman, R.J.; Kourentzes, N.; Petropoulos, F. Forecasting with temporal hierarchies. *Eur. J. Oper. Res.* **2017**, *262*, 60–74. [CrossRef]
22. Rangapuram, S.S.; Werner, L.D.; Benidis, K.; Mercado, P.; Gasthaus, J.; Januschowski, T. End-to-End Learning of Coherent Probabilistic Forecasts for Hierarchical Time Series. In Proceedings of the 38th International Conference on Machine Learning, Virtual, 18–24 July 2021; Meila, M., Zhang, T., Eds.; Proceedings of Machine Learning Research; PMLR; Volume 139, pp. 8832–8843.
23. Taieb, S.B.; Taylor, J.W.; Hyndman, R.J. Coherent Probabilistic Forecasts for Hierarchical Time Series. In Proceedings of the 34th International Conference on Machine Learning, Sydney, Australia, 6–11 August 2017; Precup, D., Teh, Y.W., Eds.; Proceedings of Machine Learning Research; PMLR; Volume 70, pp. 3348–3357.
24. Dong, M.; Grumbach, L. A Hybrid Distribution Feeder Long-Term Load Forecasting Method Based on Sequence Prediction. *IEEE Trans. Smart Grid* **2020**, *11*, 470–482. [CrossRef]
25. Yang, Y.; Zhou, H.; Wu, J.; Liu, C.J.; Wang, Y.G. A novel decompose-cluster-feedback algorithm for load forecasting with hierarchical structure. *Int. J. Electr. Power Energy Syst.* **2022**, *142*, 108249. [CrossRef]
26. Zambrano-Asanza, S.; Morales, R.E.; Montalvan, J.A.; Franco, J.F. Integrating artificial neural networks and cellular automata model for spatial-temporal load forecasting. *Int. J. Electr. Power Energy Syst.* **2023**, *148*, 108906. [CrossRef]
27. Wickramasuriya, S.L. Probabilistic Forecast Reconciliation under the Gaussian Framework. *J. Bus. Econ. Stat.* **2023**, *42*, 272–285.
28. Panagiotelis, A.; Gamakumara, P.; Athanasopoulos, G.; Hyndman, R.J. Probabilistic forecast reconciliation: Properties, evaluation and score optimisation. *Eur. J. Oper. Res.* **2023**, *306*, 693–706. [CrossRef]
29. Thangjam, A.; Jaipuria, S.; Dadabada, P.K. Time-Varying approaches for Long-Term Electric Load Forecasting under economic shocks. *Appl. Energy* **2023**, *333*, 120602. [CrossRef]
30. Gholizadeh, N.; Musilek, P. Federated learning with hyperparameter-based clustering for electrical load forecasting. *Internet Things* **2022**, *17*, 100470. [CrossRef]
31. Xiang, W.; Xu, P.; Fang, J.; Zhao, Q.; Gu, Z.; Zhang, Q. Multi-dimensional data-based medium- and long-term power-load forecasting using double-layer CatBoost. *Energy Rep.* **2022**, *8*, 8511–8522. [CrossRef]
32. Behm, C.; Nolting, L.; Praktiknjo, A. How to model European electricity load profiles using artificial neural networks. *Appl. Energy* **2020**, *277*, 115564. [CrossRef]
33. Farrag, T.A.; Elattar, E.E. Optimized Deep Stacked Long Short-Term Memory Network for Long-Term Load Forecasting. *IEEE Access* **2021**, *9*, 68511–68522. [CrossRef]
34. Li, Y.; Han, D.; Yan, Z. Long-term system load forecasting based on data-driven linear clustering method. *J. Mod. Power Syst. Clean Energy* **2018**, *6*, 306–316. [CrossRef]
35. Hamad, Z.; Abdulrahman, I. Deep learning-based load forecasting considering data reshaping using MATLAB Simulink. *Int. J. Energy Environ. Eng.* **2022**, *13*, 853–869. [CrossRef]
36. Han, X.; Su, J.; Hong, Y.; Gong, P.; Zhu, D. Mid-to Long-Term Electric Load Forecasting Based on the EMD–Isomap–Adaboost Model. *Sustainability* **2022**, *14*, 7608. [CrossRef]
37. Wen, Z.; Xie, L.; Fan, Q.; Feng, H. Long term electric load forecasting based on TS-type recurrent fuzzy neural network model. *Electr. Power Syst. Res.* **2020**, *179*, 106106. [CrossRef]
38. Jin, B.; Zeng, G.; Lu, Z.; Peng, H.; Luo, S.; Yang, X.; Zhu, H.; Liu, M. Hybrid LSTM–BPNN-to-BPNN Model Considering Multi-Source Information for Forecasting Medium- and Long-Term Electricity Peak Load. *Energies* **2022**, *15*, 7584. [CrossRef]
39. Kazemzadeh, M.R.; Amjadian, A.; Amraee, T. A hybrid data mining driven algorithm for long term electric peak load and energy demand forecasting. *Energy* **2020**, *204*, 117948. [CrossRef]
40. Hong, T.; Xie, J.; Black, J. Global energy forecasting competition 2017: Hierarchical probabilistic load forecasting. *Int. J. Forecast.* **2019**, *35*, 1389–1399. [CrossRef]
41. Zhang, H.; Chen, B.; Li, Y.; Geng, J.; Li, C.; Zhao, W.; Yan, H. Research on medium- and long-term electricity demand forecasting under climate change. *Energy Rep.* **2022**, *8*, 1585–1600. [CrossRef]
42. Olivares, K.G.; Garza, F.; Luo, D.; Challú, C.; Mergenthaler, M.; Taieb, S.B.; Wickramasuriya, S.L.; Dubrawski, A. HierarchicalForecast: A Reference Framework for Hierarchical Forecasting in Python. *arXiv* **2023**, arXiv:2207.03517.
43. Hyndman, R.; Athanasopoulos, G. *Forecasting: Principles and Practice*, 2nd ed.; OTexts: Melbourne, Australia, 2018.
44. Xu, Y.; Zheng, S.; Zhu, Q.; Wong, K.-c; Wang, X.; Lin, Q. A complementary fused method using GRU and XGBoost models for long-term solar energy hourly forecasting. *Expert Syst. Appl.* **2024**, *254*, 124286. [CrossRef]
45. Bae, D.J.; Kwon, B.S.; Song, K.B. XGBoost-Based Day-Ahead Load Forecasting Algorithm Considering Behind-the-Meter Solar PV Generation. *Energies* **2022**, *15*, 128. [CrossRef]
46. Semmelmann, L.; Henni, S.; Weinhardt, C. Load forecasting for energy communities: A novel LSTM-XGBoost hybrid model based on smart meter data. *Energy Inform.* **2022**, *5*, 24. [CrossRef]

47. Neshat, N.; Hadian, H.; Behzad, M. Nonlinear ARIMAX model for long-term sectoral demand forecasting. *Manag. Sci. Lett.* **2018**, *8*, 581–592. [CrossRef]
48. Wickramasuriya, S.L.; Athanasopoulos, G.; Hyndman, R.J. Optimal Forecast Reconciliation for Hierarchical and Grouped Time Series Through Trace Minimization. *J. Am. Stat. Assoc.* **2019**, *114*, 804–819. [CrossRef]
49. AESO. Hourly Load by Area and Region. 2022. Available online: https://www.aeso.ca/market/market-and-system-reporting/data-requests/hourly-load-by-area-and-region (accessed on 15 July 2023).
50. AESO. Annual Market Statistics Report. 2025. Available online: https://public.tableau.com/app/profile/market.analytics/viz/AnnualStatistics_16161854228350/Introduction (accessed on 15 February 2025).
51. GoA. Alberta Economic Dashboard. 2025. Available online: https://economicdashboard.alberta.ca (accessed on 15 February 2025).
52. WeatherSource. OnPoint Weather. 2023. Available online: https://weathersource.com (accessed on 2 May 2024).

MDPI AG
Grosspeteranlage 5
4052 Basel
Switzerland
Tel.: +41 61 683 77 34

Energies Editorial Office
E-mail: energies@mdpi.com
www.mdpi.com/journal/energies

www.ingramcontent.com/pod-product-compliance
Lightning Source LLC
Chambersburg PA
CBHW061839210926
51850CB00006B/925
9783725848256